The Geography of the U.S.S.R.

Originally published in 1964, and extensively illustrated with figures and charts, this volume gives an overview of both physical and human geography of the former USSR. The role that the geography of the country has played in shaping historical events and political forces is discussed, as is its role in the economy of the Soviet Union. The geography is examined by topics and regional differences explained within this framework. The book looks at some of the major problems posed by geographical conditions and how they have been tackled and as far as data allows, the success or failure of measures has been assessed.

T0225640

Geography of the U.S.S.R.

R.E.H. Mellor

Routledge
Taylor & Francis Group

First published in 1964
by Macmillan

This edition first published in 2021 by Routledge
2 Park Square, Milton Park, Abingdon, Oxon, OX14 4RN
and by Routledge
605 Third Avenue, New York, NY 10058

Routledge is an imprint of the Taylor & Francis Group, an informa business

ISBN: 978-0-367-77507-0 (hbk)
ISBN: 978-1-003-17204-8 (ebk)
ISBN: 978-0-367-77607-7 (pbk)

DOI: 10.4324/9781003172048

GEOGRAPHY
OF THE U.S.S.R.

BY

R. E. H. MELLOR

PRINTED IN GREAT BRITAIN

TO MY PARENTS

CONTENTS

ILLUSTRATIONS

(at end of text)

LIST OF MAPS

INTRODUCTION

AN understanding of the geographical character of the Soviet Union is essential to an understanding of its people and their response to the world around. Most striking is perhaps the immensity of this great continental country that overshadows man, while his habitat is made further inhospitable by an unkind climate, whose continental extremes create a rigorous seasonal control upon society.

Though limitless plains leave the country without apparent natural defences, the vulnerability is nevertheless in part cancelled by the impenetrability created by a combination of climatic adversity and distance. Yet throughout history, waves of migration and invasion have swept across the plains to leave a complex ethnic pattern. Something of the unity of the Russian lands arises from a harsh geographical environment which makes possible only certain reactions to any given circumstances. Basically, the contemporary landscape has been moulded by the long, slow, relentless spread of the Eastern Slavs from their homes in the plains of European Russia across much of Northern Asia. With 'natural' frontiers absent or extremely hard to define, the *frontier* has become a vital social force and symbol. Its expansion has incorporated into the Russian state vastly different geographical environments and peoples. The intermixture has made Russian life a strange amalgam of European and Asian, yet somehow separate from both.

Periods of stagnation and secretiveness have been broken by bursts of energy and the opening of the land to outside inspection. The Mongol period and the 1917 Revolution form two major breaks in historical continuity, while the eighteenth-century Petrine period marked the modernisation of a peculiarly archaic society. These events have left their imprint upon geographical distributions and upon the landscape of Russia. The changes wrought by the application of Marxist-Leninist doctrines since 1917 have further modified the geography of the country, but, even so, the Soviet state has been the heir to a continuum of historical development in which real problems posed by geography have not fundamentally changed. The Soviet state does not exist *in vacuo*, for a millennium of history cannot be declared irrelevant. The factors affecting the spread of the pioneer fringe of settlement and the growth of the state remain: it is man's response to them which has altered by acceptance of new concepts and by the advance of the technological armoury at his command.

This book examines some of the major problems posed by geographical conditions and how they have been tackled. As far as data allow, the success or failure of the chosen measures has been assessed. Attention is directed at the political-geographical development of the Russian state and at the geographical exploration which so often prefaced moves of the expanding frontier of settlement, whose momentum and direction altered as knowledge of terrain was pieced together. Likewise, attention is given to transport problems, so vital in binding together this vast state into an organism.

It was decided that, within the size and scope of the work, it is better to concentrate on examination of the geography by topics and to discuss regional differences within this framework. A certain dichotomy in the regional geography of the Soviet Union complicates the geographer's task. Many Russian geographers have divided their country into large regions based on the omnipotent trinity of climate, soil, and vegetation, while the Marxist-Leninist school has favoured regions based on economic and planning criteria, now used in most Soviet works, that gain their *raison d'être* from changing and changeable economic conditions. They represent more an analysis of present needs and capabilities than a synthesis of geographical distributions which forms the basis of the classical Western concept of regional geography. Despite the release of much Soviet statistical and geographical material in the last ten years, serious deficiencies in maps and local studies remain. Because an old Russian proverb says 'Every duck praises his own swamp', Soviet material has been checked and compared with non-Soviet material, while the generosity of the Carnegie Trust made possible extensive travel in the Soviet Union. Use of place names follows a modified form of the now generally accepted U.S.B.G.N.-P.C.G.N. system.

The book owes a great deal to Professor A. C. O'Dell who suggested that I embark on the task and then gave generous help and encouragement to a frequently flagging spirit. My colleague Kenneth Walton gave unstintingly of his own limited time to

Introduction

read my drafts and make many valuable suggestions. I am indebted to many people in Britain and Russia who have answered questions and given advice or welcome criticism. Any shortcomings and inaccuracies or expressions of opinion remain, however, solely mine. The maps were drawn, from my rough compilations, by Mrs. M. J. F. Barnett and much help in compilation was given by Miss Dorota Skibka. Inestimable thanks go to my wife for patience, encouragement, and help at every stage.

ROY E. H. MELLOR

THE OLD BREWERY
OLD ABERDEEN
December 1963

CHAPTER 1

The Physical Environment of Russia

ONE of the most difficult and yet fundamental problems to a proper understanding of Soviet society and affairs is to appreciate in real terms the geographical environment in which they exist. This cannot be measured in the units of time and distance commonly employed in Western Europe, a continental peninsula with a small and highly diverse regional personality. The two striking features of the Soviet world are its territorial immensity, so that despite its diversity the units which comprise it are so large as to be almost incomprehensible to the usual scales of measure in the Western European mind; and the inhospitality of the environment arising principally from climatic factors associated with its northerly position in a vast land mass, often omitted in our assessment of Russia. The atlas distribution maps give a picture of great regional diversity completely lost to the traveller across the vast, monotonously dull and relatively unchanging countryside. When in the miniscule cosmos of Western Europe, a day's journey takes the traveller across several markedly different regions, in Russia such a journey will see hardly a change in the landscape and little difference in the climate. Such changelessness is accentuated by the standardization and regimentation of Soviet life. The changes are nevertheless present in a subtle and almost imperceptible form. Russia contains both arctic and temperate desert, dense forest and barren grasslands; the wide and virtually flat immensity of the West Siberian Lowland contrasts with the high, jagged, snow-clad peaks of Caucasia; from the well-settled farming countryside of the Ukraine, with its big villages and busy country towns, the traveller can go to the sparsely-inhabited and sometimes imperfectly explored wilderness of Eastern Siberia.

The territorial vastness of the U.S.S.R. is illustrated better by saying that it is some three times the area of the U.S.A. and ninety times that of the United Kingdom than by quoting figures. An area of 8·6 million square miles, it covers one-sixth of the habitable land surface of the globe, stretching across Northern Asia and Eastern Europe, so occupying the core of the

B

I

DOI: 10.4324/9781003172048-1

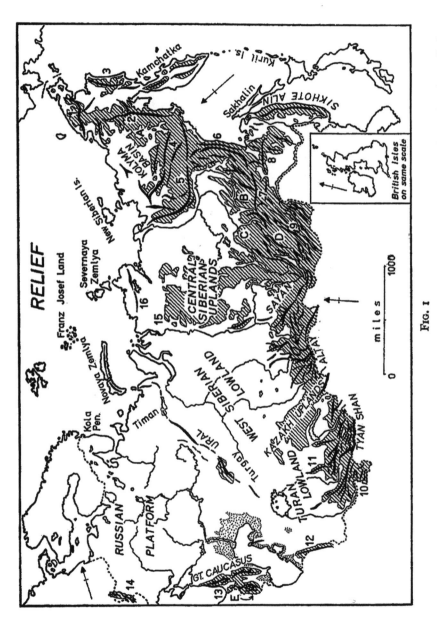

RELIEF

FIG. I

Relief. Land over 1,500 feet shaded and main trends of mountain ranges shown. *Mountains*: 1 Anadyr, 2 Gydan (Kolyma), 3 Koryak, 4 Cherskiy, 5 Verkhoyansk, 6 Dzhugdzhur, 7 Stanovoy, 8 Dzhagdy, 9 Yablonovyy, 10 Pamir, 11 Kara Tau, 12 Kopet Dag, 13 Little Caucasus, 14 Carpathians, 15 Putoran, 16 Byrranga (Taymyr). *Plateaus*: A Yukagir, B Aldan, C Patom, D Vitim, E Armenian. *Atlas Mira*, Moscow, 1954.

Eurasian continent. So great is it in longitudinal extent that the day begins eleven hours earlier in the east than in the west: when people in Vladivostok get up, Muscovites are going to bed; in fact, these two towns are farther apart than London and New York. The most northerly point of the Soviet mainland lies well within the Arctic Circle at Cape Chelyuskin, 77° 44′ N.: it is more telling to note that Moscow lies 250 miles further north than London and Leningrad is in the same latitude as the Shetland Islands. Almost half the country is affected by *permafrost*, a typical phenomenon of northerly continental latitudes, and has a winter lasting more than six months. These cold northern influences penetrate deep into the heart of Asia, across the low northern coasts and great plains which stretch to the foot of the high mountains of Central Asia and Mongolia. Yet it is a land of great climatic contrasts: in Soviet Central Asia there are deserts with intensely hot, arid summers but bitterly cold winters north of Bukhara and Fergana on the latitude of Madrid. Sheltered by the Great Caucasus, rich, humid lowlands lie around the Kolkhiz Lowland. The strong climatic rhythm of a long cold winter followed by a wet spring with great floods and then a short, warm, or even hot summer whose dust is laid in the brief, invigorating autumn, dominates the yearly cycle of Russian life.

Although Russia appears to hold a commanding central position — the nodality of a 'heartland' in the Eurasian landmass — and although its frontiers extend for over 36,000 miles, it is nevertheless surprisingly isolated by geography as well as by choice from neighbouring countries. The long southern land frontier in Transcaucasia and in Central Asia, as well as in Southern Siberia as far as the upper reaches of the Amur, is formed by rough upland and mountain country, broken by relatively few gaps or crossed by few easy passes, which rises mostly out of sparsely-inhabited steppe or forest. In the middle and lower reaches of the Amur and Ussuri, the frontier is guarded by wide marshes. Even Russia's long frontage on the Arctic and Pacific Oceans is frozen throughout most of its length for long periods each winter, while access to the Atlantic Ocean is through narrow, peripheral seas of the Baltic and Black Sea–Mediterranean, easily blocked by hostile powers. Yet in the air age, Russia lies astride many of the potentially busiest great circle routes from Europe to South-east Asia and from North America to Asia.

Does Russia belong to Asia or to Europe? It has been

customary to conceive the Russian lands as divided into a European and an Asiatic part, though the boundaries between them have been seen differently by various authorities, and the division is still used by the Russians themselves.[1] Although these terms may be useful for regional description, it is clear that the Soviet Union can only be regarded as an integral whole, and at the same time its characteristics can be described as neither fully European nor fully Asiatic: it is Eurasian, for its life and behaviour in every aspect are an amalgam moulded in the distinctive geographical and ideological environment of the Soviet state. The bulk of the population and the greater part of the nation's economic activity is concentrated in a narrow wedge of fertile black earth lands and deciduous forest which tapers eastwards into Siberia. While the centre of gravity of Soviet life tends to shift eastwards, it is unlikely that it will ever lie predominantly in Siberia or Central Asia.

STRUCTURE AND RELIEF

Fundamentally, Russia is composed of two large, ancient platforms, partly exposed and partly covered to varying depths by later deposits. On the west is the Russian Platform, and to the east the Central Siberian Platform. Between them lie the Hercynian structures of the Ural and the lowlands of Western Siberia and Central Asia. The southern and eastern periphery of the country is formed by a vast amphitheatre of mountains and uplands of more recent origin, though containing uplifted blocks of older eroded orogenic systems. The ancient platforms of crystalline rocks, such as granite and gneiss, were subjected to intense movement in Pre-Cambrian times; but later they were levelled by erosion, while fracturing and movement raised or lowered large blocks of country. Over great distances, these ancient materials are overlain by more recent sediments, frequently almost horizontal and undisturbed. In European Russia, the ancient crystalline base appears only in the shields of Karelia, a part of the Fenno-Scandian shield, and in Azov-Podolia. In Siberia, the basal structure surfaces in the Aldan and Anabar shields, although the uplifted block of Central Siberia has been deeply dissected by river erosion. Closely associated tectonically with these shields are the folds of the Caledonian orogeny in Southern Siberia.

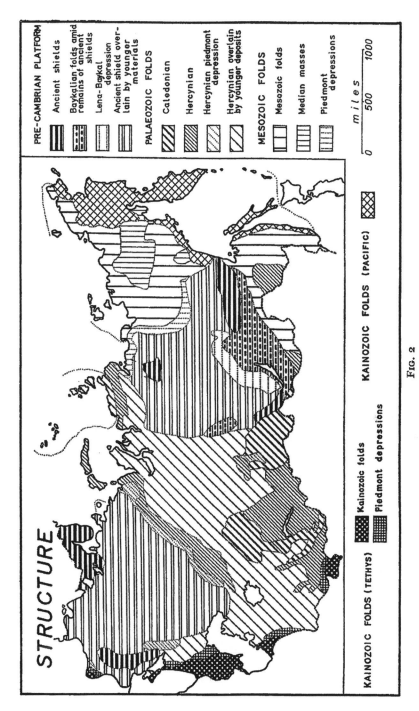

Fig. 2

Structure. Based on Shatskiy, N., Tectonic Map, 1 : 5 M, Moscow, 1956, as simplified by Saushkin, Y., in *Voprosy Geografi*, 47, Moscow, 1959. Also *Atlas SSSR*, Moscow, 1962, pl. 72-73.

5

The Palaeozoic materials, which form a broad belt between the two great platforms, were folded in Hercynian times. The folded materials outcrop in the low ranges of the Ural and in the Kazakh Uplands, which also contains some Caledonian elements. Novaya Zemlya also belongs to this group. Hercynian orogenic belts can also be found in the Altay and in the ranges of Central Asia, where they have been greatly modified by the Tertiary alpine orogeny. Epi-Hercynian platforms deeply covered by later materials form the base of the great lowlands of Western Siberia and Turan.

Mesozoic structural elements are found mostly in Eastern Siberia, including the great geosyncline in which the middle and lower Lena flows. They also include the structures associated with the Kolyma Median Mass. Kainozoic folds are divided by the Russians into two groups: the southern group (Carpathians, Crimean mountains, Caucasus, Pamirs, and Kopet Dag) of Tethys, and the eastern group of Pacific folds, including Kamchatka, Sakhalin, the Dzhugdzhur and Gydan mountains, as well as the Kuril Islands and the coastal ranges of the Sikhote Alin. The alpine movements caused structural modifications in Southern Siberia and the older Central Asian ranges, such as the formation of the many distinctive and economically important tectonic basins. The aftermath of these movements is left in the seismic and volcanic phenomena found here.

The ancient platforms form lowlands in European Russia and low but intensely dissected plateau blocks in Central Siberia. The roots of the older mountain systems are now found as low mountains of generally rounded relief, like the Ural, or as tablelands spattered by residual hills, such as the Kazakh *melkosopochnik*, or in other instances, lowlands covered by considerable depths of later deposits. The younger mountain belts have a topography of high mountain ranges and high plateaus, frequently forming serious obstacles to inter-regional movement. In the Kainozoic mountains there is great conformity between tectonic structure and morphology, which has been established over a considerable period of time; and likewise, in belts of Mesozoic folding, the relationship is less contrasted, probably as the result of extensive denudation during the Kainozoic period. The exposed roots of the ancient mountains are particularly rich in minerals, though important resources are also found in the sedimentary materials covering the platforms and lying in piedmont depressions of younger mountain ranges.[2]

Russia is, however, a country of immense rolling plains: about half the country lies below 600 feet. The European Russian Plain, an eastwards extension of the North European Plain, covers about a quarter of the whole country. The smaller West Siberian Lowland exceeds it, however, in monotony of relief. Between these two plains, the low ranges of the Ural with their easy passes do not form a barrier, and yet assume an importance in Russian thought greater than their true nature warrants. They have become the conventionally accepted divide between Europe and Asia. East of the Yenisey, uplands predominate, though they are seldom above 3,000 feet. The southern borders are formed by high mountains, including the highest elevations in the Soviet Union, in the Pamir-Alay mountains, where Communism and Lenin Peaks rise to 24,590 feet and 23,363 feet respectively. High mountain ranges also separate Central Siberia from the Pacific coast: the recently discovered Pobeda Mountain reaches 10,322 feet, while Klyuchevskaya Sopka in Kamchatka exceeds 15,000 feet. In contrast, the Caspian Depression lies below mean sea level; the present surface of the Caspian is 92 feet below sea level. The bottom of of the Karagiye Depression on the Mangyshlak Peninsula is 434 feet below sea level. Another great depression of the *graben* type is occupied by Lake Baykal, the deepest lake in the world.

Differing erosional processes have contributed to the formation of relief in Russia. During the Quaternary period, ice sheets from Scandinavia and the northern Ural spread over northern and central European Russia; Ural ice and some from the Taymyr Peninsula also covered the northern part of the West Siberian Lowland. The resultant landscapes are contrasted between the intensive ice erosion on the Karelian shield and the masses of outwash materials, moraines and other phenomena further south. Traces of widespread glaciation remain in Northeastern Siberia and parts of Central Siberia, though development of icecaps in a dry continental environment was hindered by the lack of moisture. The contemporary mountain glaciation of the mountains of Southern Siberia (the Altay and the Sayan), of Central Asia, particularly the high Pamir, and the Caucasus, is a remnant of shrunken Quaternary glaciers and icefields. The sequence of glacial advances and milder interglacial periods in Russia, however, cannot always be readily correlated with that established in Western Europe. Arctic islands still contain large glaciers and ice sheets, while over almost half the country the occurrence of permanently frozen ground exercises considerable influence upon landscape; for example, by solifluction in slope formation or the arresting of karst formation by freezing.

Throughout European Russia and in modified form in the permanently frozen ground of Siberia, normal 'humid' sub-aerial denudation is characteristic. In Central Asia, however, arid erosional processes are active in an area of inland drainage, where large sandy and clayey deserts exist. Evidence indicates that these areas were once more intensively desiccated, but that in the recent geological past they have also undergone moister periods marked by remains of old river beds, dried out lakes and broad alluvial deposits. There are also remains of former arid continental weathering in the *inselberg* (*melkosopochnik*) landscapes of the Trans-Uralian Peneplain and the Kazakh Uplands.

The deflation processes commonly associated with arid weathering have deposited around the fringe a belt of loessic materials best developed in the wide piedmont zone along the base of the dry foothills of the Central Asian mountains. The loessic deposits of southern European Russia appear to have originated in an ephemeral arid phase during the glacial epoch, either by deflation during a dry interglacial, or by wind-borne material carried from outwash deserts around the ice sheets' periphery. This may also explain some of the Asiatic deposits. These soft, friable materials have already been extensively eroded into gullies, an attack accelerated by man's destruction of the natural vegetation and by poor husbandry.

THE MAJOR PHYSICAL REGIONS [3]

The most densely peopled parts of the Soviet Union and the core of the Russian nation lie in the European Russian Plain, the largest of the lowlands, extending some 1,100 miles from the northern arctic coast of the Barents and White Seas to the almost mediterranean conditions of the Black Sea. From the foothills of the Ural it extends westwards for 1,500 miles to merge with the glacial plains of Poland. The Plain rests on a base of ancient crystalline rocks, much faulted, fractured and down-warped, overlain by almost horizontal and little-disturbed younger sediments. Only in the shields of Karelia and the southern Ukrainian

Podolsk-Azov uplands do the ancient rocks appear at the surface. Elsewhere they are brought near the surface by fracturing and warping, which exercises some influence upon the relief of the plain in a series of broad depressions between gently undulating plateaus. Where the younger strata have been eroded away from the older, Pre-Cambrian basement, low scarps occur — in the Silurian scarp of the *Glint* in Estonia and in some of the drift-mantled surfaces to the south of Leningrad — while the Donets Heights represent up-warped Carboniferous rocks of the Donets Basin resting against the Podolsk-Azov shield. The highest surfaces seldom rise above 1,000 feet and between the higher parts of the interfluvial plateaus and the bottoms of the broad valleys there is only 300-400 feet difference in elevation, so that over the vast distances of the plain the slope is indeed very gentle and the continuity of the undulating surface is seldom broken, except along the steep bluffs above the right banks of rivers and by deep erosion gullies in the southern steppe. The depressions are occupied by meandering, sluggish rivers or swamp. Settlement tends to seek higher and drier elevations, though frequently the highest surfaces are avoided, particularly in the steppe, for they are exposed to the bitter winter winds blowing from the heart of Siberia. Despite the seeming uniformity and featureless appearance of the plain, small changes in elevation or slight breaks of slope can have important military, political, or economic significance. The contrasts in the landscapes and the settlement and economic patterns of the plain are conditioned, however, more by the broad latitudinal climatic and soil-vegetation belts than by relief and physical form.

Although accurately made within the limits of scale, an atlas map can give a misleading impression of the relief of the European Russian Plain, for distances are extremely great compared to the variation in elevation over the plain. The relief features are the result of the denudation of the mantle of softer deposits which cover the crystalline base, although in some places, dislocation and fracturing of this basement has exercised an influence on the broader pattern of relief. Everywhere the action of water is evident: in the steep right banks of the rivers and their low left or 'meadow' banks, as well as in the shaping of the superficial covering of Quaternary glacial deposits, which give most of the plain its contemporary form.[4] Extending from Masuria in the west across the marchlands between Byelorussia and the Baltic republics to the Valday Hills on the east, in a broad belt of morainic hill country with many small lakes and patches of marsh, lies the main watershed of the plain, between Black Sea-Caspian drainage southwards by long, meandering, sluggish rivers, and Baltic drainage northwards by short, swifter rivers whose courses have been often modified by glacial action. Forming a drier, east-west axis of movement and a belt of portages on the north-south river routes, these uplands, mostly below 700 feet, have a key historical position in the growth of Russia and are marked by several old towns of great commercial or military significance, like Novgorod and Smolensk. The steep western face of the higher Valday Hills on the east guards the entrance to the upper Volga and the rear approaches to Moscow and Leningrad. To the north, the Baltic lowlands form wet, open plains diversified by crescentic-shaped morainic belts, with their eskers and outwash fans, best developed by the lobes of ice which occupied Riga Bay, the basin of Lake Peipus, and the country around Klaypeda.

The middle of the European Russian Plain is occupied by the Central Russian Uplands, with a general elevation of 600-800 feet but rising in places to 1,000 feet in the west. Its steeper eastern flank drops to the broad, marshy depression with its imperceptible watershed between Oka and Don drainage. On the west, it drops gently to the broad depression of Polesye, with the Pripyat marshes extending west to the Polish frontier, and the broad, left-bank terraces of the Dnepr contrasting to the steep bluffs of the right bank, well seen around Kiev. To the south, the Uplands merge into the lower eastern end of the slightly higher belt of country formed by the surface of the Podolsk-Azov shield, whose comparatively late elevation caused the ponding back of the drainage of the upper Dnepr and the formation of the wet depression of Polesye. The Dnepr crosses the shield in a series of rapids, now the site of the large Dneproges hydro-electric barrage. The shield also causes the large eastwards bend in the course of the Donets. Structures associated with the Hercynian basin of the Donets coalfield and the eastern extremity of the Podolsk-Azov shield form the Donets Heights, reaching to over 1,000 feet, in a broad upland, rolling but deeply dissected by streams and erosion gullies. Westwards, the shield is higher and its superficial deposits are also roughly dissected by gullies. The south-west corner of Soviet territory is formed by the alpine orogenic structures of the Carpathians and the rolling country of the Carpathian foreland, with its low hills and rolling dissected plateau, broken by low hills, formed by either outcrops of the underlying block or by more resistant parts of the young deposits (commonly calcareous) found in this belt.

The Volga is separated from the Oka-Don Lowlands by the Volga Heights, whose southern extension into the Yergeni Hills also separates by a mere 50 miles the drainage of the Volga and the Don. The rolling uplands of the Volga Heights rise gently eastwards to 500-600 feet and in a few places to over 1,000 feet. They overlook the Volga along its right bank in a series of steep bluffs cut in places by old river terraces and dissected by tributary streams of the great river. At their northern end, the structural anticline of the Zhiguly Hills forms an obstacle to the river, around which it curves in the great Samara Bend, one of the few places along its course where there are high banks on both sides. Traces of an earlier channel to the west, between Ulyanovsk and Syzran, which was possibly occupied to the last inter-glacial period, may still be found. The left bank of the river is low and subject to extensive flooding, but eastwards the land rises gradually into the older horst block of the Pre-Uralian or Trans-Volga Plateau, a roughly dissected block — rising to over 1,300 feet. Extensive calcareous deposits, as elsewhere in the Central Russian Uplands, have produced in places karstic phenomena. Below Volgograd (Stalingrad), the Volga breaks into a braided channel across the Caspian Lowlands, while at the foot of the Yergeni Hills, the line of the Sarpinskiye lakes marks a probable former course of the river. Widespread in the lowlands are traces of a former greater extent of the Caspian Sea in the form of relict lakes, old shore lines, fossil shells, and characteristic lacustrine clays, sands, and silts. South of the Kuma-Manych Depression rises the higher, old block of the Stavropol Plateau and the foothills of the Great Caucasus.

On the north, Volga drainage is separated by the low watershed of the slight hills of the Severnyy Uvaly from rivers draining to the Arctic basin. These rivers drain across broad, undulating plains of glacial outwash provided by both Scandinavian ice from the west and Ural ice from the east. The open lowlands are broken by occasional low morainic hills, while the low rise of the roots of ancient Caledonian mountains of the Timan Range forms a watershed between N. Dvina-Mezen drainage and the Pechora, flowing across the bare plains overlying the Pechora syncline, rich in coal. These are forested lands in the south and along the interfluves, but there are also large areas of swamp, and in the north, along the low emergent Arctic coast line, there is inhospitable tundra affected by a mediocre development of *permafrost*. On the west, the barren rocky surfaces, strewn with lakes and swamp and poor forest, of the Karelián shield rise towards the Finnish frontier to heights of 1,200 feet in the north. The Kola Peninsula, whose coast line is marked by fault cliffs, contains rounded bare hills left as monadnocks on an old erosion surface, rising in places to about 4,000 feet.

The contemporary landscape of the greatest part of the plain is the product of successive advances and retreats of the Quaternary ice sheets.[5] The extent and nature of this modification is shown in Fig. 3 where three main belts may be distinguished: a northern zone of ice-ploughing, most marked in Karelia; a broad central belt of glacial deposition, with fresh, well-defined features in the north and more indistinct features to the south where fluvio-glacial materials have modified the landscape; and beyond the maximum southward limit of the ice lobes, where the fluvio-glacial but more particularly aeolian deposits modify the surface features. The great Scandinavian ice sheets carried the superficial deposits in Karelia far to the south, leaving a bare, rocky surface of the old crystalline shield rocks, smooth and rounded by the ice. Deep troughs were gouged along lines of weakness, leaving lake and swamp-filled hollows amid bare rock or sombre forest. Turbulent, youthful streams fall in rapids to a fjörd coast, where post-glacial changes in sea level are evidenced by broad, dry valleys, wave-cut platforms and immature river profiles. The southern limit is marked between Lake Ladoga and the White Sea by a double line of morainic hills, an eastward extension of the Finnish Sälpausselkä, one of the last stages in the final retreat of the Würm ice sheets. On the Kola Peninsula, low, rounded, glaciated, residual hills and glaciated laccoliths mark the passage of the ice sheets. The northern edges of the belt of glacial deposition are marked by the belt of morainic hills extending from Masuria across the borders of the Baltic republics to the Valday Hills, the materials left by the main southward advance of the Würm glaciation. They form an undulating countryside, with swampy hollows and many small lakes, while there are also numerous traces of former large glacial lakes and their shore lines, well seen in the country around

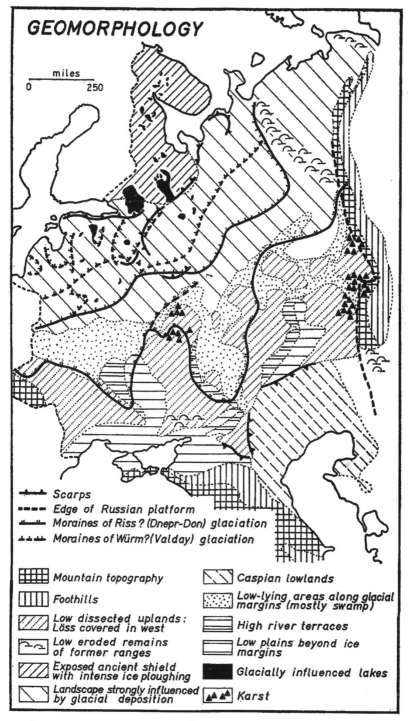

GEOMORPHOLOGY

miles
0   250

Scarps
Edge of Russian platform
Moraines of Riss? (Dnepr-Don) glaciation
Moraines of Würm?(Valday) glaciation

Mountain topography

Foothills

Low dissected uplands: Löss covered in west

Low eroded remains of former ranges

Exposed ancient shield with intense ice ploughing

Landscape strongly influenced by glacial deposition

Caspian lowlands

Low-lying areas along glacial margins (mostly swamp)

High river terraces

Low plains beyond ice margins

Glacially influenced lakes

Karst

FIG. 3

Geomorphology of European Russia. *Bolshoy Sovetskiy Atlas Mira*, Moscow, 1937, Vol. 1, pl. 100-101, augmented from Dobrynin, B. F., *Fizicheskaya Geografiya SSSR: Evropeyskaya Chast i Kavkaz*, Moscow, 1948.

13

Lake Ilmen. To the north, the wet Baltic Lowlands are lined by eskers and the moraines of former ice lobes, the largest and most distinct of which are the Peipus lobe, the Riga lobe, and the Klaypeda lobe. The main Baltic terminal moraine zone is formed by three principal parallel lines of low hills.

Immense quantities of fluvio-glacial outwash materials have greatly modified the glaciated area south and east of the limits of the Würm glaciation. It is thought that the maximum southward advance of the ice in the Riss period obliterated the earlier and less extensive Mindel advance. At the outer limits of maximum glaciation, the ice appears to have been thin and without much material in suspension, leaving poorly marked morainic belts, while the ease with which meltwater could escape led to the weak development of *urstromtäler*. The relatively thin tongues of ice extended in great lobes down the broad depressions between the interfluvial plateaus, which they did not manage to cover completely. The modern rivers have cut broad but low terraces into these fluvio-glacial outwash materials, well marked along the middle and upper Dnepr and in parts of the Polesye marshes, as well as in the Meshchera swamps and in the Vetluga and Rybinsk districts.

Across southern European Russia and into Siberia there is a broad belt of loessic materials which are usually considered to have been blown out of periglacial desert possibly at the onset of an arid interglacial period late in Würm times.[6] East of the Volga and on the Volga Heights, deposits possibly derived from loess are found, though they are stratified and appear to have been resorted by water. Aeolian deposits are also found in places on the plateau interfluves, while wind-blown sand has been deposited along the drier banks of steppe rivers. On many of these aeolian materials have developed rich and fertile soils, though overcultivation of the friable materials has resulted in extensive gullying. In the southern Russian landscape of narrow *ovragi* and broad *balki*, separated by featureless plain, yellow bluffs covered by dark soils stand above the bushes and marsh of the gully floors. East of the Don, uplift in post-Tertiary times of soft Jurassic and Cretaceous materials has started another cycle of erosion which has produced a similar gullied landscape.

In Tertiary and Quaternary times, a greatly extended Caspian Sea covered the Caspian Lowlands and has left a covering of characteristic clays, sands, and silts, besides fossil shells, old shore lines,

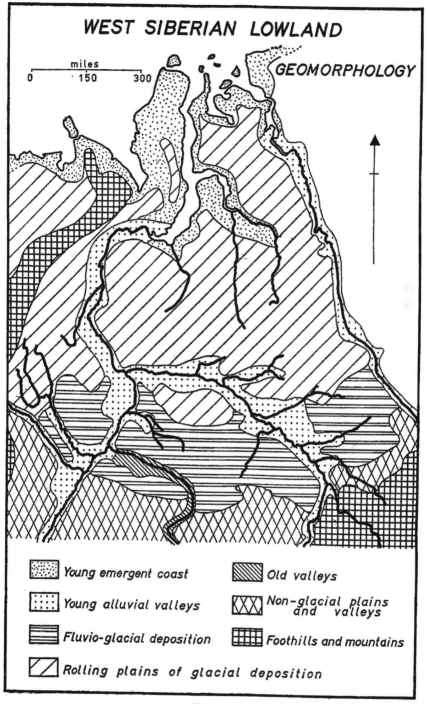

FIG. 4

Geomorphology of West Siberian Lowland. Suslov, S. P., *Fizicheskaya Geografiya SSSR: Aziatskaya Chast*, Moscow, 1956.

15

and relict lakes, while submergence along the Black Sea shore has left a typical submergent coast. In the north, changes in the relative levels of land and sea in post-glacial times have deposited marine clays on land and in parts blurred the pattern of glaciation.

The West Siberian Lowland, crossed by broad, meandering, sluggish, north-flowing rivers separated by low, marsh- and swamp-covered interfluves, is one of the more impenetrable and least developed regions and one of the most level and monotonous parts of the earth's surface. It extends some 1,250 miles from the Ural to the Yenisey and between 1,100 and 1,600 from north to south, yet maximum elevation does not exceed 400 feet and some interfluves are only 15 to 30 feet above the valley floors. Structurally, the lowland is a deep syncline filled by almost horizontal young sediments. In Quaternary times, ice sheets at their maximum extent reached south to almost 61° N., the ice converging from dispersal zones in the Ural and the Taymyr-Putoran areas. Except at their very maximum, when thin, slow-moving ice ponded back the north-flowing rivers, drainage appears to have found its way between the ice sheets north to the Arctic basin. Many relict lakes and traces of marginal waters along the one-time ice front remain in the southern part of the plain, while for a period drainage was south along the Tertiary Turgay Depression. The glacial history is imperfectly understood and appears to have differed in sequence from European Russia.[7] A marine transgression in late glacial or early post-glacial time has left typical features of submergence along the northern coast. Similarity of fish fauna suggests that the Ob and Yenisey once had a common mouth. A low east-west upland ridge, the Siberian Uvaly, at 63° N., has been suggested as a base for a dam to pond back drainage and convert the lowlands into a vast shallow sea, from which water could be drained south to irrigate Central Asia. In spring, when the upper reaches of the rivers in the south thaw, water flows north to the still frozen sections, flooding vast areas. Moreover the poor drainage from the low interfluves results in bog formation which gradually destroys the forest, well developed in the Vasyuganye swamps between the Ob and Irtysh rivers.

A broad Arctic Lowland, varying considerably in width and an eastward continuation of the West Siberian Lowland, is composed largely of Mesozoic sediments and deposits from post-Pliocene marine transgression. A southward extension may be considered to be the Mesozoic downfold of the middle and lower

Lena valley. The lowlands are also represented by the broad depression between the Byrranga Mountains on the Taymyr Peninsula and the Central Siberian Platform. This part of the lowland shows the clearest traces of the marine transgression in a series of terraces. The whole region is covered by a deep layer of *permafrost*. On the Khatanga estuary near Nordvik, salt domes with traces of petroleum have been found. The location is unfortunately one of the most inaccessible parts of the lowlands both by land and by sea, for the summer fogs and ice floes make this one of the most difficult parts of the Northern Sea Route to navigate. The lowlands are covered by open tundra and swamp.

Soviet Central Asia, apart from the southern and eastern mountain rim, is formed by the Turan Lowlands and the Kazakh Uplands, while between the depressions in which lie the Aral Sea and the Caspian Sea stands the Ust Urt Plateau. The lowlands are formed by almost horizontal Tertiary deposits covered by later materials, such as sand, loess, and clay, dropping towards the shallow central depression of the Aral Sea, to which drain the Amu Darya and Syr Darya. The area has had a long period of semi-arid and arid conditions to exercise a powerful influence on relief. Aeolian forces have produced large sand deserts, notably the Kyzyl Kum (Red Desert) and the Kara Kum (Grey Desert), probably from great quantities of material brought down by rivers during the glacial period. In the sandy deserts are found both stationary and moving dunes, with characteristic ridge dunes and *barkhans*. There are also clay deserts with scattered, flat-topped, steep-sided clay hills and *takyrs*. Stony desert is limited in extent. Around the foot of the southern rim of mountains, broad loessic piedmont plains with a general elevation between 300-900 feet stretch into the Turan Lowlands. On these, fertile grey soils have developed that produce luxuriant crops when irrigated. Between the Aral Sea and the Caspian Sea lies the wide calcareous plateau of the Ust Urt with a general elevation of 650 feet, on whose Tertiary horizontal sediments karstic phenomena have developed. The Ust Urt is devoid of surface water, though water may be obtained by digging wells.[8] The edges of the plateau are marked by a broad depression, formerly a bed of the Caspian Sea, from the older structures of the Mangyshlak Peninsula, while on the south, the gorge of the former Uzboy water course separates it from the lower and less well defined Ungus Plateau. The well-preserved channel of the Uzboy is characteristic of many similar traces of the drainage of the Turan Lowlands in more moist periods. It originates in the ancient deltaic area of the Sary-kamysh Depression, once watered by the Amu Darya, and flows into the former bed of the Caspian near the Balkhan Mountains. In exceptionally wet years, these ancient channels sometimes carry water, brought to them by contemporary rivers which usually do not flow far enough to fill them. It is unlikely that they have been regularly filled since early post-glacial times, though many classical writers and even mediaeval Persian and Turkic records suggest that the Uzboy was then still navigable. Water is the key to the modern development of this region. Ground-water is frequently brackish or saline, particularly where obtained from deep wells. The depth at which water may be encountered varies: in the Ust Urt Plateau it is found at about nine feet below sands and somewhat deeper elsewhere. Rivers draining from the melting snows and glaciers of the mountains are usually heavily charged with silt and mud, and also frequently carry much salt in solution. The rivers build up their beds between natural levees. Maximum flow is in summer, when a great deal of moisture is lost by evaporation even before it is sucked into the dry desert sands. Irrigation also draws off large and increasing quantities, for between 25-30 per cent of the rivers' water is so used. A belt of delta oases line the piedmont zone, while oasis development is also found in the desert along rivers and at groups of wells or wherever the water table is unusually high. To judge from ancient remains, irrigation was formerly more widely practised, for in many places underground reservoirs of great age are found. Several large irrigation schemes have been started in Soviet times, but these need careful control since salts detrimental to plant growth may be drawn to the surface.

The Kazakh Uplands lie between the West Siberian Lowland and the Turan Lowlands. On the west, they are separated by the Turgay Depression from the Uralian structures, and on the east they rise into the fringing mountain rim. These uplands consist of low, disconnected Palaeozoic massifs, dislocated and broken during Hercynian and later times, but also containing elements of earlier Caledonid structures. Today, they reveal complicated basal mountain structures, rich in valuable minerals. These roots of former high ranges now lie between 3,000 feet in the west and 4,800 feet around Karkaralinsk. The surface is dotted with low residual hills (*melkosopochnik*). The more arid south-eastern part contains tectonic basins, in which lie Lake Balkhash, the barren sands of the Semirechye and several other lakes. In the north-west, there is also the basin in which Lakes Tengiz and Kurgaldzhin lie. The south-western edge of the uplands is covered by the monotonous and inhospitable semi-desert and desert of the Bet-Pak-Dala, appropriately named the 'Plain of Misfortune'.

These Central Asian lands are distinguished by their unusual climatic regime of 'Egyptian summers and Siberian winters'.

The Central Siberian Uplands are an immense raised massif of Pre-Cambrian rocks resting on an ancient platform of Angara-land. The general elevation is between 900 and 1,200 feet, but the mass of flat-topped uplands has been deeply dissected to give the appearance to the traveller of rough mountain country. The ancient basal rocks appear in the north in the Anabar shield and in the south in the Aldan shield; elsewhere the surface is formed by slightly disturbed marine deposits of Cambrian to Silurian age. There are also some Carboniferous and Permian deposits with young Jurassic materials on the east, all the result of marine transgressions. Between Carboniferous and Jurassic times, extensive basalts spread over large areas, transforming some coals into valuable graphite deposits. These extrusive materials are often weathered into steep cliffs. The uplands are also rich in other minerals. There was never an ice-cap covering the whole of this vast area, though valley glaciers and small icecaps were formed on the higher parts, particularly in the moister west. In contrast to the sluggish rivers of the great plains, the rivers of the massif, like the Angara, the lower and stony Tunguska, and the upper Vilyuy, flow swiftly across rapids and through gorges, and are thus a great source of potential hydro-electric power. Economic development of the great riches of this as yet little-known region has only recently begun. Minerals are worked in the south at several sites and in the north at the foot of the Putoran Mountains near Norilsk. The rich stands of timber in the south are also beginning to be utilized.

Between the Russian Plain and the West Siberian Lowland lie the Ural Mountains.[9] A series of almost parallel ranges extend for 1,500 miles along the longitude 60° E., from the Arctic coast to the Ural river. Tectonically they are associated with the Arctic Vaygach Island and Novaya Zemlya, while in the south, the low Mugodzhar Hills form a structural extension. The general

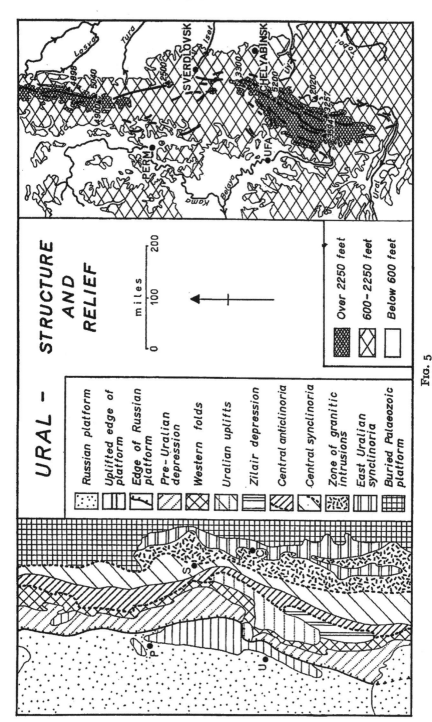

FIG. 5

Structure and relief of the Ural. *Bolshoy Savetskiy Atlas Mira*, Moscow, 1937, Vol. 1, pl. 104-105, and Komar, I. V., *Ural*, Moscow, 1959.

altitude of the Ural is between 750 feet and 3,000 feet, though some of the highest points rise to over 5,000 feet. The system is about 50 miles wide in the north, but it is much narrower and lower in the centre, while the southern ranges spread out to 140 miles. Geologically the mountains are complex. Palaeozoic sediments predominate, but metamorphic and igneous materials of various ages are common along the central axis, as in the Ural Tau. To the east, young Tertiary marine deposits associated with the West Siberian Lowland are found. The western flank is composed largely of Permian rocks. The folding of the Ural came with the Hercynian earth movements at the end of the Carboniferous period. By the Tertiary the ranges had been peneplaned, but the late Tertiary earth movements started a new cycle of erosion after uplift. The structure is also affected by Mesozoic faults and thrusts. As the roots of ancient mountains, the ranges abound in minerals, which form the basis of varied heavy industry.

The series of low and almost parallel ranges are asymmetrical in cross-section, rising gently from the western plains towards the watershed ranges, beyond which there is an abrupt drop to the rolling country which gradually dies away eastwards into the West Siberian Lowland. The lower central section sweeps in a broad crescentic bow around the Ufa horst on the western face. The western face, structurally a broad piedmont depression, consists of sedimentary Palaeozoic rocks with some quartzites (possibly Silurian); there are also lower Carboniferous coals (Kizel), salts, and bauxite. The range rises gently from the transition zone between the western tabular horst blocks to the belt of Hercynian folds through a belt of foothills. Eastwards the intensity of folding increases. A belt of crystalline schists forms the highest elevations and acts as the watershed. These schists in the northern and southern ranges impart a rougher topography, but they are poorly represented in the central part of the system, where heights seldom exceed 1,300 feet. The centre section is distinguished by the subdued relief and easy routes across it. East of the watershed, a thick band of intrusive rocks forms an important mineral zone, where there are several important industrial towns. In the north, the belt consists of gabbro, underlain by platinum-bearing dunites. Again, this belt is poorly represented in the centre section. Further east still, there is a broad zone of igneous rocks and tuffs, with some schists dating from the upper Silurian to lower Carboniferous, which drops gently below the Ob plain. The granite and gneissic masses have been localized points of intense mineralization. The Ural is a landscape of broad, open, mature forms.

The northern Ural, north of 61° N., contains the highest point of the whole range, Mount Narodnaya (6,185 feet). The system here consists of two parallel ranges cut by transverse valleys. The eastern range is the watershed. It narrows towards the north, where the Arctic Ural forms a single range (15-20 miles) with a general elevation of about 2,000 feet. The range, known as the Pay Khoy, is broken and ill-defined and drops abruptly to the barren tundra. Several stages in denudation are marked by remnants of old peneplain surfaces in the form of wide mountain benches, most pronounced at 2,400 feet and 3,000 feet. There are many traces of an extensive Quaternary glaciation, while contemporary glaciers of Mount Sablya and Mount Narodnaya lie in deep cirques, but the longest is only about three-quarters of a mile. The mountains are covered by barren tundra, though south of the Arctic Circle the lower slopes of low, rolling hills have thin, coniferous forests.

The southern Ural extends from Mount Yurma to the middle reaches of the Ural river, beyond which the ranges die away into the low, rounded Mugodzhar Hills. The ranges here are about 100 miles wide, comprising several parallel chains. The eastern range, the 3,000-feet-high Ural Tau, is the watershed. The highest point is the Yaman Tau (5,432 feet), one of the few points to rise above the snow line. Remains of the old peneplain are found but are less pronounced than in the north. South of the Belaya river, the Ural Tau and other ranges become a deeply-dissected, rolling high plain. To the south and east, a vast, rolling plain of Mesozoic denudation is broken by occasional residual hills, such as Mount Magnitnaya, a rich source of iron ore. Low hillocks and cuestas form typical *melkosopochnik*. The rolling, hilly country carries forest south into steppe country.

Elevations in the central Ural are low. The highest point, Konzhakovskiy Kamen, reaches little over 4,500 feet. The ill-defined watershed lies on the Poyasovoy Kamen ridge, though the west-flowing Chusovaya river has its source on the eastern slope. Remains of old denudation surfaces lie at 600 feet and 1,200 feet, but the low hills have once again been reduced almost to a peneplain. The whole area focuses on Sverdlovsk on the eastern slope, which lies in gently rolling wooded country, where forested hills sometimes contain granite tors.

The remaining mountain systems of the Soviet Union may be divided into three main groups. The largest consists of the mountains which border the southern fringe of the Central Siberian Uplands and extend across to the Pacific coast and also cover most of North-east Siberia. The systems which belong to this group originated during several orogenic periods. The second group comprises the complicated mountain systems, chiefly of Hercynian origin but much affected by Tertiary earth movements, in Central Asia, forming some of the highest ranges in the country; while to their west lie the alpine ranges of the Kopet Dag, the Caucasus, and the southern European Russian ranges in the Carpathians and in the Crimea. With the exception of the third, alpine group, these ranges are still imperfectly understood. Even the topographic features of the ranges of North-east Siberia are still being mapped: several major discoveries have been made during the late 1920's and some features plotted since 1945. It is also since 1945 that the first reconnaissance geological survey of much of these ranges has been completed.

Fig. 6 shows the complicated mountain systems of Southern Siberia where economic development is accelerating the detailed scientific knowledge of these ranges. One of the best-known systems is the Altay, which may be subdivided into the South Altay, a westwards extension of the Tabyn-Bogdo-Ola massif; the Inner Altay; the East Altay, which separates Ob and Yenisey drainage; and the Mongolian Altay lying mostly beyond the Soviet borders. The South Altay drops gently westwards from about 9,500 feet to 4,500 feet. The ranges are not heavily dissected and are more like broad plateaus. The northern slopes are steep but drop gently southwards into Central Asia. Passes across the ranges are difficult. The Inner Altay rises to its maximum elevation in the Katyn Belki, where the twin peaks of Mount Belukha rise to 15,154 feet in the east and to 14,563 feet in the west. The mountain carries a perennial snow cover, but it is rarely seen because it is nearly always surrounded by clouds. The ranges in this part of the Altay are separated by deep and almost impassable river gorges. The more northerly parts radiate out as low ridges (about 6,000 feet) lying everywhere below the timber line. The Eastern Altay comprises the ranges which

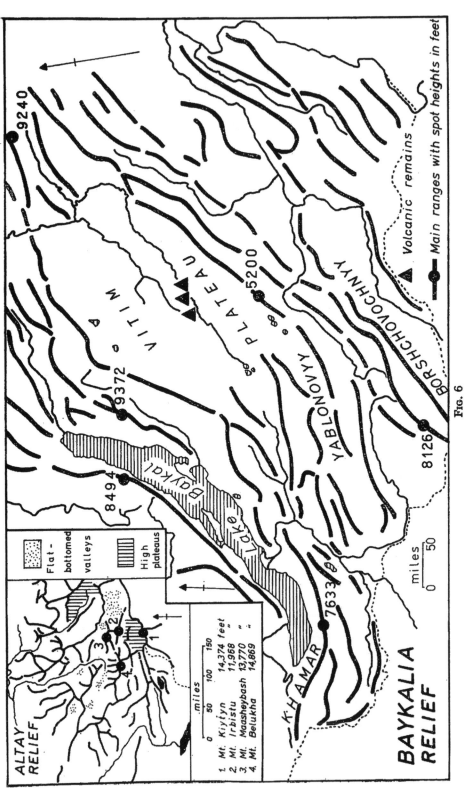

Relief of Southern Siberia. *Atlas Mira*, Moscow, 1954, augmented from Berg, L. S., *The Natural Regions of the U.S.S.R.*, New York, 1950, and Suslov, S.P. *Fizicheskaya Geografiya SSSR etc.*, Moscow, 1956.

Fig. 6

separate the Ob and Yenisey drainage. In the south, the Say-lyugem Range forms the international frontier. The ranges rise to over 9,000 feet and across them the Shapsal Pass forms a major routeway between Russia and Western Mongolia. Northwards, the Eastern Altay extends into the Gorbu and Abakan mountains, and includes the long, deep, picturesque Teletsk lake, partly tectonic and partly glacial in origin. The Eastern Altay is marked by level summits, gentle slopes, and extensive rolling plateaus, which lie between 5,000 and 8,000 feet. They resemble the high desert plateaus well-developed in Mongolia. Older maps of this part of the mountains are frequently incorrect as it is only during the later interwar years that detailed survey was made.

The Altay system, belonging mainly to the Caledonian orogeny, consists of ancient rocks with extensive granite intrusions. In the south-west, however, in the Rudnyy Altay and in the Kalbinskiy and Tarbagatay Ranges, folding took place in the Variscan period of the Hercynian orogeny. A long, quiescent, continental period resulted in peneplanation, and was followed, late in the Tertiary period and lasting into early Quaternary times, by dislocation by faulting, trending roughly east-west, creating the present mountain scenery of great plateau blocks at different elevations and producing many structural basins. The contemporary relief comprises deeply-eroded plateaus with traces of recognizable former erosion surfaces; many of the benches are so flat that bogs have formed on their surface and the streams meander across them, to fall over their precipitous edges in torrents and waterfalls. The tectonic basins are usually wide and flat-bottomed and have a dry continental climate which led early Russian settlers to call them 'steppe'. The erosional valleys are much narrower, including the impassable valley of the lower Argut. Below the old erosion surfaces extends a broad zone of rolling foot-hill relief, while above them stand higher mountains, with traces of Quaternary mountain glaciation on the Katyn and Chuya Belki.

A probable northern continuation of the Altay is formed by the Kuznetsk Basin, bordered on the west by the Salair Range and on the east by the Kuznetsk Ala Tau. The latter is an irregular mass of ancient horsts rising to over 3,000 feet, dying away gradually to the north into gently rolling plains. The Kuznetsk Basin (the Kuzbass), drained northwards by the Tom river, has an undulating relief which seldom exceeds 300 feet between the valley floors and the interfluvial crests. Its lowest point lies a little over 300 feet above sea level. The basin owes its formation to subsidence during lower Palaeozoic times: during the Devonian this was an arm of the sea, but in the lower Carboniferous the connection with the open sea closed, changing the basin into a lake around whose shores luxuriant deltaic forests provided the material from which the immense coal deposits originate. Iron ore in the Gornaya Shoriya country and excellent coking coals of the Basin form the basis of Western Siberian heavy industry.

Because of slower economic development and delayed geological survey, the Sayan is less known than the Altay, because until 1944, it lay mostly within the then nominally independent Tannu Tuva. The system forms a broad, south-facing arc between the Abakan river and the Baykalian Khamar Daban mountains. The Western Sayan, extending to 97° E., and composed almost entirely of granite and dislocated crystalline schists, rises to 8,500 feet in the west. The main ranges lie between the Bolshoy Abakan river and the Bolshoy Rapids, where the Yenisey cuts through the mountains in a deep gorge less than 200 feet wide. East of the Yenisey, the ranges are even more imperfectly known. Many parts of the Western Sayan rise above the tree line, and the crest which forms a dissected plateau has bald, rounded summits, although there is a rough, jagged crestline in the Erik Torgok Tayga. No glaciation is present, but traces of former ice action are to be found. The Eastern Sayan extends from the confluence of the rivers Mana and Yenisey to the south of Krasnoyarsk eastwards towards the Irkut gorge between Lake Baykal and Kobso Gol. The mountains are generally higher than in the west, rising to almost 10,500 feet in Mount Munku-Sardyk, from which several glaciers descend. Except where they have been eroded by ice, the mountains have rather flat summits above which rise bare dome-like peaks (*goltsy*). The formation is largely crystalline schists and limestones possibly dating from the Cambrian period, which is represented by other sedimentary deposits, including soft shales. Movement with volcanicity seems to have continued until recent geological times, for basalts are found overlying Tertiary terrestrial deposits. Extensive rockfields (*kurumy*) caused by mechanical weathering of uncertain age impart a characteristically barren aspect in many parts of the Sayan. The tectonic history of the Sayan is considered to be similar to that of the Altay.

The Minusinsk Basin, lying between the Abakan Range and the Kuznetsk Ala Tau, to the north of the Sayan, which drops to it in a steep scarp face, has a cover of loess and loess-like clays and loams, on which a rich steppe has developed, sometimes called the Granary of Siberia. It also contains lignite, coal, and iron ore. South of the concave face of the Sayan lies the Tuva Basin in the upper reaches of the Yenisey. It is a high plateau in the east, but the western and central parts are a broad depression, lying between 1,500 and 1,800 feet, much under agriculture. South of the Tannu Ola, the Soviet border crosses another tectonic depression occupied by the Ubsu Nur lying in Mongolian territory.

The structure of South-eastern Siberia is particularly complicated. Generally it consists of wide watershed plateaus dissected into dome-shaped peaks and rounded ridges. The ranges are composed largely of dislocated massive crystalline materials with a roughly east-north-east-west-south-west orientation. Of the many tectonic depressions, the best developed is the Baykal Trough, around which earthquakes are frequent and sometimes destructive. Igneous rocks are widespread, with many traces of relatively recent volcanic activity, such as cones, craters, and lava flows, notably in the Vitim Plateau. There is little agreement on the structural history, but the sequence of events seems perhaps similar to the Altay and Sayan, though the original orogeny (the Baykalian) ended by lower Palaeozoic times. The four following subdivisions may be distinguished:

1. Cisbaykalia forms a narrow belt of country along the western shore of Lake Baykal, composed of a series of nearly parallel ranges with broad, flat summits which rise above the tree line in parts. The country rises steeply from the lake shore and forms a narrow watershed between Lena and Angara drainage. The outstanding feature is the vast tectonic *graben* of Lake Baykal.

2. Western Transbaykalia is composed of lower Palaeozoic folds in Pre-Cambrian rocks, with extensive areas of granite and other igneous rocks. The south-western part is formed by four parallel ranges, the uplifted blocks of an ancient peneplain with gentle southern but steep northern faces. In several places they rise above the tree line in bare *goltsy*. The Khamar Daban Range, rising to over 6,500 feet, forms an imposing wall along the south-eastern shore of Lake Baykal. The Barguzin Range, which rises to over 8,000 feet, is more jagged, with snow-clad peaks making

an impressive eastern shore to the lake. The eastern limit of this region is set by the supposed structural anticlinorium of the Yablonovyy Range. The south-western part of the range serves as the watershed between arctic (Lena) and pacific (Amur) drainage. The average height is about 4,500 feet but it rises to over 5,000 feet in places, and there are again many bare *goltsy*; but the general impression is of broad, forested summits with much swamp. The range rises only about 1,000 feet above the valley floors. North-west of the Yablonovyy Range lies the rolling but monotonous larch forest and bog of the Vitim Plateau, with a general elevation of between 3,000 and 4,000 feet. Remains of some recently extinct volcanoes have been reported from this area. It is known to contain gold and other minerals.

3. Eastern Transbaykalia has a more varied geological composition. It is chiefly an area of Mesozoic folding, running north-east-south-west, tilting towards the north-east. Glaciation has modified relief. It is a country of deeply eroded ranges with narrow river gorges. The south-eastern part, along the Soviet Mongolian frontier, is more arid, with poor forest and even open grassland.

4. Northern Transbaykalia, the Olekma-Stanovoy region, is still only superficially explored. The western Olekma-North Baykal Highlands are an elevated peneplain, deeply dissected, with relief accentuated by Quaternary glaciation. The incompletely known Stanovoy Range forms the eastern part. Formerly it was thought to extend from Baykalia to the Bering Strait, forming a gigantic watershed, and appeared as such on many older atlas maps. It is now known that the eastern end of the range is in the Maya-Ud valleys. The Stanovoy itself is a system of plateau-like highlands bordered on the north by higher ranges: northern summits reach 8,000 feet, but on the south they seldom exceed 4,500 feet. The southern range is a series of open igneous *goltsy*.

The separate Dzhugdzhur Range forms the coastal mountain belt along the Okhotsk Sea. The ranges are formed extensively of Mesozoic extrusives. They are steep, rugged mountains, falling sharply to the coast, but fortunately, many low passes cross them. Associated with them are the mountainous Shantar Islands.

The Far East may be distinguished as an unglaciated region of widespread Mesozoic formations, though Hercynian structures — a continuation of those in Manchuria — are found in the lowlands of the Zeya and Bureya valleys. The region has a modified monsoon climate and broad-leaved forests, which cover the mountains with a rich vegetation. On the north, the region is marked by the Tukuringra Range, while the Amur valley is separated from the Bureya by the Bureya Mountains, and parallel with the coast of the Gulf of Tartary lies the Sikhote Alin, associated possibly with the edge of a submerged horst. The Sikhote Alin, formed of several parallel chains cut by *cluses* dating from the Mesozoic folding on the west and from the Kainozoic Pacific orogeny, are low, wooded, rolling mountains, with a general elevation between 2,000-3,000 feet, but summits in the south rise to 6,000 feet. There are also extensive low plains along the main river valleys, and in the south-west include the broad valley of the Ussuri and around Lake Khanka. The Bureya Mountains contain iron and coal of Jurassic-Cretaceous origin. Structurally associated with the outer ranges of the Sikhote Alin and separated by the broad depression of the Gulf of Tartary, Sakhalin is a mountainous island formed of two parallel chains with a central depression — a fault-type valley. Coal and oil of Cretaceous-Tertiary origin are found. The heavily-forested island is known for its dull misty summers.

North-eastern Siberia is the most imperfectly explored part of the Soviet Union. As late as 1926, the topographical map of the region was recast by the discoveries of the Obruchev expedition. The region is distinguished by its Mesozoic and Kainozoic foldings, which in some parts have uplifted whole blocks of country as 'median masses' (*zwischengebirge*). It is thought that the later stages of the Tertiary earth movements created these plateaus. Post-Tertiary movement is suggested by well-developed terraces at 1,000 and 1,500 feet. Glaciation has greatly modified relief and several ranges were formerly covered by icecaps. The mountain ranges form a great north-facing crescent, with the western and eastern flanks higher than the southern and more broken part, within which, dropping to the Arctic Ocean, is the large median mass forming the Yukagir and Alazey Plateaus and the Kolyma Plain, and possibly continuing northwards under the shallow, epicontinental Chukchi Sea. These structures are separated from the Kainozoic folds of Kamchatka, with its volcanoes, by the broad depression of the Parapolskiy Dol. The Kamchatkan folds and volcanic belt are continued south in the Kuril Islands, together forming what has sometimes been described as the 'primary' Pacific coast. The country in the north is barren tundra, swamp, and poor forest, with short summers and intensely cold winters, and everywhere *permafrost* is well developed. Kamchatka, with its richer vegetation, has a cold raw climate.

Soviet Central Asia is ringed on the south and east by mountains. The southern mountains, the Kopet Dag, the Pamir, the Paropamiz, and the ranges of the southern Tyan Shan, belong to the Tertiary orogeny, to which belong also the mountains of southern European Russia, and the Himalaya as well. The central ranges of the Tyan Shan, their westwards continuation in the Zeravshan, Turkestan, and Fergana Ranges, as well as the Alay, the Dzungarian Ala-Tau, and the Tarbagatay, consist principally of modified Hercynian structures. They continue eastwards in the Kwen Lun. The northern Tyan Shan may even be Caledonian. Also included in these mountains are the Mesozoic structures along the Caspian shore, standing out of the Central Asian lowlands, such as the Mangyshlak Mountains, the Krasnovodsk Plateau, and the Great Balkhan. The older mountains show considerable dislocation from the Tertiary earth movements, which were responsible for the formation of the many tectonic basins now filled by plains or lakes, such as the Fergana valley, the Balkhash-Ili Depression, Issyk-Kul and Zaysan Depression. The ranges form a majestic series of high peaks and plateaus, in which the altitudinal sequence of vegetation has contributed to the characteristic nomadic way of life. The ranges show many traces of glaciation; and in Quaternary times the snow line, now standing at 11,000 feet, descended several hundred feet below the contemporary level. Two major glaciers still exist: the Fedchenko, almost 50 miles long, and the Inilchik. Their lower parts are covered by a thick mantle of rock debris produced by intense rock fracturing under a wide diurnal temperature range. The glaciers are a source of water to the rivers, the life-blood of the arid plains. These ranges are crossed by many ancient routes which follow the tectonic basins and cross the cols by high passes, leading into China and India. The historic routeway of the Dzungarian Gates has been a key to the history of Northern Asia.

The ranges of the alpine orogeny may be divided into an eastern group of the High Pamir and Badakhshan and the western Kopet Dag, which forms the southern boundary of the Soviet Union with Persia. East of Tedzhen, it merges into the Paropamiz, of which only the foothills lie within Soviet territory.

East of the Amur Darya, the country rises eastwards to merge with the majestic Peter the First Range and the Academy of Sciences Range. Here lie the highest peaks in the Soviet Union. Pamir is the name of the mountain country which lies between the Trans-Alay on the north and Hindu Kush on the south. It forms a lofty desert plateau at 12,000-13,000 feet, crossed by broad valleys. Above the plateau, peaks (many still unclimbed) rise to 16,000-18,000 feet. The western Pamir is intensely dissected by deeply-eroded narrow river gorges. Along some of the gorges, narrow wooden platforms cling to the almost vertical sides high above treacherous waters. Described by Marco Polo, they are still in use. In the broader valleys, there are well-marked terraces used for settlement and some agriculture in an otherwise arid topography. Earthquakes are particularly common in the western Pamir. In 1911, an earthquake was associated with an immense landslide in the Bartang valley, holding back the river to form the deep Lake Sarez. This same earthquake also destroyed Alma Ata, hundreds of miles further north. The Gorno-Badakhshan district of the Pamir forms a southern projection of the Soviet Union towards former British territory in India, from which it is separated by the Wakhan strip of Afghan territory, erected as a buffer between the two powerful imperial powers in the years before the First World War.

The mountains of Caucasia may be subdivided as in Fig. 7. The high and almost parallel ranges of the Great Caucasus nearly all exceed 6,000 feet. The core of the mountains is resistant Pre-Cambrian granites and gneiss. To the north, the ranges drop away into lower Cretaceous country. At the foot of the main ranges lies the long North Caucasian depression, a broad geosyncline, disrupted in the central section by a flexure associated with the Yergeni Hills structures, so that the basins of the Kuban and the Terek are separated. To the north of this depression are uplands of the Stavropol Plateau and associated structures. To the south, the main ranges also drop away to Cretaceous and Jurassic materials. The cross-section is generally asymmetrical with steeper slopes on the south side. The present ranges date from the Tertiary orogeny, having undergone an early period of folding in Miocene times followed by erosion under the second uplift and folding in the Pliocene. With the Tertiary earth movements, there was considerable volcanic activity, seen in the great laccolith of Beshtau near Pyatigorsk and such extinct

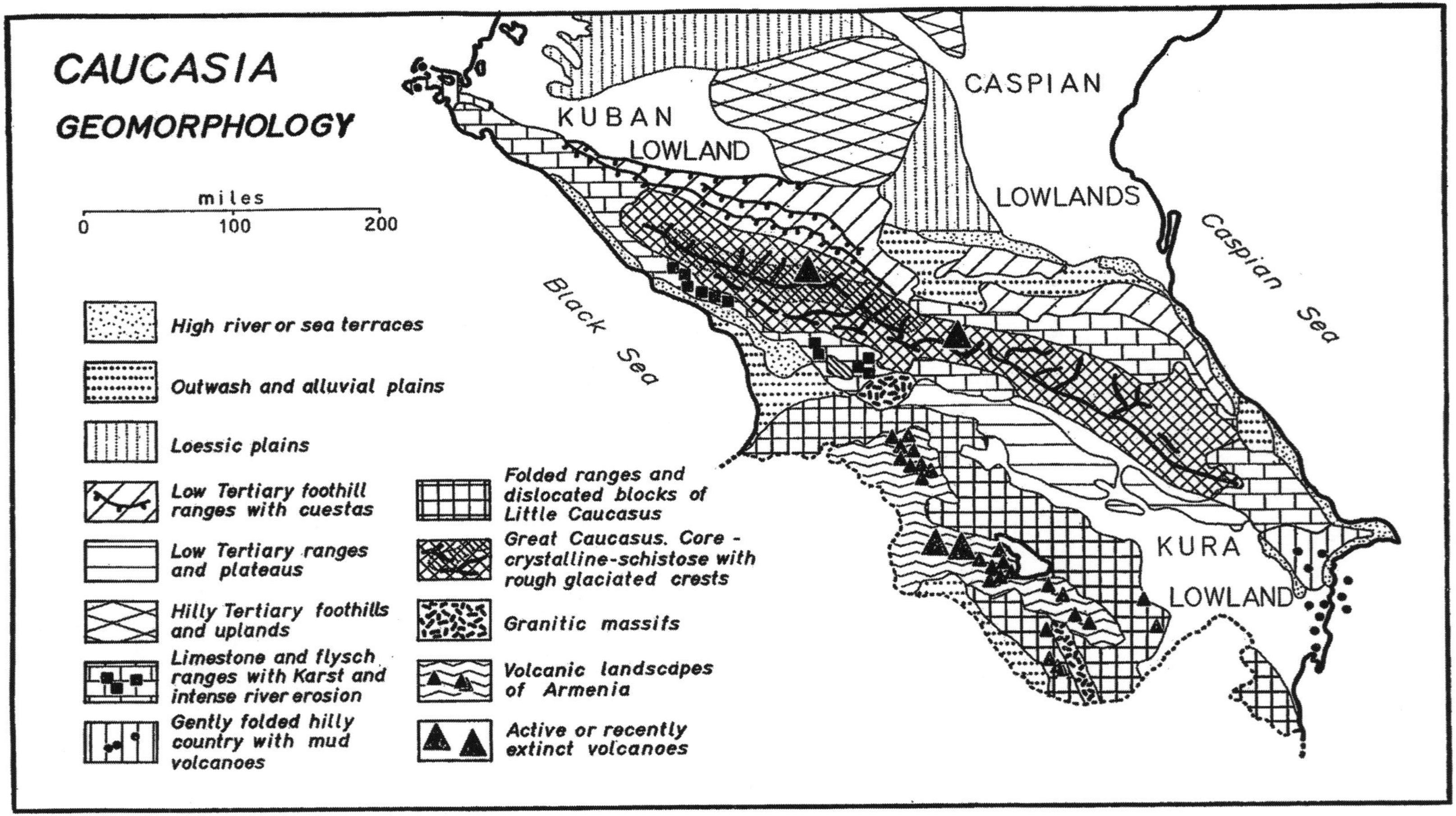

CAUCASIA
GEOMORPHOLOGY

miles
0   100   200

CASPIAN

LOWLANDS

Caspian Sea

KURA LOWLAND

KUBAN LOWLAND

Black Sea

High river or sea terraces

Outwash and alluvial plains

Loessic plains

Low Tertiary foothill ranges with cuestas

Low Tertiary ranges and plateaus

Hilly Tertiary foothills and uplands

Limestone and flysch ranges with Karst and intense river erosion

Gently folded hilly country with mud volcanoes

Folded ranges and dislocated blocks of Little Caucasus

Great Caucasus. Core – crystalline-schistose with rough glaciated crests

Granitic massifs

Volcanic landscapes of Armenia

Active or recently extinct volcanoes

FIG. 7

Geomorphology of Caucasia. *Bolshoy Sovetskiy Atlas Mira*, Moscow, 1937, Vol. 1, pl. 102-103.

volcanic peaks as Mount Elbrus, which remained active into upper Pliocene times. In some places, traces of Quaternary glaciation have been covered by younger lavas. In these mountains, the glaciation was less developed than in the Alps: the over-deepening of valleys and presence of glacial lakes is much less marked than in the Alps. There remains the Dikh Su glacier, about 9 miles long. The present snow line rises from about 8,000 feet in the west to 12,000 feet in the drier east. River terraces are well developed in many valleys; on the Black Sea coast there are also marine abrasion terraces high above present sea level, but little is known of their history. The north-east part of the Great Caucasus drops into the lower dissected mountain country of Dagestan along the western shore of the Caspian.

A broad depression separates the Great Caucasus from the Little Caucasus. In the west, this forms the Colchis (Kolkhiz) Lowlands, a humid coastal plain drained by the Inguri and Rioni. It is covered by late Tertiary materials and is known for its red, sub-tropical soils which carry a remarkably luxuriant vegetation. On the east, the larger Kura-Araks Lowlands include much steppe and semi-desert, which several large irrigation schemes have made agriculturally valuable. The lower reaches of the Kura are, however, unhealthy swamp. The two lowlands are separated by a higher depression, itself broken by the sweep of the Suram Range, a granitic massif, which links the Great Caucasus to the Little Caucasus and is crossed by the difficult Suram Pass (2,700 feet). At the western end of the Suram Pass stands Kutaisi and at the eastern end, Tbilisi. The Little Caucasus ranges seldom rise above 7,500 feet even in the highest peaks and are generally composed of younger rocks than the Great Caucasus. To the south, the Armenian Plateau is cut by the Soviet border along the Araks and Akhuryan rivers. The plateau has a general elevation of about 3,000 feet, and is composed of Tertiary and post-Tertiary volcanic materials. It contains, as in the Little Caucasus, many down-faulted troughs containing lakes, like Lake Sevan, or rivers (*e.g.*, the middle Araks). Mountains rise out of this steppe plateau, forming partly enclosed basins. The highest point, the 16,000-feet high twin peaks of Mount Ararat lie just across the Turkish boundary. The luxuri-ant, subtropical conditions of the Black Sea coast attract many tourists, and invalids visit spas such as Sukhumi, Sochi, and

Adler, while the mineral waters of Mineralnye Vody and Pyatigorsk offer health-giving properties. In many Soviet hotels, Yesentuki and Borzhomi mineral water is commonly offered. The Caucasian mountains also hold rich mineral resources, and the anticlinal structures of the piedmont depressions contain petroleum.

Although the Crimea forms a southern and almost detached portion of the steppes of the Black Sea Lowlands, the southern coast is distinguished by three parallel ranges in the west and two in the east. The southernmost is a high, dry, rolling mountain plain with pronounced karstic phenomena, known by the Tatar name of *Yaila* (summer pasture), lying at about 3,000 feet and about two or three miles wide. The southern slope drops steeply to the Black Sea, but the north falls more gently to the steppe. Along the southern coast, the combination of luxuriant vegetation, a modified mediterranean climate, and the blue sea has turned small old Tatar towns into holiday resorts. Yalta, the best known, is reached through a deep fault valley, the Baider Gate, which may be the epicentre of occasional earthquakes in this region. The sheltered valleys of the northern slope are important fruit and tobacco growers: richest is the valley around the old Tatar khanate capital of Bakhchisarai, the 'Palace of Gardens'.

Since the incorporation in 1945 of the extreme eastern districts of Czechoslovakia, the Soviet Union has included a section of the Carpathian system, the low Forest Carpathians and a marginal foothold in Danubian Plains on the vast outwash fans of the upper Tisza and its tributaries. The mountains are low and monotonous forested ridges, well below 5,000 feet, and are crossed by several low passes, like the Uzhotskiy, Veretskiy, and Tatars' Pass. The northern flysch ranges are bordered to the south by young volcanics in low, rounded hills.

The Soviet Arctic Islands

The shift of military interest to the Arctic Ocean since the end of the Second World War has greatly increased the significance of these frozen outliers of Soviet territory. The westernmost is Franz Josef Land, a group of about 800 islands lying between 79° N. and 82° N. and 42° E. and 65° E., discovered in 1873 by the Austrian Weypracht expedition and incorporated into the Soviet Union in 1926. Most of the islands are Jurassic and

Cretaceous tablelands about 3,000 feet high, covered extensively by more recent basalts. They descend abruptly to the sea and are almost completely covered by ice, though there are patches of bare tundra. They contain some lignite, but their importance is more military than economic since they guard the water approaches to western Soviet arctic waters. Novaya Zemlya consists of two large and several smaller islands. The main islands are separated by the important narrow navigable deep channel of Matochkin Shar. It represents a northern structural continuation of the Ural, with a maximum elevation of about 3,000 feet, though the islands show that they are still undergoing uplift. There is extensive glaciation, particularly in the northern island, with poor tundra vegetation elsewhere. Known since the sixteenth century, the islands were first reported by the explorer Willoughby. Nearer the Soviet coast lies the structurally associated Vaygach Island, rich in minerals, and the low Quaternary Kolguyev Island, which does not rise above 300 feet. Both are covered by poor tundra. Severnaya Zemlya, discovered in 1913 by Vilkitskiy, consists of four large and many small islands. Structurally it appears to be associated with the Byrranga Mountains of the Taymyr Peninsula from which it has been separated in postglacial times. The islands have a generally subdued relief; but as a result of uplift, the coast is usually a well-defined, raised beach backed by cliffs. Ice covers about half the 14,200 square miles of the group. Elsewhere there are mosses and lichens. The New Siberian Islands, an archipelago of over eleven islands, structurally part of the Central Siberian Plateau, was discovered in part in 1881 by the American explorer De Long. The islands have a complex but incompletely known composition. Feddeyev Island and Kotelnyy Island have between them the low raised beach of Bunge Land. Fossil ice containing Quaternary fauna, notably fossil mammoth and its ivory, is covered by later deposits. Wrangel Island and the bare rock of Herald Island were first reported in 1823, but they were explored and confirmed by Kellett in 1849. After an unsuccessful Canadian attempt to claim them, they were declared Soviet territory in 1926. Wrangel Island rises to about 2,500 feet. Warmer than average for its latitude, it has no glaciers. It is the site of a meteorological station. Herald Island is a single bare rock rising to 1,100 feet.

Rivers, Lakes, and Seas

In the vast Soviet territory, the well-developed system of rivers and waterways has played an important rôle in Russian history. Along the rivers explorers moved into the immense Siberian lands, while the rivers of both European Russia and Siberia have been important as historical routes of commerce and trade. Today, rivers still remain important as routeways and also provide growing quantities of hydro-electric power or water for new irrigation schemes in the more arid parts of the country. Russian rivers are among the longest and largest in the world. They have characteristically long courses and large basins : their flow is quantitatively great, though in relation to length and

TABLE I

DRAINAGE BASINS OF RUSSIAN RIVERS

Drainage to	Area (million square miles)
Arctic Ocean	4·4
Caspian Sea — Inland Basin	1·5
Atlantic Ocean	1·0
Pacific Ocean	0·9
Inland drainage other than to Caspian Sea	0·8
Total	8·6

basin area, relatively small. Gentle gradients and slow-moving water are also common features. Climate exerts, however, the major control on their use. The rivers of Siberia and European Russia are frozen for varying periods in winter. In spring, with the coming of the thaw, great floods hamper navigation and cause destruction and inconvenience. In the summer, navigation is frequently hampered by low water, though the rivers fill again in autumn before the frost comes. The Soviet authorities have tried to overcome the limitations of the rivers as transport arteries by constructing canals to link river basins and dams to help to control the regime, but it has been usually cheaper to provide alternative forms of transport. Lyalikov [10] defines three types of regime on Russian rivers : high water in spring, high water in the warm part

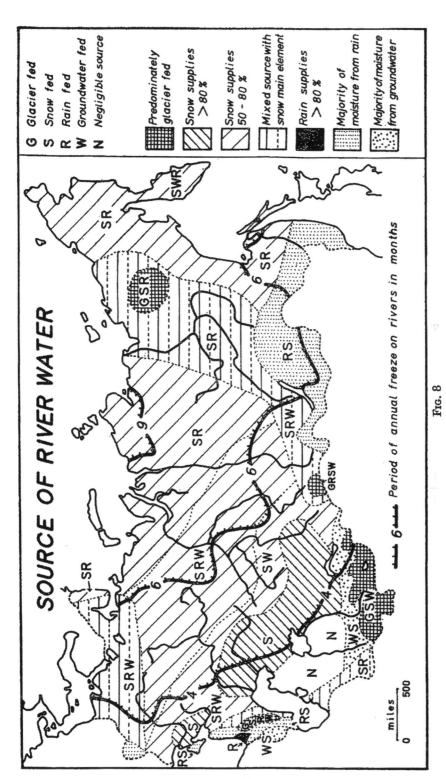

SOURCE OF RIVER WATER

G	Glacier fed
S	Snow fed
R	Rain fed
W	Groundwater fed
N	Negligible source

Predominately glacier fed

Snow supplies > 80%

Snow supplies 50 - 80%

Mixed source with snow main element

Rain supplies > 80%

Majority of moisture from rain

Majority of moisture from groundwater

━━ 6 ━━ Period of annual freeze on rivers in months

miles
0   500

Fig. 8

Source of river water.   Based on a map by Lvovich, M. I., reproduced and described in *UdSSR*, Leipzig, 1959, p. 138.

37

of the year, and an inundation regime. Some of the larger rivers incorporate more than one of these regimes into their own pattern of flow and the regimes depend upon different sources for their water, as indicated in Fig. 8. The greater part of the drainage is to the Arctic Ocean.

The rivers of European Russia flow either north or south across the almost level plain and their watersheds are often indistinct. The main watershed lies in the northern glacial morainic belt, so that the south-flowing rivers are the longer ones. The south-flowing rivers crossing the dry black earth belt gather most of their tributaries in their middle and upper reaches. The north-flowing rivers occupy in many instances channels altered by glacial action. The rivers are nearly all marked by a spring high water, often accompanied by flooding, since their main source of moisture is melting snow. Towards late summer, the north-flowing streams show a more even regime than those flowing south. The rivers are nearly all frozen for long winter periods: frost reduces the amount of water reaching the rivers, so that they have a winter low water.

The greatest system is the Volga and its tributaries. The river rises in the morainic Valday Hills and collects most of its tributaries (including the large Kama) in the reaches above the great Samara Bend, where the river flows out into the steppe. The Volga system is frozen generally from late November to early April, when the *rasputitsa* provides a sudden increase in water. The level rises slowly, carrying with it ice floes, which block the river and cause local floods. High water spreads south and the maximum is frequently marked by floods. There is usually only one maximum. The greatest volume is usually reached in the Samara Bend, where at maximum flow the volume of the river may be eighty times normal. Almost two-thirds of the annual volume passes in April–May. South of the Samara Bend, evaporation and lack of tributaries plus some seepage reduces the river's volume, though this is insufficient in spring to prevent flooding below Volgograd where the river flows into its braided channel in a 20-mile-wide flood plain. Although sudden freshets raise the summer level for short periods, low water in late summer hinders navigation, which has to face the added hazard of shifting sand-banks. Autumn *otepely* may bring additional moisture.

The Dnepr also rises in the Valday Hills, flowing some 1,400 miles to the Black Sea. It has a longer ice-free period in its lower reaches than most Russian rivers. Formerly navigation was hindered by the 40-mile-long section of the rapids near Zaporozhye, where the river cuts across the old crystalline rocks of the Podolsk-Azov shield. These have been drowned below the 120-feet-deep reservoir behind the Dneproges Dam since 1932. The level in the lower reaches has also been further raised by completion of the Kakhovka Barrage in 1956. High water is a month later than on the Volga. The short rivers which flow to the Baltic, such as the Narva, West Dvina, and the Neman, are not particularly distinguished. The Karelian and Kola rivers show immature courses, because they have not yet cut down to new base levels since post-glacial changes in sea level. They flow turbulently, with many falls and rapids, which provide sites for hydro-electric power stations, but have a relatively even flow, though tend to a minimum in winter. Because of their speed, some seldom freeze. The Caucasian rivers are also fast flowing. They have a winter low and freshets in spring, early summer, and autumn, but some dry up in the hot midsummer.

The West Siberian rivers have extremely low gradients, for instance the Ob is only 300 feet above sea level over 1,200 miles from its mouth. There is also a great deal of surface water in bogs along their interfluves and in their flood plains. They are all very slow in their middle and lower reaches. The Ob flows so slowly in the lower reaches that the water almost stagnates and fish find it difficult to live. Some rivers, like the Irtysh, have, however, fast-flowing reaches in their upper, mountainous courses. The rivers thaw in their upper reaches before the ice melts in the lower reaches. Water from upstream then flows out over vast areas of the lower reaches, causing immense floods, or is ponded back to flood the middle and upper sections. Spring high water is longer than in Europe, while in summer the rivers gain moisture from run-off from lakes and swamps. The winter frozen period also lasts much longer than in Europe, so that in general most Siberian rivers have less than 10 per cent of their flow in the winter. The East Siberian rivers have a similar regime. Thundery showers provide summer moisture, while run-off from *permafrost* areas is rapid since the frozen layer not far below the surface neutralizes the water-holding capacity of the woodland cover. May and June are the high-water period: the Yenisey at high water passes 130,000 cubic metres per second, while flow at low water at the same spot may not reach 2,500 cubic metres per second. The Yana at high water has 500 times the low-water volume! In contrast, the reservoir compensating action of Lake Baykal gives the Angara an unusually even regime. Ice floes carried downstream frequently form dams across the rivers at narrow bends or shallows, only to break suddenly when water pressure behind builds up. Floods from these bursting ice dams can cause serious damage to shipping, ports, and settlements. In the Lena delta, some channels are blocked all summer by ice floes piled high by spring floods, and large ice floes may also be stranded on the banks, to thaw slowly through the summer. The rivers are very large : the Yenisey is about 3-4 miles wide in its lower reaches, while the Lena in the middle and lower reaches exceeds 8 miles. Large ocean-going cargo ships can sail to Igarka, 450 miles up the Yenisey, though navigation from the sea into the Lena is complicated by estuarine bars.

The rivers of the Altay and Sayan mountains are usually tributaries of the major Siberian rivers, while the Yenisey rises in the Sayan. Their moisture is derived chiefly from mountain snows, which melt in spring and do not form permanent snowfields, while other moisture is derived from melting glaciers. Thundery downpours and increased melting provide freshets in autumn and in summer. The rivers have steeply-graded courses and carry down much material in suspension. Deep gorges and rapids are frequent along their courses, similar to the rivers of Central Siberian Uplands.

The regime of the rivers of the southern part of the Far East is influenced by the monsoon climate. Snow does not significantly contribute to the water resources, which are gained mainly from the summer monsoon. Winter ice does not break up into floes but melts mostly *in situ*. High water is in summer, a feature also found in some rivers of Transbaykalia and the eastern Sayan.

The rivers of Central Asia show mixed regimes. In their upper reaches, they are fed by snowfields and melting glaciers or by spring rains in their foothill courses. Winter moisture is also received from snow showers in the foothill belt. In the lower desert reaches, loss by evaporation is high and much water is also taken for irrigation. In summer, most rivers contain little water or dry up ; but throughout the year several rivers lose themselves into the desert sands. They carry down from the high mountains vast quantities of material, to form shoals in their lower courses, and along some a raised bed between natural levees is built with this material, so that conditions for dangerous floods in exceptionally wet years are created.

The Soviet Union contains some of the world's largest inland waters, truly described as 'seas', but with the exception of the Caspian Sea they do not play a major rôle in the economy and cannot be compared with that of the American Great Lakes. The Caspian Sea whose surface area exceeds that of the United Kingdom, and whose present surface lies 92 feet below mean sea level, consists of a shallow northern basin, which freezes over in winter, and two deeper basins, with the greatest depths reached in the southernmost basin. The water is brackish, though the salt content is kept down by the movement of Caspian water into the almost totally enclosed shallow 'evaporating pan' of the Kara Bogaz Gol, whose water is highly saline and which is covered by a 6-feet layer of salt at the bottom. It is estimated that 23·5 cubic kilometres of water are annually evaporated from the Kara Bogaz Gol. In the northern basin, Caspian water has a salinity of 2-4 pro mille and between 12 and 15 pro mille in the south. It also contains less sodium chloride but more sulphates than ocean water. Most water is brought to the Caspian by rivers, of which by far the greater volume is from the Volga. Berg estimates annual evaporation amounts to a water layer 1 metre deep, but this is offset by rainfall sufficient to cover the sea with a layer of water 20 centimetres deep and the yearly inflow of Volga waters, some 60 centimetres deep, as well as other rivers (Emba, Ural, Terek, etc.). The level of the sea shows long-term fluctuations and is at present falling so that the outline of the shores has already changed greatly in the last half-century, while there is plenty of evidence in marine deposits and fossils to show that in glacial times the area covered by its waters was very much greater.[11]

The Aral Sea, from the Kazakh meaning 'island sea', is much smaller and generally only about 60 feet deep. The greatest depth does not exceed 200 feet. It lies in a gentle downfold bordered on the west by the scarp edge of the Ust Urt Plateau. It is fed entirely by the Syr Darya and Amu Darya, and its water, rich in sulphates but poor in chlorides, is brackish with about 10 pro mille salinity. Like the Caspian, the northern part freezes for about three months in winter.

Most of the lakes of Northern European Russia, like Ladoga, Chud (Peipus), Onega, and others, are glacial in origin. Ladoga and Onega belong to the type produced by ice-gouging, while

Chud, Pskov, and Ilmen and Beloye are moraine-dammed. They are frozen in winter. The frozen connection between lakes Chud and Pskov was the scene of Alexander Nevskiy's defeat of the Teutonic Knights in 1224, while the railway tracks and roads laid by the Russians across the frozen Lake Ladoga at the siege of Leningrad in 1941 were important in the relief of the city. Several large artificial lakes have been created in European Russia by hydro-electric power station construction and canal building: examples are at Rybinsk, the Moscow Sea, the Tsimlyansk reservoir and behind the barrages at Kakhovka and Kuybyshev as well as Volgograd.

Lake Baykal [12] in Siberia, one of the most interesting lakes, is the deepest in the world (3,939 feet deep). It is entirely fresh water and lies in a deep *graben* of probably mid-Tertiary origin. Over 400 miles long, it is only 50 miles wide, bordered by high mountains. The Mongolian Selenga river and the upper Angara supply 85 per cent of its water, which drains off in the turbulent and clear Angara. The water temperature in the lake is very low all the year, while below 800 feet it remains almost constant at 37·6° F., and there is a three-feet-deep ice cover in winter but an ice-free period lasts for between 110 and 248 days per annum. The lake markedly modifies local climate. It has a distinctive fauna, containing unique forms of ancient origin with some relation to arctic species. Lakes lying in tectonic depressions and in troughs are common throughout the mountain structures of Southern Siberia, Central Asia, and even in Transcaucasia. Balkhash is thought to be of Quaternary formation and once to have extended over a greater area and to have perhaps included the lakes of the Dzungarian Gates. It is fed by the Ili river which reaches it across a broad belt of sand (*Semirechye*). The eastern part is brackish and the western end fresh. Its level fluctuates widely, but it is generally shallow. The lake is frozen for five months. Similar fault basins and depressions underlie Issyk Kul (Warm Sea) and Kara Kul (Black Sea). The former is over 2,000 feet deep, slightly saline and does not freeze. The latter is brackish. Lake Sevan in Transcaucasia is also tectonic, consisting of a deep western and a shallow eastern basin. Its level has been falling slowly. Plans are prepared to drain the shallow basin for agriculture and to use the lake's water to supply the Sevan-Zanga hydro-electric cascade.

Soviet Seas [13]

The land frontier of the Soviet Union measures 10,550 miles, but the coast line is over 26,700 miles long. The Soviets claim a twelve-mile limit to territorial waters. Unfortunately, nearly all the coast line freezes for varying periods each year, so that the search for 'warm-water' ports has been politically and economically desirable; but the Russians have also become skilled in the art of navigation in ice. Several of the seas on which the Soviet Union has a frontage have narrow entrances easily commanded by foreign and possibly hostile powers; for example, the Baltic narrows and the Dardanelles entrance to the Black Sea, while even the Atlantic and Pacific sea approaches to the Soviet Arctic coast, can be reasonably easily commanded, as can the entrances to Russia's frontage on the Sea of Japan.

The Arctic coast line is the longest. Along it has been developed the Northern Sea Route from European to Far Eastern ports of the U.S.S.R. Between the coast and the deep basin of the Arctic Ocean lies a broad shelf on which lie numerous islands and island-systems between 70° N. and 80° N. These island groups help to divide the coastal waters into sectors which bear the names of seas. Russia commands the sector of the Arctic Ocean between 32° E. and 169° W., where Soviet scientists have made an important contribution to the understanding of the northern seas. One of the most outstanding discoveries has been the charting of the submarine Lomonosov Ridge, which divides the Arctic Ocean into two distinct basins. The relatively shallow Barents Sea is bordered on the south-west by the inhospitable, high Norwegian and Soviet Murman coasts; on the north by the islands of the Spitsbergen-Franz Josef groups; and on the east by the northern Uralian extension of Novaya Zemlya. On the west, it is open to the warming influence of the North Atlantic Drift, which brings in warm, salty, oceanic water, and helps to keep the western sea open for navigation in winter. The southern coast is generally low and smooth east of the narrow fault-controlled neck of water to the White Sea, frozen from November to May. For its size the White Sea is deep, reaching 1,050 feet in the Kandalaksha Gulf. It is fairly fresh water. In the southern part of the sea lies Solovetsk Island, site of an historically famous monastery.

The Kara Sea overlies a sunken part of the West Siberian Lowland. It is sheltered from warm Atlantic influences behind

Novaya Zemlya on the west. The deepest part of the sea lies immediately offshore from Novaya Zemlya, reaching 1,500 feet. The sea is filled by fresh warm water brought by the Ob and Yenisey which helps to thaw the southern shore by June, though open water is later in the north. Ice frequently persists well into summer in the narrow approaches to the sea from the west and is also held by the many islands of the north-eastern part of the sea, the Nordenskiöld Archipelago. The most difficult waters to navigate along the Northern Sea Route are those between the Kara Sea and the Laptev Sea, because ships have to pass between the islands of Severnaya Zemlya and the mainland through the shallow Vilkitskiy Strait frequently strewn by grounded ice floes, while equal danger is present in trying to sail north round the islands. Ice is present all the year in these waters, while in bad years the Vilkitskiy Strait may remain closed throughout the year. The Laptev Sea is open to navigation in August and most of September, but fog makes navigation hazardous. Shallowness, low salinity, lack of tides, and the low annual temperatures, all contribute to rapid ice formation in the waters of the East Siberian Sea and the Chukchi Sea, though the latter is warmer in the south owing to penetration of Pacific water through the Bering Strait. These waters are open from July to September. The Chukchi Sea is sometimes thought to lie over a sunken massif. The coasts of the Siberian seas show evidence of uplift in raised beaches and driftwood deposits high above present sea level. The map of this part of the Russian coasts changes rapidly with better survey and as a result of these natural movements between the levels of land and sea : a good example is the creation of the Zemlya Bunge as land in the New Siberian Islands. With uplift, new islands and shoals are constantly appearing, though some imperfectly sur-veyed islands have disappeared, possibly owing to false reports or to them having been formed on ice masses which have thawed away.

The northern and eastern shores of the Black Sea and the almost totally enclosed Sea of Azov are Soviet territory. The Black Sea is geologically young and shows many traces of recent submergence : for example, loessic deposits with terrestrial mole holes and *kurgany* are found below water, as well as the many drowned river estuaries (*limany*). The northern part of the sea is a broad, shallow shelf, on to which rivers empty large quantities of fresh water, so that it freezes in winter. Spits and lagoons have been formed here, though the currents in the sea are generally weak. To this shallow northern shelf belongs the even shallower Sea of Azov, connected to the Black Sea by the narrow Strait of Kerch, and whose water is even fresher, brought down by the Don, Kuban, and other rivers, while the Sivash Sea, lying behind the well-developed spit of the Arabatskaya Strelka, acts as an evaporating pan. Spits are also well developed along the Azov coast, and to landward of them are frequently found broad coastal marshes, like those of the lower Kuban. The winter freeze in the Sea of Azov is more severe than in the Black Sea. The southern basin of the Black Sea drops steeply to over 6,000 feet. Below 600 feet, no life is found, as the water is contaminated by sulphuretted hydrogen and is deficient in oxygen. The muds found in the deep basin contain 25 per cent organic matter, of which 10 per cent has already been changed to heavy hydrocarbons soluble in benzene, regarded as incipient petroleum formation. The surface water has a salinity of 18 pro mille, less than half that of ocean water, but below 600 feet, the salinity rises to 22 pro mille. Surface water flows out into the Mediterranean through the Bosphorus, to be replaced in part by an undercurrent of saline Mediterranean water.

Since the incorporation of the Baltic republics and the former German East Prussia (now Kaliningrad *oblast*), the Soviet frontage on the Baltic has spread south-westwards. The Russians now command the southern shore of the Gulf of Finland and the broad bays, spits, and offshore islands to the Vistula delta. The Soviet sector of the coast is frozen in winter, though the Prussian ports are easily kept open. The sea is almost tideless and has relatively fresh surface water, though salt water enters below the surface from the North Sea.

The Russian Pacific coast lies behind an outer belt of islands and faces on to the epi-continental Sea of Japan, Okhotsk Sea, and Bering Sea. Submarine relief allows only the surface waters generally to be interchanged with Pacific basin water. The Okhotsk Sea, with its indented coast line in the south-west and north-east, drops towards the Kuril Islands, where a deep trench lies to both flanks of the islands. The northern part of the sea is a shallower shelf, about 3,000 feet deep. Saline Pacific water enters through the gaps between the Kuril Islands, while less saline Okhotsk water runs on the surface to the Pacific. There is a marked tidal movement, with tides of over 30 feet in narrow bays and gulfs. The sea freezes in winter and ice is present from December to June. The Bering Sea is separated from the Pacific by the Aleutian Islands. There is, however, a 200-mile-wide channel between the Aleutians and the Commander Islands on the Soviet shore, which is very deep and allows Pacific water to enter. The northern basin is about 600 feet deep, but the southern basin falls to about 9,000 feet. The complex system of currents causes an up-welling of cold water off the Soviet coasts, producing fogs and accentuating icing. Warm Pacific water does, however, manage to penetrate the 20-mile-wide but shallow Bering Strait into the Arctic basin, where it modifies conditions in the Chukchi Sea. The Bering Sea, despite its flanking continental landmasses, receives warm Pacific water, which keeps its surface temperature above local air temperature all the year. It remains navigable longer than the Okhotsk Sea. The Sea of Japan is linked to outside waters by the Straits of Korea, the La Pérouse Strait on the north-east and the Tsugaru and Shimonoseki Straits on the east. The basin drops to over 6,000 feet on the eastern, Japanese, side. A warm current, a branch of the Kuro Siwo, comes up from the south and runs along the Japanese shore; but unfortunately cold currents wash down from the north on the Russian shore, entering through La Pérouse Strait; ice therefore extends along the Russian shore in winter. Again, fog is a hindrance to navigation on the Russian shore.

NOTES ON CHAPTER 1

1. Parker, W. H., 'Europe: How Far?' *Geographical Journal*, vol. 126, 1961.
Sumner, B. H., 'Russia and Europe', *Oxford Slavonic Papers*, vol. II, 1951.
Spengler, O., *The Decline of the West* (Trans.), New York, 1928.
2. *Tectonic Map of the U.S.S.R.*, *1:5 M*, 1956. Edited by N. S. Shatskiy, with accompanying text.
Nalivkhin, D., *Geology of the U.S.S.R.* London, 1960.
3. Dobrynin, B. F., *Fizicheskaya Geografiya SSSR: Evropeyskaya Chast i Kavkaz.* Moscow, 1948.
Suslov, S. P., *Fizicheskaya Geografiya SSSR: Aziatskaya Chast.* Moscow, 1956.
Mikhaylov, N., *Sibir.* Moscow, 1956.
4. The undercutting of the right bank has been postulated in Baer's Law, though this has not been received universally as the explanation.
5. Maximum extent of the ice sheets possibly came in the penultimate glaciation. (Equivalent to the Saale moraines in Germany.) Discussion in Charlesworth, J. K., *The Quaternary Era*, vol. II, pp. 954-962, London, 1957.

6. The origin of loess is discussed in Charlesworth, J. K., *op. cit.* vol. I, chapter xxvi.

7. Thiel, E., 'Die Eiszeit in Siberien', *Erdkunde*, no. 5, 1951.

8. Thiel, E., 'Die Grundwasserverhältnisse Turkmeniens und ihre Beziehungen zur Morphologie und Geologie des Landes', *Petermanns Mitteilungen*, vol. 95, 1951.

9. Komar, I. V., *Ural*. Moscow, 1959.

10. Lyalikov, N. I. (Ed.), *Geografiya SSSR*. Moscow, 1955.

11. Taskin, G. A., 'The Falling Level of the Caspian Sea', *Geographical Review*, vol. 44, 1954.

12. *Po Baykalu-Putevoditel*. Moscow, 1955.

13. *Morskoy Atlas*, vols. I and II. Moscow-Leningrad, 1955.

Admiralty Pilot — relevant volumes on Russian waters. London, various dates.

For readers of Russian, a rich collection of articles on special topics can be found in the *Bulletin of the Academy of Sciences of the U.S.S.R., Geographical Series, The Bulletin of the All-Union Geographical Society of the U.S.S.R., Geography in Schools*, as well as the irregular *Problems of Geography*. In English, there is *Soviet Geography: Review and Translation*, published by the American Geographical Society in New York.

CHAPTER 2

Climate, Soils, and Vegetation

Most of Russia has an extreme continental climate, with long, cold winters and short, warm summers. About three-quarters of the country lies more than 250 miles from the tempering influence of the sea; but the seas which surround the country are also limited in their influence, either because they are comparatively small in area or they are frozen for long periods each year, while cold currents running off the coast also help to limit maritime penetration inland. Warmer air from lower latitudes is excluded by the high mountain barriers which lie across Central Asia, but the low Arctic coast allows a deep penetration across the northern plains of cold, northern air. Temperate, moist Atlantic air, when it penetrates into Russia, has been modified in crossing the intervening European landmass, while Pacific air seldom penetrates deeply into the interior because of the high, concordant coastal ranges and the cold currents running in Pacific waters along the Russian coast. Great winter cold is a feature of almost all the country: even the desert areas of Central Asia whose summer temperatures compare with other hot deserts have abnormally cold winters. Regularity of seasonal conditions suggest that Russia has climate but little weather.[1]

Winter is dominated by an intense high-pressure system developed in a large, polar continental air-mass over Mongolia and Central Siberia. From this system, intensely cold air streams out as the Asian winter monsoon. Over most of Siberia, the bright, but intensely cold, still weather is broken only when storms develop as an occasional cyclone passes along the edge of the polar anticyclone. From this high-pressure system, a ridge extends westwards along the 50° N. line across Southern Russia and Central Europe, forming a barometric divide with westerly circulation to the north, so that cyclonic development along the atlantic polar and arctic fronts brings occasional winter warm spells to North-west Russia. To the south, easterly circulation keeps much of Southern Russia and Trans-Caspia under the bitter

DOI: 10.4324/9781003172048-2   48

cold of the Siberian interior. Cyclonic movements from the Mediterranean low pressure send a few infrequent weak cyclones into Southern Russia and Caucasia, but these have little influence on the broader pattern.

As the sun moves north, the Central Asian high pressure disappears with the rapid heating of the land. It is replaced by an intense low-pressure system over Baluchistan, extending northwards over Soviet territory. This draws in warm, moist air from the southern oceans, but the strong monsoonal conditions which develop over India do not penetrate north of the great Himalayan barrier. Russia remains dominated by warm, dry, tropical, continental air. A monsoonal effect develops over Eastern Siberia, however, but its penetration inland is restricted by the surface configuration. It brings cool and unpleasantly damp conditions to the Eastern Siberian littoral and islands. Unpleasant, overcast weather marks the weak front between arctic and polar continental air along the northern coasts of Siberia.

Thus in winter the country is in the grip of intensely cold air. Only the occasional penetration of cyclones from either the Atlantic or Mediterranean low-pressure systems breaks the monotony by violent blizzards. In the continental heart of Siberia, bright, cold, dry weather prevails, with some of the lowest

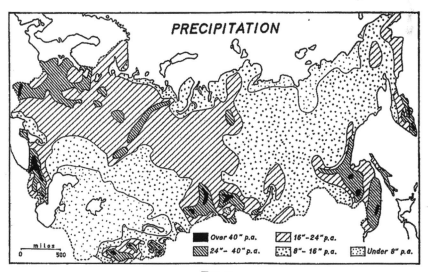

Fig. 9

Precipitation. Compiled from climatic maps in Soviet atlases.

temperatures recorded on earth; but in European Russia the skies are overcast and the humidity greater, so that it feels raw. Towards April, the winter pressure system begins to break down, the earth warms up and the ground thaws, marking the change from winter to summer in the unpleasant, cool, raw weather of the *rasputitsa*. Summer is short but usually hot, with thundery showers or periods of dry and dusty conditions. Within four or five short weeks of autumn, everywhere is again in the icy grip of winter, although occasional reversions to warmer conditions mark the first half of the European Russian winter. In the high arctic latitudes, the long summer days with almost no night give way to short winter days with intensely cold nights and little solar heating.

The lack of strong relief over much of the country is reflected in the uniformity of precipitation. Fig. 9 shows that precipitation decreases eastwards as well as to the south and to the north away from the central belt of moderate fall (*circa* 20 ins. p.a.). Falls are particularly small in North-east Siberia and in Central Asia around the Aral Sea. The somewhat heavier fall in Amuria compared to the rest of Siberia comes from the local south-east monsoon. North of 45° N., much precipitation, particularly in uplands and mountains, arrives in the form of snow. Throughout most of the country, summer is the rainiest season. In winter, penetration of moisture is prevented by the strong winds blowing from the winter anticyclone in Siberia. Over European Russia and Siberia, July and August are usually the wettest months, with rain brought by thunderstorms. In the Amur and Ussuri basins, the south-east monsoon brings summer rain, with fine drizzle along the Okhotsk coast where the precipitation is reduced and the monsoonal effect weaker. The extreme west and north of European Russia have autumn rain brought by westerly depressions, which are most active at this time and are not yet excluded by the strength of the winter high-pressure system. Although there is no true dry season along the Crimean and Caucasian Black Sea coasts, late autumn and winter rain is the main precipitation : for this reason, these are sometimes classified as having a quasi-mediterranean climate. An autumn-winter maximum is also characteristic along the western shore of the Caspian Sea ; though in the Armenian Plateau, late spring is the maximum. Central Asia has its precipitation mostly in March and April ; in the north, along the steppe fringe, May and June

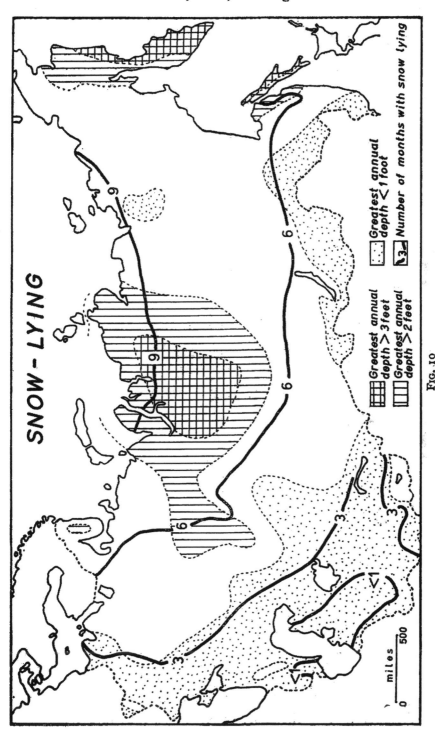

SNOW-LYING

Greatest annual depth > 3 feet

Greatest annual depth > 2 feet

Greatest annual depth < 1 foot

3— Number of months with snow lying

FIG. 10

Snow. Compiled from climatic maps in Soviet atlases and augmented from Alisov, B. P., *Klimaty SSSR*, Moscow, 1956.

are the wettest months, but much rain comes from copious falls accompanied by thunder, so that loss to run-off is high. Winter precipitation is mostly snow, though it is frequently blown away by strong winds. European Russia has snow between mid-November and March, but in the south snow is far less frequent. In Arctic Russia, snow is common between mid-September and early June. Fig. 10 shows the average depth of snow cover, which reaches the maximum in the tayga belt where depressions bring considerable moisture and the forest prevents powdery snow being blown away. Although it has a longer period when snow is likely, the cover in Siberia is lighter, since moisture for its formation is not present in large amounts. As an insulating layer to keep the ground warm, snow has an important effect on vegetation and the depth of cover also bears some relationship to the development of *permafrost* intensity.

OCCURRENCE OF PERMANENTLY FROZEN GROUND [2]

The severity and length of winter over much of the Soviet Union is closely related to the widespread occurrence of *permafrost*, which covers 47 per cent of its area, notably in the north and east and extending over most of Siberia. It is an important agent, in association with related phenomena, in the landscape formation of the tundra and northern forests, marked by four zones of intensity (Fig. 11). *Permafrost* occurs as a continuous cover or with islands of thawed ground, while around the edge, islands of frozen ground lie amid unfrozen country; but whatever form it takes, it influences drainage, erosion, and vegetation as well as human utilization of the land. It may be found at depths down to 2,000 feet, but this is exceptional. More usually it is a broad layer extending down to several feet and thawing in summer in the upper, so-called active layer. In coniferous forest, this layer may be 6-7 feet deep, but in marshy ground only 3 feet and peat bogs thaw only in the top 18 inches. *Permafrost* is particularly well developed in cold, damp clays and in peat as a poor conductor of heat. It is less well marked in sands. *Permafrost* forms most readily where the snow cover is thin and frost can easily reach the ground or where there is prolonged drainage of cold air into hollows. The clear winter skies allow intense heat radiation from the earth, while the short summer has cold nights despite

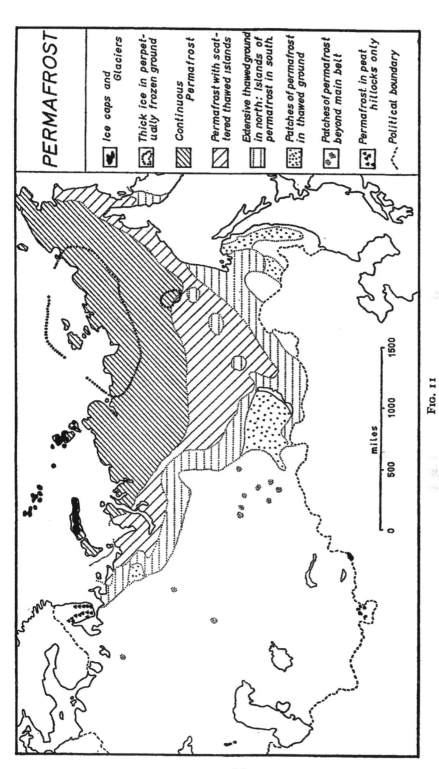

PERMAFROST

Legend:
- Ice caps and Glaciers
- Thick ice in perpetually frozen ground
- Continuous Permafrost
- Permafrost with scattered thawed islands
- Extensive thawed ground in north: Islands of permafrost in south.
- Patches of permafrost in thawed ground
- Patches of permafrost beyond main belt
- Permafrost in peat hillocks only
- Political boundary

miles

0   500   1000   1500

FIG. 11

Permafrost.   Compiled from Berg, L. S., *The Natural Regions of the U.S.S.R.*, New York, 1950, and Suslov, S. P., *Fizicheskaya Geografya SSSR*, etc., Moscow, 1956, augmented from pl. 70 in *Atlas SSSR*, Moscow, 1962.

high midday temperatures. Buried ice is commonly found in *permafrost*, caused by several different reasons. In Northern Siberia, there are deposits of fossil ice overlain by more recent marine or terrestrial deposits which often contain remains of Pleistocene fauna.

By preventing drainage, water-logging is a feature of *permafrost*. Flowing water prevents formation, so that along rivers there are islands of unfrozen ground, frequently marked in the northern forest and wooded tundra by well-grown clumps of trees. Rivers tend to incise their valleys deeply into the unfrozen ground along their course. It is generally more usually found on north-facing slopes, so that asymmetrical valleys are common. On slopes, it may cause extensive solifluction; in North-eastern Siberia, mud avalanches are sometimes caused by *permafrost*. Up-welling of ground water through *permafrost* layers or the formation of ice lenses may cause the growth of ice blisters (*pingoes*) of great size, the *bulgunnyakhi* common in Yakutia. When these collapse, they leave small ponds and some areas are littered by myriads of such hollows. In places, hummocky bog, string bog, or polygonal tundra develops. Fields of ice covered by water and slush are also common and covered by small ice mounds out of which the ground water pours: such *naledy* fields may cover large areas and are virtually impassable. Trees growing on *permafrost* are often bent and tilted.

In construction work, as in nature, slight changes in the thermal balance can produce the development of *permafrost* with consequent earth movement, which may have unfortunate results on foundations and the embankments or cuttings of roads and railways and the pillars of bridges. Opening up *permafrost* country is expensive and requires great skill.

MAJOR CLIMATIC BELTS

The major climatic belts form remarkably regular latitudinal zones across the country, again an indication of the monotony of relief over vast distances. Despite the variety of climatic regimes, the transition between belts is almost imperceptible to the traveller, so that the border lines between the different regions are therefore exceptionally arbitrary: the authorities agree only on the broad general division and on the nature of the regimes shown in Fig. 12.

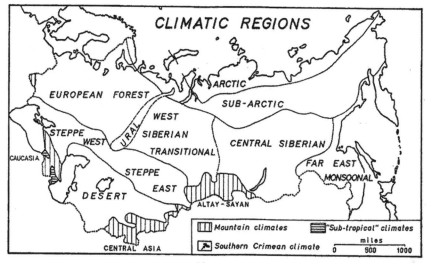

FIG. 12

Climatic regions. It should be remembered that the boundaries between regions are approximate and represent in reality broad zones of transition between climatic types.

Arctic

1. Arctic climates may be defined as extending south to the 50° F. isotherm for the warmest month. A further distinction may be made between tundra climate and perpetual frost climate along the isotherm of 32° F. for the warmest month. This line is also the divide between the permanent and non-permanent settlement. The arctic climates are distinguished more by their cool summers than by the intensity of winter cold. In these latitudes, length of day and consequent solar radiation is an important factor in climate. The strongest continental influences are felt in the central section between the Yenisey and Lena rivers. In spring, temperature rises slowly, so that autumn is generally warmer than spring. Summer warmth is sufficient to thaw only the surface layers for two or three months, while widespread *permafrost* tends to cool the lower layers of the air. The intense cold of winter is bearable so long as the air remains still and the skies clear. Snow, the main form of precipitation, is brought by blizzards, when life in the open is almost impossible. Drifting snow makes it hard to estimate the actual depth of cover. Along the coast, misty and cloudy conditions of summer also bring some moisture, but make conditions most unpleasantly raw.

55

Sub-arctic

2. There is a gradual transition to sub-arctic climate, whose southern limit is marked by the southern boundary of continuous *permafrost*. In the west, this regime is associated with tundra vegetation; but in Central and Eastern Siberia there is wooded tundra and even poor forest. The western sector, which extends to the River Taz, has more varied and warmer winters. The belt is noted for its great winter cold, which is frequently more intense than in the Arctic. Temperatures are particularly low in North-east Siberia where cold air settles at the bottom of valleys. Verkhoyansk ($-59°$ F.) and Oymyakan are within the world's 'cold pole'. Anticyclonic conditions give bright, clear weather which makes the intense cold bearable. Despite the lack of personal hygiene in such intense cold, the air is healthy and germ-free. It is only during the occasional *buran*, a violent cyclonic storm, that conditions become trying for man and beast, and health is endangered when humans and animals huddle together in huts to seek protection from its fury. Summers are warm, though plagues of flies and midges spoil man's enjoyment of it: up to $100°$ F. has been recorded, but the diurnal range is very great, with daytime temperatures of $95°$ F. dropping to $41°$ F. at night. The contrast between the west and east of the belt in summer is less than in winter.

European forest

3. European forest climates are characterized by winter cyclonic activity. Eastwards there is a gradual increase in continentality, while there is a progressive decrease in temperature from south to north. The belt has warm summers and long cold winters; but cyclonic influence in the north and north-west gives a more varied winter than in the south and south-east. North-west and Central European Russia frequently have the winter cold broken by invasions of warmer, moister atlantic air which cause warm spells (*otepely*), particularly during November and December when temperatures may rise to over $45°$ F. Snow lies for 80 days in the south and up to 200 days in the north between October and April, and about a third of annual precipitation comes as snow brought by moist atlantic air. Rivers freeze between late December and mid-January and remain ice-

bound until March. The winter is gloomy and dull with overcast skies and because it is moister than in the bright anticyclonic
conditions in Siberia, it is more trying and raw, so that where
people crowd together respiratory complaints are common. In
April, when the thaw begins, there is much surface water, roads
turn to muddy quagmires, and rivers flood. Movement is difficult
and uncertain, so that it is well-named *rasputitsa*. The floods
subside and the ground dries out as the short but warm summer
approaches. Temperature differences between north and south
are smaller in summer than in winter. Over a third of annual
precipitation comes in the months June to August, when there is
a good deal of cloudy, overcast weather and frequent thunder or
thundery showers. The thunderstorms are often intense, with
great displays of lightning and torrential rain, and are preceded
by hot, dusty winds. The showers are welcomed to settle the dust,
much of which is blown in from the steppes. Western travellers
sometimes find the wide summer diurnal range of temperature
unpleasant. Autumn is short but frequently a bright and pleasant
period. Along the Baltic coast, maritime influence tempers both
summer warmth and winter cold. The Ural shows a notably
heavier precipitation, particularly on the western slopes where
snow is deeper and lies longer, and brings northerly influences
far to the south as the result of increased altitude.

West Siberian transitional and Central Siberian

4. The Western Siberian Lowland forms a transition between
the forest climates of Europe and the true Central Siberian tayga
climate east of the Yenisey. Winter cold is intensified under the
clear skies of the Mongolian-Siberian winter high pressure. In
North-east Siberia, available records are for low-lying stations,
where drainage of cold air may tend to exaggerate the intensity
of winter cold. Despite the short winter days, clear skies allow a
high degree of insolation, while stillness of the air makes the cold
bearable. Between 110° E. and 160° E., the anticyclonic winter
conditions are most markedly developed. The low annual precipitation and consequent light snow cover leaves the ground open
to contact with the intensely cold air, so that *permafrost* phenomena
are common and the frost produces strange rumblings in the
ground and other curious effects. Again, winter blizzards can
hazard life outside. Spring is moist and raw and comes later

towards the north and east. The summer is much warmer than expected for the latitude, but it lasts only about seventy days in the north and a hundred days in the south, with a wide diurnal temperature range. Over half the light annual rainfall comes in late summer when the low-pressure system is most strongly developed. A frequent, unpleasant, dusty wind from the north and north-west spoils the summer by bringing plagues of mosquitoes, while the acrid smoke of forest fires hangs in the air. Autumn is short: by September, night frost occurs, and everywhere is bound in an iron frost by mid October. Slope and position in this belt cause important local variations in climate.

Far East monsoonal

5. East of the Stanovoy Range and the mountains of the Okhotsk coast, the influence of the south-east monsoon is felt: this is the Far East monsoonal climate. The winter is similar to Central Siberia, with temperatures unusually low for the latitude. Summer is moist and warm, tempered by an inflow of pacific maritime air. Monsoonal rain is brought by winds off the sea in May and lasts until September, with the maximum in July. Away from the coast, amounts rapidly decline. Annual precipitation exceeds 40 ins. in parts of the Sikhote Alin. The coasts have misty, cool summers. North of the Amur estuary, inland penetration of the monsoon is restricted by the high, concordant, coastal ranges which flank the Okhotsk coast, where cold winters are followed by raw and misty summers. The Okhotsk coast, the Kuril Islands, and Kamchatka are notorious for their unpleasant moist and misty summers.

Steppe

6. Steppe climate is transitional between the forest and the desert. It has cold winters but hot summers, with a spring maximum rainfall. The European steppe has warmer winters and cooler summers than the Asiatic steppe. Dryness increases towards the east and south, except in the Black Sea littoral. Precipitation, which is not reliable, varies from 8 to 16 inches a year, but this is low for agricultural purposes because the great heat and dryness of summer cause big losses by evaporation. Rain falls mostly in heavy showers so that much is lost by run-off

from the parched ground. Gullying is a serious erosional problem in the steppelands. The sparse winter snow cover exposes the earth to severe winter frost, reducing the friable soil to dust and providing raw material for dust storms. A little winter snowfall followed by a dry and windy spring can produce, as in 1960, disastrous dust storms and the blowing away of valuable topsoil and winter-sown crops. One year in five may prove a crop failure. Over the hard dry ground in winter movement is easy, except when conditions are made trying by violent north-east blizzards along the northern edge of the steppe belt, particularly in the southern Ural. Spring comes slowly, with frequent reversions to winter, and the thaw turns the friable soil into a sea of mud, making the short season unpleasant. Rainfall reaches a maximum in late spring, when heavy showers and run-off cause great damage to young crops. The summer mornings begin bright and cool, but rapidly the temperature rises and clouds gather, the air becoming stuffy and oppressive, with perhaps a late afternoon thundery rain shower, after which the air again clears and temperatures drop rapidly in the evening. Vegetation withers towards the end of the hot and dry summer, when local low-pressure systems off the Black Sea lift and carry dust over great distances, to cause bad visibility at sea and disease-bringing dust storms in the forest belt to the north. The dry *sukhovey* wind blowing along the periphery of anticyclones withers grain crops and is another summer hazard of the farmer.

Desert

7. To the south, the traveller passes into the desert climate, which may be defined by an annual precipitation of less than 10 inches, though it is extremely unreliable and the actual position of the isohyet varies greatly from year to year. The transition is a gradual one, through ever-drier steppe into the semi-desert and into the true deserts around the Aral Sea. Summers are intensely hot and dry, after a brief spring in which most of the scanty precipitation falls, a short period of life when plants blossom before they wither under the hot, dry winds from the south, such as the *Afganets*. Summer is a period of intensely clear skies and great diurnal range of temperature. A brief autumn, and the cold of winter rapidly increases under anticyclonic influence from the cold heart of Siberia. Rivers freeze. There may be a light

snow cover lying for up to sixty days in the north, though it is rare in the south. It mostly comes from weak cyclones from Mediterranean low-pressure systems which manage to penetrate beyond the Caspian Sea. Again, the change to summer is a rapid one.

Transcaucasian

8. Transcaucasia has a climatic regime not found elsewhere in Russia. In the lowlands there is a warm, moist climate with heavy rainfall and much cloud. No true dry season is noted, but a maximum occurs in autumn and winter. Moist, cool air flows from sea to land in summer; but in winter the movement is reversed, so that dry, warm air flows out from the interior, producing a weak, monsoonal effect. For Russia, the diurnal range of temperature is small. In the Armenian Plateau there is a continental steppe regime. The winter is cold with a good deal of snow, though maximum precipitation comes in spring and early summer. In the later part of the summer, great evaporation results in a withering of vegetation. Two small areas on the Black Sea coast, around Alushta and Yalta in the southern Crimea, and between Sochi and Sukhumi in Transcaucasia, have conditions closely akin to mediterranean climate. Protected from cold northerly influences, winters are mild and summers hot and dry. The rich vegetation, the blue skies and sea, make these favourite resorts for people from the harsh climate of the interior.

Mountain

9. Altitude produces special local climatic regimes, frequently having characteristics of more northerly climates, but the detailed study of these is hampered by the lack of recording stations. In these mountain climates, location and aspect are important factors. The Great Caucasus, rising to over 6,000 feet, has a western face which is wetter and warmer than the eastern face. Near Tbilisi, local *föhn* effects temper winter cold but bring damaging hailstorms. The Dagestan part of the mountains is particularly dry and sunny in winter. With an increase in altitude, precipitation increases but temperature decreases, and in the highest parts snowfields and glaciers are found. The mountains of Central Asia have what may be described as a continental

mountain climate. Able to trap high-altitude moist air currents, many western and northern faces carry dense forest between 10,000 feet and 17,000 feet, where precipitation is deposited by occluded depressions passing over Central Asia. The snow line varies with aspect from 11,000 feet on the north-facing slopes to 18,000 feet on southern faces. On the lower slopes, desert conditions prevail far into the mountains. Although winter on these slopes is raw and foggy, *föhn* winds occasionally bring very warm conditions. The high plateaus, particularly in the Pamir, are dry and have a rare atmosphere. Here snow does not lie below 16,000 feet, but some islands of *permafrost* are found. Nilotic conditions are found in many of the sheltered, lower mountain basins. A continental regime is also found in the mountains of Southern Siberia: the winters are more severe than might be expected for the latitude, but nevertheless the mountains stand as islands of warmth surrounded by much cooler lowlands. In summer, however, the mountains are cooler than the lowlands. In the high mountains, snow cover is light, so that *permafrost* is widespread. Many of the lower, sheltered valleys have only light precipitation and form patches of steppe grassland. Climatic conditions on the whole are harsher in the Sayan than in the Altay, with colder winters and snow on high ground from October to June. The extreme southern slopes of the Sayan have a Mongolian steppe climate with relatively little snow and yet great cold in winter and frost becomes a major factor in weathering. Mountains in North-east Siberia have an extreme sub-arctic climate, though again they stand as warm islands in winter amid very cold lowlands. They are, however, almost a complete blank on the climatic map, since so few recording stations exist among them.

SOIL AND VEGETATION [3]

The remarkable uniformity of relief over much of the Soviet Union is reflected, as in climatic distribution, by the broad latitudinal belts of soil and vegetation types. The close association between the three and the limited relationship between soil type and underlying geological formations has led Russian soil scientists to favour a climatic theory of soil formation. It has also attracted Russian geographers as a suitable means of regional

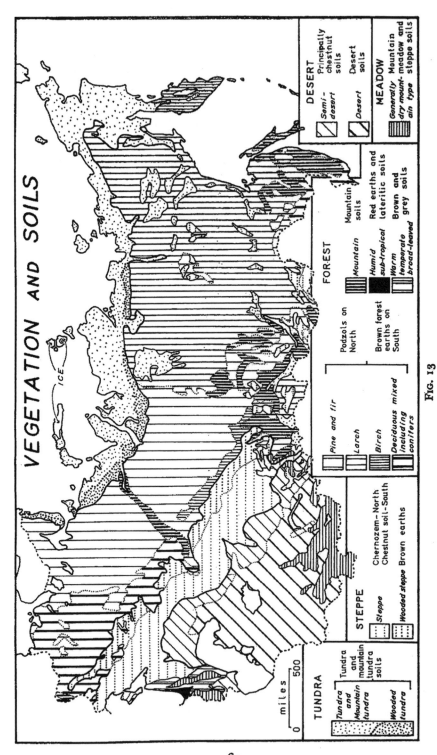

VEGETATION AND SOILS

TUNDRA

Tundra and Mountain tundra | Tundra and mountain tundra soils

Wooded tundra

STEPPE

Steppe | Chernozem–North Chestnut soil–South

Wooded steppe Brown earths

FOREST

Pine and fir

Larch

Birch

Deciduous mixed including conifers

Podzols on North

Brown forest earths on South

Mountain | Mountain soils

Humid sub-tropical | Red earths and lateritic soils

Warm temperate broad-leaved | Brown and grey soils

DESERT

Semi-desert | Principally chestnut soils

Desert | Desert soils

MEADOW

Generally dry mountain type | Mountain meadow and steppe soils

miles
0   500

ICE

Fig. 13

Vegetation and soils.   Simplified from *Atlas SSSR*, Moscow, 1962, pl. 84-85 and 88-89.

62

division of the country, best expressed in L. S. Berg's great work, *Priroda SSSR*. Unlike peninsular Europe, human action has so far had less impress on the natural condition of soil and vegetation in the Soviet Union. The relation between climate and vegetation is, however, usually closer than between vegetation and soil, since vegetation has responded to climatic change since the last glaciation more quickly than soil. Each of the major vegetational belts has also its characteristic fauna, despite important seasonal migrations and the influence of man in modifying patterns by hunting and trapping, particularly among the fur-bearing animals of the tayga and the wide-ranging species of the steppe.

Tundra

The northernmost belt, corresponding to the arctic climates, is the tundra. Here the soil lies mostly on a permanently frozen substratum with poor drainage, so that despite the low annual precipitation, there is consequently much surface water and the ground is constantly saturated. Under such conditions, micro-organisms are not very active and the decay of organic material is slow. The soils generally have a thin, peaty layer above a dull grey or black layer of partly decomposed organic material. Poor podzol is found on sandy, better-drained areas, while spotty tundra (patches of vegetation between bare ground) develops on waterlogged clays. Peat is widespread, but because of the reasons indicated above it is seldom deep. The tundra is not a continuous bog, although swampy conditions are most common and do cover large areas. Peat mound bogs, 15 to 75 feet high, occur in sheltered basins and depressions along the southern boundary. Vegetation is retarded by the short growing season and comprises relatively few species. The true tundra is treeless: the open, windswept ground is covered by hardy, woody plants, many of them evergreen perennials. There are also many lichens, including the reindeer moss, and small, brightly flowering plants. Upward growth is restricted by strong winds, while downward growth is hindered by the underlying *permafrost*. Towards the south, stunted trees and low bushes begin to appear, gradually passing into clumps of trees or scattered individual trees which form the wooded tundra. *Permafrost* inhibits tree growth or produces patches of trees standing at crazy angles. The trees increase in numbers southwards to form the northern limits of the great forest belt, though its northern edges are indeed poor, but nevertheless provide wild reindeer and other animals with winter shelter and pasture. The tundra landscapes are bare, grey, and monotonous, broken only in summer by the bright flowers of a carpet of small plants, mostly on south-facing slopes, while snow lies all the year on the north-facing slopes. On clays, great clusters of colour stand out in summer amid bare ground; but an altogether richer vegetation is found on sands. In the long, dark winter, when the ground is frozen or covered by frozen snow, the tundra is almost without animals, which retire to the shelter of the forest or fly away to warmer latitudes. Occasional polar bears, inhabitants of the northern ice floes, sometimes visit the coast and an arctic fox or the much rarer blue fox is seen. Ptarmigan come in from further north. The summer brings wild reindeer from the forest to search for food and to try to escape the swarms of gnats and mosquitoes, but they are followed as prey by wolves. Foxes and furry animals of the tayga also move north, so that ermine, ferret, lemming, and the snowshoe rabbit may be seen. Flocks of aquatic birds come to open water and to swamps, which are covered by great numbers ('bazaars') of a few species. Insects and reptiles are poorly represented, except for flies, gnats, and mosquitoes.

Tayga

Across Russia's middle latitudes extends a broad belt of forest, composed predominantly of the coniferous tayga, associated with the ashen-coloured podzol, which forms under forest in temperate latitudes where there is sufficient moisture and relatively moist summers. It is a poor, acid soil with a thin, peaty layer of raw, acid humus originating from the coniferous vegetation. There are three clear horizons: an upper mineral layer of greyish-white, highly siliceous, leached acid soil poor in plant food; a central layer, brownish at the top, rich in humus and material washed from above, and yellow or rust-coloured at the bottom, stained by iron hydroxide; at the base is a reddish-brown layer containing material washed from nearer the surface. Hard pan is often found beneath the top layer, while bog iron occurs in nodules, and can facilitate swamp development by water-logging. The degree of podzolization varies greatly between sandy and clayey areas: leaching is usually greater in the higher elevations than in the valley bottoms. Towards the south, where there is greater summer evaporation, podzolization is also usually less complete. North of 60° N., bog soils are common. The rougher terrain of Eastern Siberia allows better drainage than in the west, so again the soils show less marked podzolization. In Southern Siberia and in parts of Central European Russia, grey forest soils, usually found associated with wooded steppe, occur under proper forest, and suggest encroachment of forest into the steppe. In the middle Lena around Yakutsk, on river terraces above the flood plain, chernozem-like soils have developed on a carbonated, loess-like clay loam. These have formed as far north as 62° N. under a dry climate with little precipitation and warm summers. Underlying *permafrost* now prevents the removal of salts from these soils.

The tayga is coniferous forest interspersed by bog. It contains about a third of the world's forest, extending almost unbroken for 3,000 miles across Russia in a belt about 600 miles wide in north to south extent. It reaches far into the north along gulfs of warmth formed by the great Siberian rivers. The vegetative period lasts between three and four months, limiting the species to those with low transpiration such as conifers. The snow cover is deeper than in the tundra, because the forest prevents drifting, and it provides an insulating layer for small plants against the winter cold. The forest is composed of extensive stands of a few species, with rather small trees, seldom exceeding 3 feet in girth. They grow close together, so the forest floor is often covered by a tangle of decaying wood. Where the trees are wider apart, there are small shrubs and flowering plants, but otherwise the forests are gloomy, mournful, silent, and forbidding: it is said that to venture inside them is to know fear. Bogs break the forest cover, while along the rivers there are coarse meadows. Encroachment of bog, notably on interfluves, causes large areas of decaying forest, while in places *permafrost* is marked by 'drunken forest'. Notably in Eastern Siberia, where the country is drier, vast forest fires burning for long periods leave bare country to regenerate naturally, and various stages of colonization can be found. Almost all this vast forest is untended and in its primordial state. In the European tayga, spruce and fir are found on moist clays, while stands of pine (*bory*) mark sandy patches. The cold, acid, and badly-aerated soils are poor in fertility and difficult for agriculture, though much land has been cleared and improved.

F

In the Ural mountains, the vegetation is spruce and larch in the north, giving place to pine, birch, and larch in the south. The forest is poor in the higher parts of the range and many of the higher northern ranges are covered by wooded tundra. Altitude is thus sufficient to carry northern vegetation belts well to the south of their usual latitude. Western Siberia has spruce in the north, since its superficial root system grows well on *permafrost.* Larch is also common. Bogs are widespread on the low inter-fluves between the West Siberian rivers, the result of water-logging of dry land by sphagnum. To the south, dense thickets of trees with patches of bog (*urmany*) occur, with fir predominant. The rivers are lined by coarse meadow (*yelan*). South-west Siberia, notably in the Altay, grows fir and larch. Aspen, alders, and willow bring some diversity to many of the forests. In the southern forests, more undergrowth and many ferns and berry-bearing plants, as well as the beautiful Daurian rhododendron, give the forests a cheerful aspect, with more open patches. Larch forest is common in Eastern Siberia with its drier lands and the Daurian larch, which grows well on *permafrost*, is a hard, decay-resistant wood valued for constructional work. In Trans-baykalia, many vast expanses of virgin forest contain rich pine forests in the valleys. Bogs are less common because of the better drainage and drier climate, but the dryness of the summers creates a grave fire danger. Travellers have frequently remarked on the smell of burning which hangs in the summer air of Eastern Siberia. Dry summers are also a factor in the development of grassland and parkland in sheltered valleys of Southern Siberia and the Vilyuy-Lena lowlands. The birch forests of Kamchatka distinguish it from the mainland of Siberia. They are thin and park-like, while spruce and larch also occur on low ground.

The gloomy tayga is poor in animal life, whose absence in-tensifies the oppressive stillness. The animals tend to live in the fringe of the forest around clearings and open patches. Wolf, bear, fox, and lynx, and many small fur-bearing animals and carnivores are found. The valuable fur-bearing animals such as the silver fox, ermine, squirrel, and the Siberian sable, attracted early Russian settlers into this country, but hunting and trapping took such numbers of the prized furs that the animals have been driven into the more inaccessible regions. The sable, for instance, was so reduced in numbers by hunting that it became almost extinct at the end of the nineteenth century. It has been under protection since 1913. The beaver, also protected, is now rare. Some species of deer have been exterminated in the European forest, though they still occur in Siberia. Elk is widespread and maral is found widely in Siberia. There are many birds of passage as well as the ptarmigan and the capercaillie. Several poisonous snakes, including the viper, and other harmless reptiles occur. The prolific insect life of flies, gnats, and mosquitoes in summer greatly discomforts man and many animals in the tayga belt and has proved a hindrance to colonization.

Mixed forest

Between the tayga and the grasslands lies a wedge of mixed forest in which broad-leaved trees predominate in the south. It is much wider in European Russia than in Siberia, petering out at the foot of the Altay to reappear again in Amuria, and lies on modified, less acid podzol, where summer warmth produces more decomposition of organic material and consequently a richer humus content, with more active bacteria and earthworms. Along the southern edge, darker soils suggest invasion of forest into the steppe, while outliers of plants of the steppe are found in clearings in the forest. The forest, with more openings and more light, is less forbidding than the tayga. The presence of oak and spruce together is characteristic : oak decreases to the north and spruce to the south. In the northern forests, oak grows mostly on clay loams and avoids intensely-developed podzol. There is some fir. The southern parts have chiefly hornbeam with comparatively little spruce. Other trees include pine, ash, elm, linden, and even maple. Swamp is extensive in parts, but a lot of clearance and drainage has been undertaken. The better soils are well suited to agriculture, so that much of the area has been cultivated for a long period, though the early penetration of man has reduced the fauna, which includes many of the animals found in the tayga belt. In the nature reserve of the Belovezhskaya Pushcha near Brest is preserved the last herd of European bison.

Broad-leaved forests

Broad-leaved forests of the Far East grow on varied soils. There are also areas of a dark-coloured pseudo-chernozem in Amuria on which grows grass. The forests have usually a park-like character, particularly stands of larch. Southwards, the luxuriance of the forest increases: in the Ussuri valley, the vegetation becomes almost completely Manchurian in form. Species include the Mongolian oak, Daurian birch, Asiatic white birch, and many Japanese and Manchurian trees. The legendary and mysterious plant of the Chinese, the ginseng, is found in secluded valleys of the Sikhote Alin. Wild pear and crabapple provide fruit: the growth of imported fruit trees is inhibited by the late spring, frequent frost, and damp monsoonal summers. Wild Amur grapes are gathered for preserves and even wine. Fauna is composed of both northern and southern Asiatic forms: sable, squirrel, and lemming of the north live among Manchurian tiger, wild boar, leopard, Amur wild cat, and the racoon dog of the south. Many of the birds are Japanese and Manchurian. The forests are rich in colourful butterflies.

Grasslands

The grasslands which extend across Southern European Russia, Southern Siberia, and in isolated patches into the Baykal lands, form a transition from forest to desert. They change from wooded steppe on the north to poor steppe and semi-desert on the south. Historically, these open grasslands have been a routeway from the heart of Asia into Europe. Today they form the granary of Russia. The soils vary from modified forest soils to the rich black earth and the poorer chestnut and brown soils of the desert fringe. Most important is the black earth, the chernozem, the most fertile soil of the Soviet Union, which covers about 12 per cent of the area. The surface layers are loose and crumbly, black or chocolate in colour. The humus layer produced by a grassy vegetation is deep: the humus content varies from 10 per cent in the north to 20 per cent in the south-east. Much of the chernozem overlies loess, which allows easy drainage. It is no longer considered that chernozem may only form on loess, since it has been found overlying several parent materials. Southwards, the chernozem passes into chestnut and brown soils, which have their greatest extent in Asiatic Russia, where they have formed under strongly continental conditions and are poorer in humus. The soluble materials, however, are retained and are drawn to the surface in the summer, so that there is a tendency for the surface to become saline, notably in depressions where there is a high water table. In arid areas, where salts are left in the surface soil by evaporation, there is *solonchak*, a soil with a loose texture but no definite structure. Sometimes the salt concentration is so high that it leaves a white encrustation on the surface (often called *sor* or *shor*). In general, sodium salts predominate. A fall in the water table, increased irrigation or rainfall produces leaching of the soluble salts, except sodium, leaving an alkali soil, called *solonets*, a sticky, unworkable material when wet and an iron-hard surface when dry. Further leaching leaves *solod*, a soil poor in humus and minerals. All three raise difficulties for cultivation. The neutral chernozem with its ability for natural rejuvenation of its fertility is the most sought.

Wooded steppe

As the continuous forest cover thins and becomes more open, the traveller passes into the wooded steppe, with either patches of woodland or trees standing apart like parkland, while 'gallery forest' is also found in some places. The trees in Europe are mostly oak and mostly birch in Siberia. Other trees of the broad-leaved forests are also present. Aspen grows chiefly on *solonchak* or *solonets*. Beech is found in the extreme west, in Podolia. Pine grows on sands, chalk, or limestone. In the Siberian Kulunda steppe, pine grows along lines of dunes, forming ribbons of forest. Gradually, southwards, trees become fewer and poorer, so that in the true steppe trees are absent, except along gullies where salts have been leached or along river banks. The biting winds of winter and the dryness of summer are the enemy of trees, while deposition at the base of the humus layer in chernozem of minerals detrimental to trees also limits growth. There has been much planting of trees for shelter belts and moisture conservation in the steppe lands.

Steppe

The true steppe varies from a thick, close cover of grass in the north to more scattered tussocks of grass with bare ground between them on the southern edge. Towards the north, the bare patches are covered by moss, but they become a marked feature of the semi-desert fringe. Little of the grassy northern steppe remains in its virgin condition : the most northerly belt might be described as 'meadow steppe', which lies mostly in the wooded steppe zone, and to the south are found the narrow-leaved varieties of grass which form the true 'grassy steppe', while along the semi-desert southern fringe there is 'polyn steppe', a form of wormwood. The aspect of the steppe changes rapidly with the seasons. In spring, during the *rasputitsa*, the steppe is lifeless and easily turned to a muddy quagmire by passing animals and vehicles. With the rains of late spring and early summer, it is transformed to a mass of brightly-coloured flowers: in Europe, these include red and yellow tulips, buttercups, purple hyacinths, and other colourful flowers. By midsummer, the steppe is a waving sea of silvery, feathery grasses; but towards the late summer, the vegetation withers in the heat and dryness, leaving rolling, brown and lifeless plains, with a dust pall hanging in the air. A few flowers appear briefly in the moister autumn before they are killed by winter cold.

Ease of hunting the herds of steppe animals by man has led to the extermination of some of the original species. These include the wild horse (tarpan), the saiga antelope, and several species of deer. Species of ground-nesting birds have also disappeared with the ploughing of the steppe. The many small rodents are important in bringing new soil to the surface; but the small suslik, one of the most common forms, is sometimes a nuisance to farmers because of the size and number of its warrens and its incredible ability to reproduce which often far outstrips attempts at its extermination. Crops are also occasionally damaged by invasions of grasshopper and locust, while flies and mosquitoes, breeding in stagnant water in gullies, are an unpleasant summer feature.

Deserts

Towards the south, the steppe grows poorer and the brown soils turn to grey. The parent rock becomes a more important factor in soil formation. The typical grey soil is alkaline and clayey, with only a thin layer of humus. The intense evaporation of summer draws many salts to the surface to form widespread *solonchak* and *solonets*. On the southern fringe of the desert in Central Asia, some of the piedmont and foothill soils appear to be of loessic origin, which are rich and fertile when properly watered. There are also large sandy and clayey deserts, notably to the south of the Aral Sea, while the floors of undrained desert basins become salt pans, known in Russia as *takyr*.

The vegetation is scant and poor, with a continuous plant cover at no season, so that bare clayey and sandy patches give the landscape an unfriendly appearance. Moisture changes the aspect quickly: ground moisture or a sudden downpour can change the drab, parched country into a brilliantly-coloured wilderness. Many plants are specially adapted to resist drought, loss of moisture and swamping by sand, while others have long roots to draw up moisture. Trees are absent, except in the oases where a remarkable fertility is shown. Luxuriant and tangled jungle develops along rivers and in their deltas. A common tree of the moister parts is the strange *saksaul*, which grows in comparatively open thickets offering no protection from the intense sun, so that within, where wind is excluded, it is hotter than in the open country. The black *saksaul* has a peculiar dead appearance and stillness, with weirdly contorted forms. The fauna is composed of animals adapted to resist aridity and heat and able to move quickly over sand and stone. Many lie dormant for long periods each year. Protective colouring is well developed. Small rodents like jerboas and susliks are common. Foxes and hares are also found, besides wild cats (such as the rare Pallas cat). Antelopes, gazelle, and the wild ass (kiang) are now rare. The *chink* and other escarpments have mountain sheep. In the riverine *tugai* jungles, with their reeds, rushes, and tamarisks, live wild boars, hares, and jackals, while in the jungles of the Amu Darya delta and the Semirechye tigers abound. Birds are few, though aquatic forms live in swamps and oases and there are sparrows and storks, while the villages are renowned for the cooing of their doves. Lizards and snakes are frequently met, and there is a rich insect life. Scorpions are troublesome.

Transcaucasion lowlands

In great contrast to the desert are the regions described by Russian geographers as the 'humid sub-tropical lowlands' of Transcaucasia; on the west, the Kolkhiz Lowland, and on the east, the Talysh Lowland. The Kolkhiz Lowland has hot summers and very mild winters which allow plants to grow all the year, while the heavy precipitation is fairly evenly distributed without any marked dry season. Cloudiness also increases the relative humidity. The winds have a monsoonal character; warm and dry in winter and cool and moist in summer, though dry *föhn*-like winds sometimes shrivel vegetation. Soils are mainly types found in association with boggy conditions, though brown and alluvial soils are found on the higher surfaces, while red soils occur in the foothills. There is a luxuriant vegetation, with many evergreens, vines, and ferns, showing rapid growth and a great diversity of species. The Talysh Lowlands, around Lenkoran, have a climate similar to Kolkhiz, though the rainfall is less even and there is a marked dry period in June. Winter is sometimes extreme, destroying aquatic birds when rivers and swamps are frozen. Red and brown soils again predominate and there is a similar luxuriant vegetation, though evergreens are absent. The most common tree is the Persian parrotia, which has an especially durable wood. The forest, as in Kolkhiz, has been much cut over. There is a rich fauna of birds and even jungle cat, tiger, and the porcupine.

Diversity of soils and vegetation characterize the Soviet mountain regions, where aspect and altitude greatly influence the floral composition, and altitude often carries more northerly floral and fauna associations far to the south. To botanists, mountain forests, particularly in the south-east, are interesting as relics of Tertiary vegetation.

Mountains

The lower parts of the mountains of Central Asia are chiefly desert, with desert soils and areas of *solonchak*. Where piedmont loess occurs, irrigation may convert the country to rich farming land. In the foothills, desert quickly changes to steppe, though subject to frequent drought. As altitude increases, the steppe grows more luxuriant and the soils become richer in organic matter, while *solonchak* and *solonets* become less frequent and then absent entirely. Between 2,100 feet and 2,700 feet in the Tyan Shan, unirrigated land begins to prevail over irrigated. At about 3,500 feet, the forest steppe appears and on wetter slopes there is true forest, which thins into alpine meadows below the permanent snow line, whose lower border lies at about 8,000 feet. Pine and oak trees do not occur in the forests, but there is much juniper. Wild fruit trees are common: many species of our own fruit trees originated here. Coniferous forest grows in some of the eastern ranges, while deciduous forest is found in limited patches in the

Tyan Shan. The fauna is similar to the vegetational zones of the lowlands, though mountain sheep and goats are common, and there are the ibex and snow leopard, but in the Pamir fauna is similar to that found in the Himalaya and Tibet. The fauna also includes western Asiatic forms, such as found in the Kopet Dag.

The Great Caucasus rises from the steppe which gradually turns to forest steppe in which oak, hornbeam, elm, maple, pear, and apple are found. The trees thicken with altitude to become true forest, with beech common. In lower parts, cut-over forest has been replaced by thorn and scrub. The wetter western slopes carry a thicker and more luxuriant forest; while on the lower slopes around Novorossiysk, stunted woods and thickets of xerophytic species occur, to form a type of mediterranean vegetation transitional between forest and steppe. On the northern slopes, forest occurs directly in succession to steppe, in contrast to the western slopes where there is the intervening belt of evergreen mediterranean forms. Eastern Caucasia is much drier, so that steppe extends to higher altitudes and the forest is thinner. Owing to the late thaw and the development of plants in a moist soil at high temperatures, tall, herbaceous vegetation is common. Between 6,000 to 9,000 feet, the alpine zone begins, mostly poor grassy vegetation with few bogs, though again richer on the moister west. Caucasian ibex, chamois, and the snow pheasant are common.

Armenia contains much mountain steppe on the plateau, though summers are cooler than the true steppe and there is a deep winter snow cover. Summer precipitation is low and drought withers vegetation in late summer. The soils are chernozem-like, developed on highly calcareous residuals of igneous weathering. On the mountains of Armenia, xerophytic oak forest and juniper occur. The fauna contains forms of the suslik, jerboa, hamster, and fox found in Asia Minor. Patches of semi-desert occur in lower basins and valleys, such as along the middle Araks, while steppe along the middle Kura degenerates to semi-desert downstream. Both can be extremely fertile once irrigated.

In the mountains of the Southern Crimea, the calcareous materials produce meadow broken by rocky outcrops and patches of 'limestone desert', though there are forests, forest steppe, and steppe on the lower slopes. On south-facing slopes along the coast, where the climate is almost mediterranean, there is a very

TABLE 2

SELECTED DATA FOR REPRESENTATIVE CLIMATIC STATIONS

Region and Station	Altitude (ft.)	Months below 32° F.	Average Temp. °F.		Precipitation	
			Jan.	July	Total (ins.)	Main Months
Polar						
Tikhiya Bay *	—	10	− 2	34	12.0	n.a.
Wrangel I.	10	9	− 11	36	5.5	Ju.A.S.
Tundra						
Anadyr	75	8	− 19	53	7.5	S.O.
Dikson I.	66	8	− 12	41	6.4	Ju.A.S.
Kola (Murmansk)	23	7	11	55	16.1	A.S.O.
Sagastyr	11	9	− 34	41	3.3	A.S.
Northern Coniferous Forest						
West						
Arkhangelsk	50	6	8	59	16.8	Ju.A.S.
Berezovo	98	7	− 13	59	13.1	J.Ju.A.
Perm	535	6	12	64	23.5	J.Ju.A.
Solovets	56	6	14	54	15.4	A.S.O.
Tomsk	399	7	− 6	63	19.9	J.Ju.A.
Yeniseysk	260	7	− 12	64	17.2	Ju.A.S.
Zlatoust	1,502	5	4	61	20.4	J.Ju.A.
East						
Irkutsk	1,532	5	− 5	65	14.5	J.Ju.A.
Olekminsk	495	7	− 26	67	9.8	J.Ju.A.S.
Verkhoyansk †	330	7	− 59	59	3.9	J.Ju.A.
Yakutsk	300	7	− 46	66	13.7	J.Ju.A.
European Mixed Forest						
Kaliningrad	23	3	27	64	26.4	Ju.A.S.
Kazan	250	5	7	67	15.4	J.Ju.A.
Leningrad	16	5	17	64	19.2	J.Ju.A.S.
Lvov	930	3	25	66	27.2	J.Ju.A.
Minsk	660	3	20	65	23.9	J.Ju.A.
Moscow	480	5	14	66	21.0	J.Ju.A.
Wooded Steppe						
Chita	2,218	7	− 18	63	12.3	J.Ju.A.
Kiev	590	4	21	67	21.1	J.Ju.A.
Omsk	295	5	− 3	68	12.1	J.Ju.A.
Ulan Ude	1,527	7	− 15	68	9.5	J.Ju.A.
Voronezh	360	4	14	67	22.0	M.J.Ju.

* Franz Josef Land.
† Oymyakon has probably the coldest weather in Asia: − 70° C.

TABLE 2—*continued*

Region and Station	Altitude (ft.)	Months below 32° F	Average Temp.		Precipitation	
			Jan.	July	Total (ins.)	Main Months
			° F.			
Steppe						
Dnepropetrovsk	259	4	20	71	19.4	M.J.Ju.
Minusinsk	836	5	-4	68	12.2	J.Ju.A.
Odessa	210	3	25	73	16.1	M.J.Ju.
Orenburg	360	5	4	72	15.2	M.J.Ju.
Tselinograd	1,148	5	2	70	13.1	J.Ju.A.
Desert and Semi-desert						
Astrakhan	-46	3	19	77	6.4	Ap.M.J.
Irgiz	422	5	4	77	6.9	Ap.M.J.S.O.
Kazalinsk	205	5	15	79	4.8	O.N.D.
Mary	755	1	34	86	7.5	Ja.F.Ma.Ap.
Tashkent	1,596	1	30	80	14.6	Ja.F.Ma.Ap.
Turt-Kul	307	3	23	82	3.2	Ja.F.Ma.Ap.
Far Eastern						
Blagoveshchensk	466	5	-9	71	19.8	J.Ju.A.S.
Khabarovsk	246	5	-9	68	22.3	J.Ju.A.
Kirovskoye ‡	375	6	10	61	23.0	Ju.A.S.O.
Okhotsk	30	7	-13	53	7.5	Ju.A.S.
Vladivostok	50	5	5	69	14.7	J.Ju.A.S.
Humid Caucasia						
Batumi	20	0	43	74	93.3	S.O.N.D.Ja.
Black Sea Coast						
Yalta	13	0	39	76	19.9	N.D.Ja.F.
Mountain						
Gaudan §	4,455	1	29	72	8.6	Ma.Ap.M.
Kislovodsk	2,481	2	24	65	21.0	A.M.J.
Murgab	11,985	6	1	56	2.3	J.Ju.A.
Tbilisi	1,350	1	32	76	21.4	M.J.Ju.
Krestovyy Pass	7,735	3	11	53	66.7	A.M.J.

‡ Sakhalin.   § In Kopet Dag.

Key to Months

Ja. = January	Ju. = July
F. = February	A. = August
Ma. = March	S. = September
Ap. = April	O. = October
M. = May	N. = November
J. = June	D. = December

rich flora: 1,400 species have been found here compared to some 3,500 known for the whole of European Russia. Mediterranean, Japanese, and Chinese plants introduced artificially as well as native forms grow in profusion. The fauna shows an absence of forest forms and there is a poor bird life, with species foreign to the Black Sea steppes but found in the Balkans and Asia Minor.

The Ural ranges carry northern vegetation, soil, and fauna far to the south. In the extreme north, tundra covers the mountains, but gradually trees appear in the lower slopes. As one passes further south, the forest cover grows denser and tundra is found only in the higher elevations. In the central Ural, spruce and fir forest stands above the forest steppe to either flank, while the alpine zone, which is reached only on the highest summits, comprises arctic forms. The southern Ural carries a forest cover south to 52° N., mostly pine and larch, but forest steppe is also common. In some southern valleys, temperature inversion produces an inversion of vegetation: the middle slopes having a vegetation of more southern character than the valley floors. Few points in the southern Ural rise above the forest zone. The eastern slopes carry birch-forest steppe of Siberia and the broad-leaved trees of the west are absent. The ranges have a mixed fauna and do not form a zoogeographical divide as once believed.

The Altay ranges show marked variation in soils and vegetation depending on elevation and aspect. Steppe, tayga, light larch forest, and vast alpine meadows occur. Broad-leaved species found in the Caucasus and Tyan Shan are, however, absent. Fauna is composed of Mongolian species in the south-east, but Siberian tayga forms in the north-east Altay. These include bears, sable, polecat, lynx, gazelle, red deer, ibex, and reindeer, as well as the common bobac and suslik. The Sayans have steppe in the lower foothills and sheltered basins, above which stands Siberian tayga. Between 4,500 and 6,000 feet, the forest gives way to sub-alpine vegetation and immense areas covered by stones and almost without plant life. Siberian tayga forms are strongly represented in the fauna. The mountains of North-eastern Siberia have poor tayga, with Daurian larch, Mongolian poplar, and occasional patches of pine. At about 2,750 feet there is wooded tundra giving way to arctic vegetation. East of the Kolyma, tundra predominates over forest, which is almost absent. Mountain sheep, musk deer, and the hibernating suslik and bobac are found. Fox, squirrel, and ermine are the important fur-bearing species. They are bare and desolate mountains.

NOTES ON CHAPTER 2

1. Alisov, B. P., *Klimaty SSSR*. Moscow, 1956.
   Borisov, A., *Klimaty SSSR*. Moscow, 1948.
   Kendrew, W. G., *Climates of the Continents*. Oxford, 1955.
2. Suslov, S. P., *op. cit.* chapter iv.
   Cressey, G., 'Frozen Ground in Siberia', *Journal of Geology*, vol. 47, 1939.
   Charlesworth, J. K., *op. cit.* vol. I, chapter xxvii.
   Thiel, E., *The Soviet Far East*. London, 1957.
3. Berg, L. S., *Natural Regions of the U.S.S.R.* New York, 1950.
   Geobotanical Map of the U.S.S.R., *1 : 3.75M*. Moscow, 1956, with accompanying text.

CHAPTER 3

The Growth of the Russian State and the Russian Contribution to Geographical Exploration

THE expansion of the Eastern Slavs across northern Eurasia was part of a fundamental urge to expand and to push a wedge of agricultural colonization against the sparsely-settled forest and steppe. Their nationhood first grew on the vast and uniform Russian Platform, where natural defences were difficult to find, yet which enjoyed a certain impenetrability. The search for natural defences often conditioned expansion and movement, while impenetrability protected rising Slav nationhood from being swamped by more dynamic neighbours. Its growth was controlled by what has been termed 'the inevitable logic of geography which lies at the basis of history'.[1] Later this movement was coupled to the expansionist policy of the tsars.

Changing geographical values of terrain have changed the patterns of expansion. Westwards, the bounds of the Russian state have been limited by resistance of neighbouring nation states, their position reflecting the relative politico-military strengths of Russia and its neighbours. Southwards, the limits were influenced up to the end of the eighteenth century by the balance of power between the forest Slavs and the steppe nomads : in some periods, nomad power was strong enough to extend effective control deep into the forest, particularly in mediaeval times under the great universalist Mongol empire, whose organizational and jurisdictional concepts have been partly absorbed by the Russian state. Nomad decline allowed the Russians to push their boundaries in the eighteenth and nineteenth centuries to the Black Sea littoral and deep into Central Asia, until they impinged on the national or imperial interests of other powers. Thus restricted in their spread to west and south, the great expansion of the Russians moved north and east into vast expanses of *terra*

DOI: 10.4324/9781003172048-3   **78**

incognita, where they contributed to geographical knowledge by their explorations. From the seventeenth century, they have held a foothold on the shores of the northern Pacific and turned the emptiness of northern Eurasia into a Russian land.

The Russian Platform is crossed by great north- and south-flowing rivers, linked by easy portages across low watersheds. In early historical times, their interfluves were broad marchlands of forest and marsh. Whoever was able to control one or more of these great river routeways commanded great political and commercial wealth. Of particular importance, the routeway along the Dnepr, the Lovat, and the Volkhov gave control of the western Russian lands, with the important settlements of Kiev at the junction of forest and steppe, and Smolensk on its drier island commanding the northern portages; while further east, Moscow, lying in a commanding position in the nodal 'mesopotamia' between the Oka, Dnepr, and Volga basins, used its geographical advantages coupled with wise and foresighted diplomacy to rise to the political and spiritual leadership of the Russians. Likewise, the drier morainic lands offered defensive possibilities and were often chosen as the centres of small principalities, surrounded by wetter, marshy, impenetrable ground. The great morainic wall left in north-western Russia was an important military factor in the struggle between Swedish, German, and Baltic groups to deny Russia access to the sea itself. The transitional zone between the steppe and the forest in mediaeval times became the cockpit of the struggle of Russian nationhood to exert itself over the steppe nomads.

EARLY RUSSIA

The Slavs

Long before any Russian state appeared, the Slavs lived in the marshy valleys of the middle Vistula, the Pripyat, and the upper Dnepr.[2] Trappers, collectors of forest wealth, and primitive agriculturalists — they formed large family groups which held all property rights and from which arose later tribal political organisms. By the sixth century, their outward spread already split them into distinct western, southern, and eastern groups, living not only in the forest but also in the steppe fringe, where they were sometimes, like the Antes, under a non-Slavonic aristocracy. Between the fifth and ninth century, the Slavs spread northwards as far as Lake Ladoga and the upper Volga and Oka. The Finno-Ugrian tribes in the valleys of northern rivers, separated by forest and swamp, were unable to form allegiances against the superior organization of the invaders, who conquered and absorbed them, or pushed them back into the tundra fringe. The Slavs also pressed against the older Baltic stocks. As assaults from the steppe grew, particularly after the eighth century, along the edge of the forest, the movement of Slavs northwards increased.

The Nomads

The first historical record is of the Iranian Scythians holding the Pontic steppe in the seventh century B.C.; they replaced the earlier and enigmatical Cimmerians. Among them were possibly also Turkic and Mongol elements. The Scythian royal clan occupied the Dnepr bend from which all outsiders were excluded and where they left their most imposing burial mounds (*kurgany*). The Greek colonies established from the fifth century B.C., like Olbia (Odessa), Chersonesus (Sevastopol), and Panticapaeum (Kerch), traded with the steppe in grain and cattle, raised by Slav tribes subjugated by the Scythians, and with the northern forest in beeswax and Baltic amber, as well as in cloth and metalware brought from Central Asia. From the colonies hellenistic ideas filtered northwards. In the second century B.C., the Sarmatians started to occupy the eastern steppe and slowly conquer the Scythians. Now began a procession of invasions which continued until the thirteenth century A.D. From the north-west, Germanic Goths invaded and conquered the Sarmatians but rapidly accepted their customs and material culture. They reached their greatest power under Hermanaric in the fourth century, but were shortly afterwards defeated by the invasion of a Mongolian aristocracy leading Turkic-Mongol Hunnic tribes displaced westwards after their defeat by the Chinese. The Huns, however, established their main base in the strategically commanding plains of Pannonia, and in the fifth century A.D., under Attila, they were able to control most of Central and Eastern Europe. From east of the Volga in the mid-sixth century appeared the Turkic Avars who, like the Huns, recruited many Slavs into their service, as they spread rapidly across the steppe and into the Balkans, where they attacked Constantinople in 626. The Avars were crushed by the Franks. The Eastern Slavs now fell under the Turkic Khazars who had settled between the Dnepr and Volga as well as in Northern Caucasia from the seventh century. From their capital near Astrakhan, the Khazars began to build a great trading empire. Jews had begun to settle in the Greek Black Sea colonies, which slowly lost their Greek character, from the first century onwards, and from them the Khazars accepted the 'neutral' Judaism rather than either of the Christian concepts of Rome and Byzantium. At this time, a small kaganate of Irano-Slav As and Rus tribes also existed in the easily-defended Taman Peninsula in the Azov region.[3]

The Varangians and Kiev Rus

In the seventh century, Norse parties began to appear at the mouths of the rivers emptying into the Baltic, first seeking plunder and later trade.[4] By the eighth century, the Norse were moving in the upper Oka and Volga : shortly afterwards they defeated the Magyars in the upper Donets-Oskol region and, as vassals of the Khazars, established control over the As and Rus. Danes began exploring the Ladoga region, where, in A.D. 856, Rurik established Staraya Ladoga and shortly after moved to Novgorod, which led them down the Dnepr routeway. In 878, the Norwegian-Danish Varangians under Oleg, having taken Smolensk, captured Kiev from the Magyars and their lieutenants by ruse, shifting the geographical pivot from the Don-Azov region to the Dnepr basin. The Norse, numbering at most a few thousand, with superior capacity to organize and conduct military operations, were able to command the allegiance of Slav tribes and weld them into the first Russian state, Kiev Rus, where the Norse aristocracy was soon absorbed by a Slav majority. This state depended primarily on agriculture, but it also traded extensively and tried to establish contact with Central Asia, the Arabs, and particularly Byzantium across the steppes.

Early Russian Princedoms

The Russian lands comprised many small principalities, developed mainly in the basins of the river routeways, separated

FIG. 14

Early Russia. *Atlas Istorii SSSR*, Moscow, 1954, Vol. I, pl. 10.

from each other by wide marchlands of forest and swamp. They competed for superiority in an incessant internecine warfare, broken only when some prince stronger than the others for a short time forced unity upon them. Under Svyatoslav (964–972), Kiev Rus had a short-lived expansion, when he captured the Khazar towns of Sarkel and Itel and the 'Russian' kaganate of Tmutorakan as well as subduing the Danube Bulgars. Destruction of

Khazar power, however, opened the vital lower Volga-Don gateway to the fierce, nomadic, Turkic Pechenegs, who quickly destroyed the Russian garrisons and besieged Kiev. The situation was saved by Vladimir of Kiev, who brought a brief unity to the Russian princes and introduced Byzantine Christianity (986), though it was a century before it was generally accepted. Again, from 1019 to 1054, another strong prince, Yaroslav, exploited the weakness of the Pechenegs to extend briefly Russian territory to the Black Sea littoral, though on his death it was again taken by a new invasion of the Polovtsy (Cumans).[5]

From the tenth century, mounting nomad pressure along the steppe fringe forced Russian princes to build defensive walls to prevent their raids, but these were little use against an enemy enjoying the mobility of the horse and when the defenders lacked unity. To escape the nomad terror, large numbers of Slavs from the wooded steppe trekked northwards into the forests, where new principalities arose. In the north, Novgorod, with democratic government, began to develop as a commercial power with links outside Russia through the Hansa merchants; on the west, the united principality of Galicia-Volhynia was involved in the struggle between Roman Catholic Poland and the Orthodox principalities; in the south, Kiev lay as a great cultural centre and outpost on the steppe fringe; while in the north-east, Vladimir-Suzdal developed as the forerunner of Moscow. The Orthodox principalities were closely involved in the intrigues of Byzantine policy in its struggle with Rome, which brought new dangers to the Russian lands.

The Rise of Lithuania and the Baltic Germans

In the twelfth century, Germans appeared on the Baltic coast to trade with the Baltic people and to convert them to Christianity. Their reluctance to abandon paganism brought support for the missionaries from German knights, who were as ready to convert Orthodox Christians as pagans to Roman Catholicism. In the unsuccessful struggle with the Germans, Letts and Lithuanians moved eastwards into the thinly-settled marchlands of Russia. The growing power of the Lithuanian principality sought defendable, fortified towns and a healthy commercial backing, so that the spread became an agreement with Russian nobles rather than a conquest, for Polotsk, threatened by the Germans, passed to Lithuania, as did Pinsk in the south and the devastated Galician principality, both victims of the Mongol invasion. Despite the strong Russian influence, however, Lithuania fell under Polish culture, which separated the Roman Catholic nobility from the Orthodox peasants. Through Polish contacts many German ideas were introduced into municipal concepts and into land tenure. Whereas at first Lithuania had been a protector against the Mongols, later it was the Mongols who began to protect Russian princes against the powerful Lithuanians.

The Mongol Period

The Mongol invasion became one of the most significant formative periods in Russian nationhood.[6] After the election of Genghis Khan in 1206 to lead the dynamic expansion, the Mongol tribes invaded China before they attacked Khorezm and invaded Iran and Transcaucasia. A Mongol scouting party was defeated by a Russian-Polovtsian army at the Kalka river in 1223. Fifteen years later the Mongols appeared again, having subdued the Volga Bulgars and concentrated an army of 120,000-150,000 horsemen on the Volga. To secure their flanks, they attacked north-east Russia first, raiding Ryazan, Moscow, Vladimir, and Tver, riding on to Novgorod, but sixty miles from the town they retreated, fearing snow and floods. The Mongol-Tatar armies then settled in the key area between the Don and Volga before attacking Kiev and raiding Poland and Austria. The death of Udegey Khan in 1241 and the withdrawal of Batu Khan's armies saved Western Europe from their depredations.

The unity of the great and rapidly created 'universalist' empire quickly broke, so that the control of the Russian lands passed to the Western or Golden Horde settled at Sarai on the Volga. The Mongol-Tatar core of the state was surrounded by a much greater 'rimland' of subjugated peoples, so that the ratio between the Mongols and their subjects was never less than 1 : 100, becoming a symbiosis between the nomad and the subject agriculturalists and employing each group in its special capacity. The stability of the *Pax Tatarica* encouraged trade between Europe and Asia. Among the Russian princes, Mongol rule created a dichotomy. The western princes tried to free themselves of Mongol yoke by alliance with Lithuania, ready to recognize the authority of Rome. The eastern princes saw a policy of loyal submission to the Mongol khans as a means of defending Russian lands and Orthodoxy against western attacks and of strengthening their own internal authority. From the mid-fourteenth century, Mongol power began to decline, partly because its dynamism had passed and partly because of the cultural uncertainty of the khans, but the choice of Islam by the Mongol-Tatars separated them from the tenacious Orthodoxy of the Russian princes.

The impress of Mongol rule was deeper in north-eastern Russia than in the south-west. The old democratic ideas of the early social order were swept away. In their place came the widespread enforcement of Mongol cultural and legal ideas by the now dominant Moscow, which had taken over, as the khans weakened, the Mongol concept of universal submission and service to the state, the autocratic power of the prince requiring unqualified submission. Such ideas encouraged serfdom, though this was not enforced until after the main victories over the Mongols and Lithuanians. Even in the mid-sixteenth century, Russian peasants were still free to move from one allodial landowner to another. But there was a rapid change to the *pomestie* land-owning system conditional on military service to the tsar, while at the same time a shift from the old predatory, shifting cultivation to a three-field rotation did less to raise the status of the peasants than to increase their obligations and indebtedness to the landlords, who saw in them useful labour to supply the growing urban corn markets. Under such pressure, peasants began to escape to the frontier regions and beyond, so that in 1649 serfdom was made absolute. To escape the severity of Mongol rule in the fourteenth and fifteenth centuries, many men had sought refuge in monasteries, which multiplied in number and wealth, becoming an important colonizing force in northern Russia. Their influence was further enhanced by the final step to absolute power, after the fall to the Turks of Constantinople in 1453, when Moscow, exercising its political power against the claims of Kiev, became the centre of the Byzantine church.

The Rise of Moscow

The central position of Moscow placed it well to expand and unite other Russian principalities. From 1328 until abandoned in 1480, it held the right to collect tribute for the Tatars and had transferred to it the allegiance of several Tatar princes. By the early sixteenth century, its westwards expansion began to feel the stiffening resistance of Poland-Lithuania, but the capture of the dry morainic lands of the Smolensk Gateway put Moscow in a key strategical position. In 1478, Moscow had subdued in the north the rich commercial Novgorod, greatly adding to its material resources and opening the route into Siberia, as well as bringing Muscovite power close to the Baltic, where the strength of the Livonian Knights was waning, though a new danger appeared with the growing Swedish interest in Baltic domination.

Eastwards, Moscow started to absorb Tatar princedoms. With the help of the Kasimov Tatars, Ivan the Terrible defeated and annexed the Kazan Tatars in 1552, followed by annexation of the Astrakhan khanate in 1556. Now Russian control spread into the Don basin to include as suzerains the Don Cossack frontiersmen, later subjugating the Terek Cossacks and making the Nogay Tatars of the Ural river suzerains. Fortified towns were built to hold these districts (*e.g.*, Samara, 1586, Saratov, 1590, and the strategically key Tsaritsyn, 1589) and defended lines constructed to repulse nomad raids. As the ground was consolidated, Russians flowed in to settle the black earth lands, which became a new pioneering belt amid which rose towns such as Orel (1564), Voronezh (1586), Belgorod (1593), Valuyki (1593), while from Ufa (1586) the Bashkirs were drawn into the Russian state. Nevertheless, much of the southern steppe remained dominated by the Crimean Tatars and their overlords the Turks, while the remains of the Golden Horde in the steppe-desert fringe still controlled the southern road into Siberia. It was not until the eighteenth century that Russia was able to extend control to the shores of the Sea of Azov and the Black Sea.

The Move into Siberia

By the eleventh century, Novgorodians had explored the Ural and crossed the mountains to the mouth of the Ob in Siberia. Perm was annexed by Moscow in 1473 and Vyatka in 1489, becoming starting-points for the road to Siberia. The brunt of the exploration and conquest was borne by Cossacks, who built fortified settlements as they penetrated along the rivers (Fig. 15). Although the Siberian Khan paid tribute to Moscow from 1555, when the tsar described himself as 'Ruler of Siberia', it was not until 1584 that annexation began with Yermak's adventurous

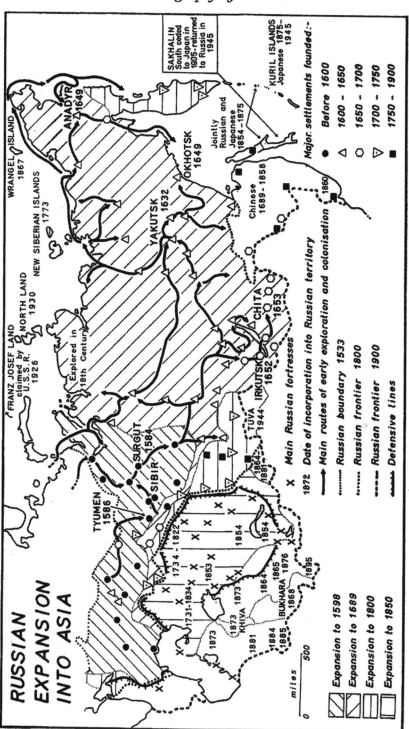

FIG. 15

Russian expansion into Asia. *Atlas Istorii SSSR*, Moscow, 1954-1955, Vol. I, pl. 19 and 28, Vol. II, pl. 5, 13 and 18.

campaign; but government regulation was first defined in the *Sibirskiy Prikaz* of 1637. By the mid-seventeenth century, Russia claimed all Siberia to the Pacific coast, though on the south there were clashes with warlike Buryats and Chinese outposts. Cossack penetration into Amuria was halted by the rising power of the Manchus; and in the Treaty of Nerchinsk (1689) Russia withdrew, laying a foundation for comparatively amicable Sino-Russian relations.[7]

THE DEVELOPMENT OF MODERN RUSSIA

By the seventeenth century, the long struggle against the Moslem Mongol-Tatars and the Roman Catholic Poles had isolated Russia from the outside world, while the self-sufficiency developed in isolation and the conservative attitudes of the Orthodox church acted against penetration of new ideas and outlook. The gulf between Russia and the rest of Europe therefore grew wider, but was to be bridged by the dynamic personality of the enlightened despot, Peter I (1689–1725), who turned still mediaeval Muscovy towards europeanization, so laying the foundation of modern Russia.[8]

On the south, Russia was isolated by the Crimean khanate and the Turks who controlled the Black Sea littoral as well as western Transcaucasia; the west was still blocked by a strong, but weakening, Polish-Lithuanian Commonwealth, who supported the decomposing Livonian Order to prevent Russia getting a foothold on the Baltic, from which it was also excluded by Swedish control of Estonia, Ingria, and Finland that had sealed the ultimate decline of the Novgorod trading empire. Against these adversaries only small territorial gains were made after long, exhausting campaigns in Peter's reign, but they taxed Russia's enemies more than Russia itself. They were offset by the encouraged expansion into Siberia, with penetration into the far north-east and the rich Altay country. The old search for furs was replaced by a new interest in metals, silver, copper, and gold. Peter's interest in maritime activity brought encouragement for exploration in the north Pacific along the Siberian and Alaskan coasts. Trade and diplomatic relations with China and Japan were strengthened. Although failing to attain its objectives, an expansionist policy into Central Asia began (Fig. 15). Unsuccessful campaigns to take Khiva and Bukhara roused fear in Persia, and a brief war left Russia with temporary control of the western and southern Caspian shore, though the aim of opening trade with India did not materialize. After Peter's death, the Central Asian advance continued, with growing Russian pressure among the Kazakh hordes, exercised from Orenburg (founded 1737) and other fortified posts. To contain the nomads and secure the Russian occupation, defensive lines were built along the Ural river, then between the Ural and Omsk, and later along the Irtysh to Semipalatinsk, each defended by a Cossack army.

The Partition of Poland

The mid-eighteenth century shifted interest to the southern and western frontiers. In the west, the greatest territorial advances were made after the partition of Poland between Russia, Prussia, and Austria : in 1772 areas around Polotsk, Vitebsk, and Mogilev fell to Russia ; further internal disorders gave Russia an opportunity to take large areas around Minsk and parts of Volhynia and Podolia in 1793 ; but the rebellion against the occupation in 1795 brought the final elimination of the Polish state, when Lithuania, Kurland, and Western Volhynia also passed to Russia (Fig. 16). Finally, in 1815 at the Congress of Vienna, Russia received the Grand Duchy of Warsaw created in 1807 with only slight modification. A few years earlier (1809), Sweden surrendered Finland to Russia, though it retained a measure of autonomy.

The Advance to the Black Sea and into Transcaucasia

In the late eighteenth century, it became imperative to secure the steppe frontier if Southern Russia was to be guarded against the threat of Turkish power and raids from the Crimea. Defensive lines were built as the frontier moved south ; between 1731–1735 the Ukrainian Barrier had been built between the Dnepr and Don, while a fortress was erected at Rostov-na-Donu. In 1736, Russia won from Turkey the bend of the Dnepr, the home of the Zaporozhye Cossacks, and the area between the Southern Bug and Taganrog. Here extensive colonization was conducted and 16,000 Serbs were planted on the right bank of the Dnepr and on the Lugan and Bakhmut. Another Turkish defeat in 1775 awarded Russia the mouths of the Dnepr and the Southern Bug, as well as the Don estuary and the Kerch Strait. The Crimean and Azov Tatars ceased to be Turkish vassals. In 1783, the Crimea was stormed, while later the Kuban was settled with

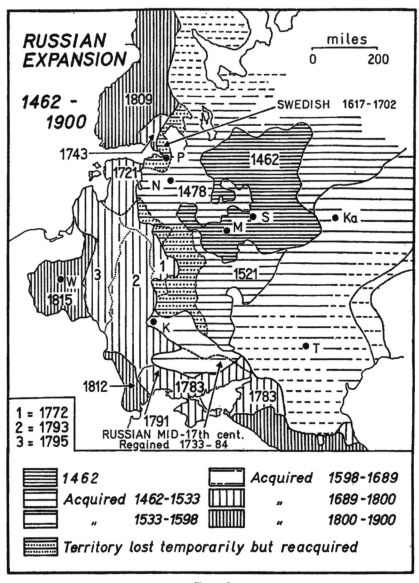

FIG. 16

Russian expansion in Europe, 1462–1900. *Atlas Istorii SSSR*, Moscow, 1954–1960, three volumes; *Westermanns Atlas zur Weltgeschichte*, Brunswick, 1956; *Putzgers Historischer Weltatlas*, Bielefeld, 1961. K: Kiev; Ka: Kazan; M: Moscow; N: Novgorod; P: S. Peterburg; S: Suzdal; T: Tsaritsyn; W: Warsaw.

Cossacks. In 1791, Russia received land between the Southern Bug and the Dnestr, Turkish since 1526. In 1812, Bessarabia passed to Russia, while for a time (1829–1856) even the Danube delta was held.[9]

At the beginning of the nineteenth century, the advance into Transcaucasia began from the defensive lines built at the foot of the Great Caucasus between 1763 and 1797 (Fig. 17). The main base for the advance was Vladikavkaz, a fortress built in 1784 to guard the upper Terek and the Krestovyy Pass. By 1783, the Russians had established a protectorate in Eastern Georgia, which when threatened by Persia in 1801 was ceded by its king to the tsar. In 1803–1804, treaties were made with Western Georgian princes, while between 1804 and 1813 the Azerbaydzhani princedoms were incorporated. In 1828, a war with Persia led to inclusion of Eastern Armenia, the khanates of Yerevan and Nakhichevan. Later additions were made from Turkey. The mountains were absorbed much later (Fig. 17): Kabardinia had been associated with Russia since 1761 and incorporated in 1825, though the main annexation period was

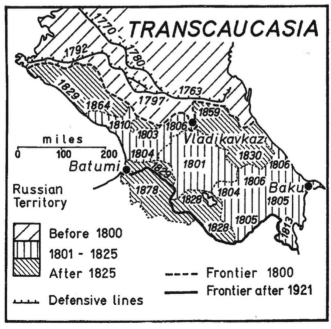

FIG. 17

Russian expansion into Transcaucasia. Sources as for Fig. 16.

between 1859, with incorporation of the Chechen, and 1864, with the final absorption of the Adegey. From 1878–1921, Kars and Ardahan were also held from Turkey. Russia was now in a strong position to command the Black Sea and the Caspian and to interfere in Middle Eastern affairs.

The Conquest of Central Asia

In the nineteenth century, as Russian control in Central Asia spread southwards towards the Aral Sea, so British control in India also extended.[10] From the middle of the century, British interest grew in Afghanistan and in Baluchistan, while raids by Khivans on Russian outposts from 1840 had kept Russia militarily active against the khanate with consequent territorial acquisitions. The Russian frontier moved south from Orenburg to the key fortress of Aralskoye in 1847, guarding the Syr Darya and the northern Aral Sea, while Fort Alexandrovskiy (Shevchenko) held the north-east Caspian shore (Fig. 15). On the eastern flank, Russia advanced to the Lake Balkhash area, with Sergiopol (Ayaguz) and Kopal, both erected 1847, as forts and a defensive line between them guarding the Dzungarian Gates. For further security, the Kokand fortress of Akmechet (Perovsk) was also taken. The Cossacks slowly pushed south : in 1854, Verniy (Alma Ata) was built as part of a new defensive line, while in 1865 the country between the Chu and Syr Darya was consolidated and Tashkent passed to Russia, followed in 1868 by Samarkand, with the Kokand khanate reduced to a vassal but dissolved after a revolt in 1876. Bukhara was also made a vassal. Pressure against Khiva could now be increased from these more suitable bases in the rich, loessic, piedmont belt, while the establishment on the south-east Caspian shore of Krasnovodsk in 1869 enabled Russia to enclose the Khivans and Turkmen in a pincer.

The struggle between the imperial interests of Russia and Britain now began to cause friction. Fearing the growing strength of Russia in the oasis belt and possible tsarist aims for a warm-water frontage on the Indian Ocean, feelers were started by Britain for a neutral zone between their respective possessions. Russia showed little interest in an offer to recognize expansion to the Amu Darya in return for acknowledgement of British interests in Afghanistan, hoping perhaps to wring a concession to maintain a fleet in the Black Sea again, lost in 1855 after the Crimean War.

British fears grew in 1871 after Russia annexed Kuldzha when a revolt the Chinese could not quell threatened their territory, though it was later returned to China. Further alarm came when the Khivan khanate was subdued in 1873 in order to stop raids against Russian outposts. Between 1880 and 1884, Russia conquered the Turkmen Akkal Teke and the Murgab oases, notably Merv (Mary). A critical clash between imperial aims arose in 1885 when Russia defeated Afghan troops at Kushka. A decade later, the occupation of Gorno-Badakhshan brought tsardom to the borders of India, so that Britain pressed for accurate frontier delimitation and the creation of the narrow 'buffer' of the Afghan Wakhan strip. Russia also increased political influence in Western Sinkiang and Tibet, where Chinese power was failing. The collapse of Russian prestige after the defeat by Japan in 1905 gave Britain an opportunity to regulate affairs. In 1907, Russia recognized British interests in Afghanistan, while Persia was divided between a northern Russian sphere of influence and a British one in the Persian Gulf littoral, and Tibet was accepted as neutral and nominally independent territory.

Russian Encroachment on China and Japan

Whereas in the seventeenth century, rising Manchu power had checked Russian spread into Amuria, expeditions visiting the area in the 1840's reported that it seemed unoccupied, but there was little diplomatic interest until Nevelskoy in 1849 showed that Sakhalin was an island, when a new strategic significance for the Amur mouth was seen. Pressure applied to China led to the surrender of the north bank of the Amur in 1858 at the Treaty of Aygun, while in 1860 the east bank of the Ussuri and the Sikhote Alin passed to Russia, who founded Vladivostok as a port and naval base. In 1854, agreements were made with Japan about the joint control of Sakhalin and the Kuril Islands, but the latter were exchanged in 1875 for full Russian control of Sakhalin. Increased pressure on China, who was supported in the 1894 war against Japan, led to the signing of a treaty allowing Russia to build a railway across Manchuria to Vladivostok to avoid the difficult terrain in Amuria. In 1898, to guarantee financial interests in the railway and the development of Manchuria, Russia took a twenty-five-year lease of the Liaotung Peninsula. At this time, Harbin grew into one of the largest

Russian towns in Asia, while troops occupied the northern railway zone in the Boxer rebellion. Offended by the Russian leasehold on Liaotung and worried about their intentions in the Yalu valley and Korea, Japan unexpectedly attacked the Russians a Port Arthur in 1904. Inadequately prepared for fighting so far from their main bases and depending on a weak railway link with them, as well as without real public interest at home and hampered by inferior commanders, the Russian forces were quickly and decisively defeated. The 1905 Treaty of Portsmouth awarded Liaotung to Japan as well as southern Sakhalin, while Russia withdrew from the Chinese Eastern Railway zone.

TERRITORIAL SPREAD AND ETHNIC DIVERSITY

In its limited spread on the west and north-west, Russia incorporated territory already well settled by Slav (mostly Ukrainian and Byelorussian) and Baltic peasantry dominated by Polish Lithuanian, or German landlords. The Baltic Germans (along with Germans from the west) formed an important element in tsarist administration and armies. On the north, Slav settlement assimilated or russified the less developed Finno-Ugrians. In the southern European steppe, Russian colonization swamped a sparse Tatar population including some Greeks and Jews.

On the Volga, a mostly Russian squirearchy was imposed on Tatar and other non-Slav peasantry, while Russian settlers occupied the best lands and mixed freely with the natives. The rapid spread of Russian power across Siberia added a diverse collection of primitive tribes scattered thinly across its vastness The conquest owed more to vodka, European diseases, and intermarriage with the Russian settlers than to guns. At first, Siberia filled with adventurers, Cossacks, and the banished, but later came a sedentary colonization by Great Russian and Ukrainian peasants who filled the best farming country. Although the Orthodox church never conducted great missionary campaigns russification was spread widely, particularly among tribes with weak national development. Yet some, like the Yakuts, managed to alter the customs and speech of Russian settlers and under Russian power, Komi and Yakut have become the lingua franca over great areas. Such an amalgam has produced the distinctive habits and dialect of the true Siberian, the *Sibiryak*.[11]

Spread of Russian political control into Central Asia did not bring a major Slavonic colonization until Soviet times. It introduced, however, the Cossacks, loyal allies of the tsar wherever booty was to be had along expanding frontiers. Some Russian peasants were settled in tsarist times, however, in the Golodnaya Steppe, and Siberian settlers usurped nomads along the northern steppe fringe. The southern oasis lands were already well settled by ancient oasis civilizations, now despotic, degenerate, feudal societies, who had never fully recovered from Mongol depredations. They tended to maintain their old Islamic traditions and remained apart from Russian settlements, so that Russian ideas were not accepted. The spread of commercial capitalism brought by the Russians passed, however, into the hands of Bukharan Jews. The uncompromising attitude of tsarist policy towards these people made little use of their cultural contacts beyond Russia's borders and may have contributed to some of its failures in adjacent countries.

Incorporation of such varied groups could have important repercussions on external relations. Russian behaviour in Bessarabia towards Moldavians, who regarded themselves as attached to Rome and anti-Slav, brought one of the greatest failures of imperial policy among the related Romanians. The conquest of Transcaucasia drew in the Armenians, who were confirmed as the dominant commercial element, spreading into the Black Sea littoral and at times dominating trade in Southern Russia. Their religious centre, the Echmiadzin monastery, seat of the *Katholikos*, lay in Russian territory, where a liberal policy led many to look to the tsar as protector, a powerful instrument in relations with the *Porte*. The acceptance by the Georgians of Russian protection and many cultural facets of Russian life further strengthened imperial control in this strategic bridge between Europe and Asia. Likewise, Russian interest in the Mongols in Southern Siberia was conditioned by the existence of a Pan-Mongolism generally anti-Chinese.

The widening territorial horizons of the eighteenth century broadened the opportunity for trade and economic development, to which the westernizing reforms of Peter the Great gave added impetus. Industrialization on a rising scale and the influence of commercial capitalism on agriculture in the nineteenth century were hampered by backward social conditions, notably continuation of feudal customs, only abolished in 1861, after which the spread of industry, the growth of commercial agriculture, and the building of roads and railways, developed quickly in close association. The population rose from 15 million in the eighteenth century to 40 million early in the nineteenth century and to over 100 million in 1897.

RUSSIAN TERRITORIAL DEVELOPMENT SINCE THE REVOLUTION

The tensions created by social and economic changes culminated in the Revolution of 1917, as big a divide as the Mongol period. It seriously weakened the country, so that many neighbouring states took the chance to seize territory, while other groups sought independence outside Russia's boundaries. The Baltic Letts, Lithuanians, and Estonians formed their own republics outside the new Soviet state, while Finland also asserted its independence. A war between the newly formed Polish republic and the Soviet state turned to the former's favour and was settled at the Treaty of Riga in 1920 with big territorial awards to the Poles east of the line suggested as a suitable boundary by Lord Curzon thus leaving in Poland large Byelorussian and Ukrainian minorities. Romania seized Bessarabia, the western Moldavian districts; and Turkey retook Armenian districts around Kars and Ardahan. There were also short-lived attempts at independence in Transcaucasia, among the Leningrad Ingrians and by non-Bolshevik Russian groups, particularly in very confused circumstances in the Ukraine. The only territorial acquisitions made by Russia were the unoccupied but strategically important arctic Franz Josef Land and Severnaya Zemlya.[12]

In 1939, in association with the German invasion, Soviet forces retook eastern Polish districts roughly to the Curzon Line, though including Lvov and Białystok. In 1940, the Baltic republics were annexed and pressure applied to Romania to surrender the Bukovina and Bessarabia, while after a winter war Russia took from Finland the northern shore of Lake Ladoga, which made Leningrad less vulnerable, and in Northern Karelia widened the former narrow corridor between Finnish territory and the Gulf of Kandalaksha, securing the defence of the Leningrad-Murmansk railway. Later, and most important, the richly metalliferous Petsamo (Pechenga) district was added and a common Russo-Norwegian frontier created. Russia received the Hankö Peninsula as a naval base, exchanged after 1945 for Porkkala (returned to Finland in 1955), both commanding positions in the Baltic.

The Soviet Union returned to these territories after their temporary loss in the German invasion as a victor in a peculiarly strong position. Without hesitation, Estonia, Latvia, and Lithuania were reincorporated and Memel (Klaypeda) given to the latter. The annexed portion of former German East Prussia, including the naval ports of Königsberg (Kaliningrad) and Pillau (Baltiysk), possibly because of its strategic significance, became a detached *oblast* of the R.S.F.S.R. Bukovina and Bessarabia taken from Romania were made into an enlarged Moldavian S.S.R., while a portion of the low forested Carpathians, including the important Uzhotskiy, Veretskiy and Tatars' passes and the out-wash fan of the upper Tisza, was placed in the Ukraine, which also received control of Ismail on the Danube delta. These were areas with substantial Ukrainian populations.

The Poles made powerful representation to sympathetic allies for a postwar return to the pre-1939 frontier, while the Russians were equally anxious to see a line settled more in accordance with Curzon's suggestions, with modifications to their advantage in the rich loessic and petrol-bearing Galicia and Volhynia, in the strategic Białystok-Suwałki area and in the Pripyat marshes. To achieve this, Russia supported a westward shift of Poland's boundaries to include German territory long regarded as a spring-board for attacks against the East.

Russia thus now commands in former East Prussia an important military position in the south-east Baltic, balanced on the south by the strong forward position on the Tisza fan, a gateway to Danubia and the Balkans (Fig. 18). These two positions are further strengthened by the right to maintain troops in Central Germany, a hundred miles from a main Western defence line along the Rhine and a mere thirty miles from the vital lower Elbe. From such a position of power, diplomatic and political means have incorporated most of Central Europe and the Balkans into the Communist system, with a stronghold on the Adriatic and Mediterranean in Albania, the importance of which has risen with the failure to infiltrate Greece and since the part defection from the Soviet *bloc* of Jugoslavia.[13]

Between the wars Russia was contained by non-Communist powers in its European and South-west Asian boundaries, and

FIG. 18

Russia's western frontier, 1900–1955. Sources as for Fig. 16.

developed its long-standing interest in the vast but potentially rich Sinkiang, which contains several peoples with ethnic relations in Soviet Central Asia. Manchu control was never popular, though the Chinese had consolidated their hold after 1882 in fear of Russian aims, and the fall of the Manchus led to increasing orientation towards Russia. A large Kazakh minority tried to establish a Soviet-type republic, but while the Russians were fighting Germany the Chinese reasserted their authority. In 1945, in the Sino-Soviet treaty, Russia declared no intention of interfering in Chinese internal affairs; but in the 1950 treaty, joint Russo-Chinese companies were formed to develop its wealth and Russia has taken part in building the Lanchow-Aktogay railway. In 1911, Russians occupied the upper basin of the Yenisey in the Sayan mountains, Tannu Tuva, a titular part of China, making it a protectorate. The basin around the main settlement, Kyzyl, controls several important routes into Western Mongolia. In 1921, the protectorate ceased and the country became nominally independent; but in the 1945 Sino-Soviet treaty, China abandoned claim to the area, which was incorporated into Russia, possibly because of its control of the Yenisey headwaters and its potential mineral wealth. At the fall of the Manchus in 1911, Russia also penetrated into Outer Mongolia, when it became an autonomous province within China. During 1921, it was declared a Soviet-type republic. Increasing Japanese activity in Manchuria in the thirties roused Russian interest in the country, as its eastern steppe was a gateway from Manchuria into Transbaykalia coveted by Japan. In 1945, China renounced claims made since the seventeenth century to Outer Mongolia, which is more closely associated with Siberia, for its richest and most thickly-settled part lies north of the great void of the Gobi, isolating it from China. The Khalka Mongols of the Republic have ethnic affinities with the Buryats of Soviet Baykalia. After attacking the Japanese in Manchuria in 1945, Soviet forces remained for a time and held a concession until 1955 in the Liaotung Peninsula. The Soviet retreat from this potentially rich and strategically placed marchland was influenced by the good surface relations with China, which has a powerful *de jure* claim, and the strength of its new Communist regime. The Soviet position was, however, strengthened in Mongolia by the 1945 treaty and in the north-east by the control of southern Sakhalin and the Kuril islands, which turned the Okhotsk Sea into a Soviet lake.

Russian troops entered Persian Azerbaydzhan during the war to handle Lend-Lease supplies from Persian Gulf ports and remained in the strategically commanding route centre of Tabriz after the official date of withdrawal. Hopes of establishing a puppet regime and separatist movement among the Kurds failed, but might have seriously threatened eastern Turkey and northern Iraq as well as giving Russia ultimate control of a Persian Gulf port.[15]

THE GEOGRAPHICAL EXPLORATION OF RUSSIA AND THE RUSSIAN CONTRIBUTION TO GEOGRAPHICAL EXPLORATION

Territorial expansion has carried Russians — both adventurers and scientists — into little-known regions, adding substantially to European knowledge of the world. At the same time, Russians have explored outside their own territory, just as foreigners have explored within Russia.[16] The earliest knowledge of the interior came from Greek colonies along the Black Sea littoral in classical times : Herodotus had a good knowledge of the Scythians of the steppes and of the Volga and Caspian lands.[17] The Alexandrian conquest, likewise, temporarily opened the way into Central Asia, while Greeks in later Roman service appear to have established land contact with Mary and with China through Kashgar. The Roman period also led to exploration in the Caucasus and along the Caspian littoral, so that by the time of Ptolemy (A.D. 150) geographical knowledge spread from the Aegean to the borders of China, but much knowledge gathered in classical times was later lost.[18]

Besides Norse parties from the Baltic coast, in the tenth and eleventh centuries, Christian missionaries and Byzantine diplomats also travelled in Russia, while in 1106–1108 the Russian Daniil went to Damascus and Jerusalem via the Dnepr and Byzantium. In the thirteenth and fourteenth centuries, diplomatic missions visited Russia to enlist Mongol support for the West against rising Islamic and Turkish power. The first envoy, the Franciscan Carpini, went in 1245 on a mission for the Pope to the Mongol Grand Khan, travelling through Kraków, Kiev, and down the Dnepr to the Sea of Azov, across the Caspian-Aral depression to the Syr Darya and along the western foot of the

Tyan Shan to Karakorum. He returned with a remarkably accurate knowledge of the country and its inhabitants. In 1252, Rubruck set out from the Sixth Crusade at Acre, followed a similar route but returned across the steppe north of Lake Balkhash and then through eastern Caucasia and Armenia. The unity of the steppes brought by the Mongols made it possible to 're-discover' Asia, with as great an impact on mediaeval commerce as the discovery of America on Renaissance commerce. The greatest journey was by the Polos, though they did not see much of later Russian territory. Arabs also travelled extensively: in 921–922, Ibn Fadlan had visited Urgench and the Volga as far as Bulgar near modern Kazan. Later in the same century, Ibn Dulaf went along the Volga and the Oka and returned along the Dnepr, while in the fourteenth century, the greatest Arab voyager, Ibn Battuta, visited the lower Volga, the Caspian, and reached the fringe of Siberia.[20]

In the fifteenth century, Turkish power spread across the gateways between Europe and Asia, already blocked by the enmity between Islam and Christendom, isolating Europe from trade by land with the East. By sea, the Portuguese dominated trade with the East via Africa and the western Atlantic was divided between Spain and Portugal, so that exploration to find a north-west or a north-east passage began. Although two Russians had reputedly sailed from Tallin to the White Sea along the Norwegian coast in 1497, the main voyage was by Chancelor and Willoughby from Deptford in 1553 to find a north-east passage to Cathay. Of three ships, only Chancelor reached Russia, landing near modern Arkhangelsk and then continuing to Moscow, where his friendly reception led to trade opening between Russia and England through the Muscovy Company. In 1557, Jenkinson, by the same route, travelled to Moscow and down the Volga to Bukhara, later (1561) visiting the western shore of the Caspian and Kazvin in Persia. The first Russians to visit Britain are supposed to have returned in Chancelor's ship.[21] Russian 'pomory' are claimed to have reached Novaya Zemlya and east of the Ob estuary by the late fifteenth century, but the principal voyages in the White Sea and adjacent waters came in the sixteenth century, when Vaygach Island and the lower Pechora were explored. In 1584, English servants of the Muscovy Company reached the Ob mouth and found Matochkin Shar, while about the same time a Dutchman visited Siberia overland and sailed in the Ob estuary. In 1596,

Barents, sailing a more westerly course than usual, reached Bear Island and part of Svalbard, wintering on Novaya Zemlya. Towards the end of the century, an apparently warmer period opened otherwise ice-bound seas.

First adventurers into Siberia were possibly Novgorod merchants in the eleventh century, and a wedge of Novgorod territory extended across the northern Ural in the twelfth century. By the sixteenth century, Russia was established on the eastern slope of the Ural. The exploration of the Ob opened the conquest of Siberia, though the name had first appeared on a Russian map in 1367. Entry was made through the northern river routes, for the steppe nomads still dominated the steppe route. From the capture in 1583 of the Sibir khanate by the Cossack Yermak, a rapid spread across almost empty forests to the Pacific shore was completed in less than sixty years, though filling in detail was to last many decades (Fig. 19). The impetus carried Russians into Alaska and made their contribution to exploration of the Pacific Ocean.[22]

Using the great rivers and the easy portages between them, Russians reached the Tom by 1600, crossing the Yenisey to Lake Baykal and the middle Lena. In 1620, Penda travelled through Central Siberia and returned along the Angara; in 1630, using the Lower Tunguska and the Vilyuy, Vasilyev reached the fertile middle Lena basin. From here, Moskvitin went up the Aldan and the Maya and over the Dzhugdzhur mountains to the Pacific shore near Okhotsk in 1639. In 1643, Ivanov visited Lake Baykal, more fully explored by Galkin in 1648–1650. Stadukhin and Rebrov explored the northern Siberian coast east of the Lena estuary and by 1644 had travelled along the Indigirka and the Kolyma. For two years Stadukhin lived in Nizhne Kolymsk, gathering information about the Chukchi, who told of mounds of ivory waiting collection far to the east. Leaving Yakutsk in 1648 to reach the fabled Pogikha river east of the Kolyma, the Cossack Dezhnev unwittingly sailed between Asia and North America through the strait later named after Bering and reached the lower Anadyr. Some authorities disbelieve the story and claim that he crossed to the Anadyr by land along the upper Anyuy as Stadukhin did in 1650. Some of Dezhnev's party are supposed to have eventually reached Kamchatka, though its exploration is usually credited to Atlasov between 1697 and 1699. Dezhnev's report handed to the Governor of Yakutsk in 1655

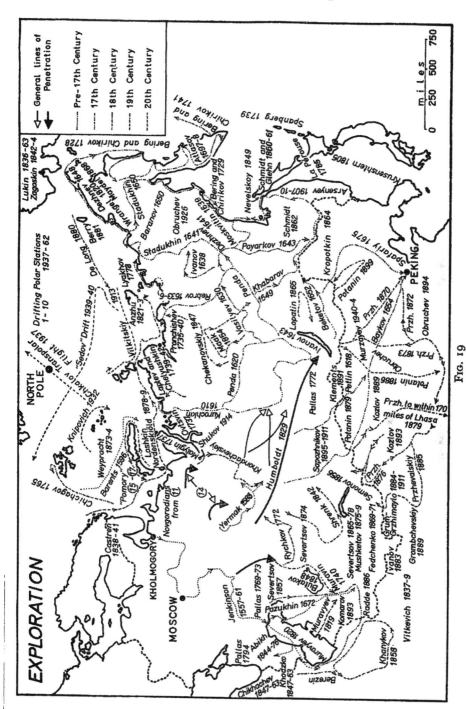

Exploration. *Atlas Istorii Geograficheskikh Otkrytii i Issledovanii*, Moscow, 1959.

Fig. 19

was not known in S. Peterburg until over eighty years later, eight years after Bering sailed from Kamchatka to find the mysterious Strait of Anian through water supposedly visited by Dezhnev.

With the outline of Siberia known and the Arctic still unpromising and inhospitable, rumours of riches across the water roused Russian interest in the North Pacific, which was to extend south along the North American Pacific coast as far as modern California. Initial encouragement was given by Peter the Great, though the main phase came after his death. In 1728, Bering sailed through the strait between Asia and America, but as he had not reached a point definitely known to be on the Siberian coast, he had not proved conclusively that the continents were separated. The American coast was probably first sighted about 1732; but Bering's second expedition in 1741 saw Mount Elias and Kayak Island when in search of the imaginary Gama Land in latitude 47° N. Chirikov, who lost contact with Bering at the start of the expedition, reached the northern tip of Queen Charlotte Islands and sailed along the coast to about 58° N. Other important voyages were made by Shelekov and Pribylov in the 1780's. The furs brought back by these voyagers encouraged the foundation of the Russo-American Company in 1798. By 1804, the Russians had thirteen colonies between Kodiak and Sitka, their main base of Novoarkhangelsk. By 1812, they had founded Fort Ross near San Francisco, held until 1841. The principal exploration of the Alaskan interior was by Zagoskin and Lukin in 1836–1863.

Difficulty of overland communication with Eastern Siberia began a search for sea routes from European Russian ports to the Northern Pacific, leading to exploration within the Pacific basin and to the great Russian circumnavigations. Between 1803 and 1806, Krusenshtern and Lisyanskiy sailed round the world, discovering several Pacific islands and visiting Hawaii, which the Russo-American Company tried to secure in 1813, abandoning the claim a year later.[23] About the same time, Lasarev discovered the coral Suvorov Islands and Rumyantsev found islands in the Tuamotu and Marshall groups, though Russia never appeared to consider annexation. From 1823 to 1826, Kotzebue sailed round the world carrying the oceanographer Lenz and was followed within three years by Litke (Lütke) who surveyed several island groups in the Pacific. In coastal waters, Krashenninikov (1737–1741) explored Kamchatka and Spanberg sailed along the Japanese coast in 1739. These waters were also visited by the Frenchman La Pérouse (1785–1788) as they had been by Cook in 1780.

Russian voyages in the seventeenth and eighteenth centuries added much knowledge about the Arctic. The search was for warm, ice-free water believed to exist east of the Ob estuary, first reported by Cherry in 1587, while others sought a narrow neck of land supposed to link Asia with Greenland, though natives reported otherwise. In 1620, Russians had approached the Taymyr Peninsula from the Yenisey and there had been exploration a little later eastwards from the Lena delta. Thorough exploration of the Taymyr Peninsula was made by Chelyushkin and Laptev (1739–1742) and in 1773, Lyakhov reached the New Siberian Islands, but these were not properly explored until 1881 by de Long during voyages in the East Siberian and Chukchi Seas, after Anzhu (Anjou) had sailed round them in 1820–1824. About 1760, Russians sailed round Novaya Zemlya, though detailed investigation came only in the first thirty years of the nineteenth century. Wrangel Island was reported to Wrangel by natives in 1823, but it was first seen by Kellett in 1849 and visited by Berry in 1881. Weyprecht's expedition of 1873 visited Franz Josef Land and it was not until 1913 that Vilkitskiy discovered Severnaya Zemlya. The end of the nineteenth century brought the important journey of Nordenskiöld (1878–1879) from Tromsö to the Pacific along the northern Siberian coast, so completing after three hundred years the search for the north-east passage. In contrast, Russians under Bellinghausen in 1819–1820 made one voyage to the Antarctic.[24]

Russian explorations in Caucasia began early, though records are few. In the fifteenth century, Nikitin of Tver travelled down the Volga and along the western Caspian shore to Persia and to Indian ports on the Arabian Sea, returning through Armenia and the Black Sea. Several Russian envoys crossed the Great Caucasus into Georgia during the sixteenth and seventeenth centuries. In 1620, a Russian envoy journeyed to Khiva and in 1669–1672 Pazukhin crossed the Ust-Urt from Astrakhan to Khiva, visited Bukhara, and returned through Persia and the Caspian. In Central Asia, Baykov travelled along the Irtysh and crossed Mongolia to Peking between 1654 and 1658, a more westerly course than that followed in 1618 by Petlin, who went from Tomsk via Ubsu Nur to Peking. In the eighteenth century, journeys into the Central Asian oases were hazardous because of the hostility of the Khivans, who killed many Russian agents and explorers, though Muravin's journey in 1741 along the eastern shore of the Aral Sea to Khiva marked a success. Gmelin travelled between 1763 and 1773 along the Caspian into Persia, only to die in confinement in Caucasia. Along the northern steppe-fringe between Southern Russia and the Amur, Pallas made important observations on fauna between 1768 and 1774.[25]

The nineteenth century brought great activity in all parts of Russia and carried Russian explorers into adjacent as well as distant lands. The detailed exploration of the interior of Siberia continued and scientific journeys in European Russia became common. Topographical survey started, notably in the second half of the century, with the development of railways. There was the exploration of the Amur-Ussuri lands where superficial exploration began in 1643–1648 under Poyarkov and later Khabarov (1649–1651), but Chinese opposition prevented their continuation. In 1805, on a mission to China, Golovin saw more of the region, followed in 1832 by Ladizhinskiy and the unsuccessful Muravyev expedition of 1848. In 1850, Russians reached the mouth of the Amur, where a new importance for the river was seen after Nevelskoy in 1849 had proved Sakhalin to be an island, visited and described by Schmidt and Glehn in 1860–1861.[26]

Elsewhere explorers filled in detail. Yerman studied the northern Ural in 1828 and made voyages on the lower reaches of the Siberian rivers. Humboldt, on his last journey, accompanied by Rose and Ehrenburg, travelled through Perm and Tobolsk to the Ob and then south to the Altay, returning through Omsk and the Volga to the Caspian and back to Moscow along the Don. He produced a hasty but brilliant account of the area's geology and mineralogy. In North-eastern Siberia, Middendorf in 1843 from Turukhansk visited the lower Yenisey, the Pyasina, and lower Khatanga, following the Laptev Coast into the Taymyr Peninsula. Later he crossed the Dzhugdzhur mountains from Yakutsk and went south over the Stanovoy and into Amuria before returning to Lake Baykal. Siberian ethnography was studied between 1842 and 1849 by Castrén and by Ahlquist in 1853–1858, followed in the Altay by Radlov (1868). After 1850, greater interest was taken in Baykalia and in Central Siberia, notably in the Angara, Vitim, and Olekma basins by Lopatin (1865) and Chekanovskiy (1873–1876). The desolate swamps of the Ob-Irtysh lowlands were explored between 1877 and 1881 by Khondachevskiy.

In Central Asia towards the latter part of the nineteenth century, exploration was given a stimulus by the clash of Russian and British imperial interests.[27] Russian control along the Irtysh brought exploration of Eastern Kazakhstan, when in 1833 Fedorov reached Lake Balkhash at the mouth of the Lepsa river, while between 1840 and 1842 the Semirechye was explored and ten years later Kuldzha, in the upper Ili, was reached. After the building of Alma Ata in 1854, Russians began to explore around Issyk Kul and in the Chu valley, preparing the ground for explorations in the Tyan Shan and later in Dzungaria. In 1819, Muravyev explored from Balkhan Bay to Khiva, opening Russian interest in Turkmenia. Khiva was visited by several Russian and British travellers, of whom Stoddart and Conolly perished horribly in the Emir's vermin pit. In 1848 Butakov had systematically explored the Aral Sea and later reached Chinaz by boat along the Syr Darya. In 1856–1858, Semonov explored the Tyan Shan to test Humboldt's theory of their volcanic origin, and between 1869 and 1871 Fedchenko climbed in the Pamir and adjacent ranges, while Valikhanov crossed from Issyk Kul to the Naryn valley and then on to Kashgar.

In Western China and Mongolia, British explorers had come from the Chinese coast in the late 1850's, while later penetration from India was made. Few records of early Russian investigations exist; but a Russian party travelled from Peking to the Altay country in 1868, Radlov visited Kobdo in 1870, and in 1875 an expedition went to China seeking the best road from Semipalatinsk to Lanchow. Though quite numerous, Russian (like other) expeditions had no important geographical results. The exception was Przhevalskiy who had explored in the Ussuri region between 1867 and 1869 before starting his four great journeys in Central Asia between 1871 and 1888, doing much to complete the map of Mongolia and Western China.[28] In 1871, he set out from Kyakhta to Ulan Bator and crossed the Gobi to Kalgan, visiting the Ordos country and north-eastern Mongolia, returning eventually to Peking, which he left in 1872 to travel through Kansu to Kuku Nor and the Tsaidam. He intended to reach Lhasa but failed, so returned along the edge of the Ala Shan and across the widest part of the Gobi to Ulan Bator, bringing a rich collection of fauna and flora as well as news of wooded country standing amid arid lands in the Kansu mountains north of the Hwang Ho and east of Kuku Nor. Before Przhevalskiy's second voyage, Padarin explored northern Mongolia trying to identify the site of Karakorum, and Sosnovskiy in 1872 visited the Black Irtysh and Dzungaria. At this time the British from India were making important journeys and Lhasa was becoming an international focus. The second journey begun in 1876 aimed at reaching Lhasa from the north, and Przhevalskiy, starting from Kuldzha, crossed the Tyan Shan to the Tarim river and then Lop Nor. He discovered the Altyn Tag range but was unable to find a way through as winter set in. Exploring the foot of the mountains, he concluded it formed an outer buttress to the Tibetan Plateau. Before Przhevalskiy's third voyage, Russian explorers made a great advance in knowledge, bringing the first modern account of Hami and Dzungaria and exploring in the Pamir. Detailed information about the Tyan Shan, the Alay, and the Trans-Alay, as well as the Tannu Ola and the Bogdo Ola, was collected, while the Valley of the Lakes in Mongolia was explored. Turfan was also reached. Przhevalskiy left Zaysan on his third voyage in 1879, crossing Dzungaria to Hami and then across the Altyn Tag and east of Tsaidam, over the Tang La range, to within 170 miles of Lhasa. He was refused permission to visit the town and returned by a new route across Mongolia to Kyakhta. In 1883–1884, he made a fourth and last journey, examining more country in Tibet and Eastern Turkestan in an attempt to reach Lhasa, which was again foiled. He crossed the Gobi again and explored the headwaters of the Hwang Ho. Reaching Khotan, he turned north across the dreaded Takla Makan to Aksu, crossing the Tyan Shan to Issyk Kul, where he died. On his grave a bronze eagle spreads its wings above a map of Asia. Tsybikov reached Lhasa in 1889 and other companions of Przhevalskiy continued to fill in the gaps, among them Potanin, Kozlov, Bogdanovich, and Roborovskiy. Between 1889 and 1890, the brothers Grum-Grzhimaylo travelled in the Tyan Shan away from accepted routes along the northern flank and discovered a part of Turfan below sea level. Grombchevskiy explored the Pamir, while Younghusband was doing so on the British flank. In Mongolia, Radlov, Klements, and Obruchev filled in detail and in the Altay-Sayan country, Sapozhnikov worked between 1895 and 1911 and the Englishman Carruthers was active in 1910–1911.

Contribution to Geographical Exploration

Russian political spread into Caucasia was also accompanied by exploration. Abikh for thirty-three years studied the geology and topography of the region, particularly in Dagestan, later continued by Radde. Russian officers under Khodzko surveyed the provinces between 1847 and 1863. The mountains of Caucasia were first described in detail by the Englishman Freshfield after 1868. Chikhachev, an attaché at the Russian embassy in Istanbul, made extensive journeys in Asia Minor and Armenia between 1847 and 1863, while Berezin (1842–1844) and Khanykov (1858) travelled widely in northern Persia.

The Russian contribution to geographical knowledge outside Eurasia is generally little known. In Africa, Barskiy visited Sinai in 1727 and Norov travelled along the Nile in 1834, followed later by Kovalevskiy. Junker's travels in Africa in 1875–1886 are also claimed by the Russians. Yefremov in 1774–1782 visited India and Ceylon and was followed by many Russians in the late nineteenth century. Between 1871 and 1883, Mikhlukho-Maklay travelled widely in the East Indies and New Guinea, though he began his career as an explorer in the Canary Islands and the Red Sea coast. In South America, the most noted Russian traveller was Langsdorff and his Brazil expedition in 1821–1828, while the geographer Voyekov travelled in the U.S.A. in 1873 and Yeshchenko visited Australia in 1903.

Detailed scientific investigation, geological and topographical survey have been intensified by the Soviet authorities and great encouragement given by the Academy of Sciences. Unfortunately, the international tension has prevented publication of a great deal of the results, which are regarded as state secrets.[29] All maps on scales greater than 1 : 500,000 are secret, though in 1946 the first 1 : 1M map of the country was completed and several important thematic maps have been published. The scope of Soviet geographical knowledge is mirrored in the magnificent volumes of the *Atlas Mira* and the *Morskoy Atlas*, based on a new ellipsoid measured from Pulkovo Observatory to a point near Svobodnyy in Eastern Siberia, showing a greater radius than Bessel's ellipsoid of 1848. The Krasovskiy ellipsoid has been standard use since 1946. Arctic exploration in connection with the operation of the Northern Sea Route has been actively pursued, particularly in the deeper waters of the Central Arctic Basin. In 1930, the exploration of the ice-breaker *Sedov* confirmed the existence of Vise Island, while many small islands have since been plotted in the Kara Sea. Detailed plotting of the northern lands began with Urantsev's expedition to Severnaya Zemlya in 1930–1932, when a detailed survey of the northern Siberian coast revealed many changes in outline, notably in Gydan Bay and Taz Bay. In 1943, a scientific survey of the Taymyr region was made. Navigation in high latitudes has been studied and a warmer period has eased movement in Arctic waters, so that a west-to-east passage was made in 1932 by the *Sibiryakov* and the unsuccessful but scientifically interesting voyage of the *Chelyushkin* added further knowledge. In 1934, an east-to-west voyage was made by the *Litke*. Research into arctic conditions has also been conducted from a series of bases drifting on polar ice. The first was laid down in 1937 and since then well over half a dozen stations have been maintained for varying periods, from which the Russians have made a detailed study of submarine topography, leading to the plotting of the Lomonosov (Harris) Ridge, as well as gaining additional knowledge of earth magnetism. In the late thirties, they also pioneered trans-polar flight. During the late fifties, Soviet scientists maintained stations on the Antarctic icecap, mostly between 80° and 110° E. in Queen Mary Land on the Shackleton Ice Shelf, where Mirniy has been the base. At the same time, Russian vessels have conducted oceanographic exploration, notably the *Vityaz* in the Pacific Basin.

The Soviet period has also completed the detailed investigation of relief in Central Asia and in Siberia, where major discoveries were made by the Obruchev expedition in the North-east in 1926.[30]

NOTES ON CHAPTER 3

Many excellent histories of Russia exist. Two very readable accounts are Vernadsky, G., *A History of Russia*, 4th revised edition, Oxford, 1954, and Sumner, B. H., *A Survey of Russian History*, London, 1944. Three historical atlases presenting ancient, mediaeval, and modern Russian history are available : *Atlas Istorii SSSR dlya sredney Shkoly*. Moscow, 1954, 1955, 1960.

1. Vernadsky, G., *A History of Russia*, 4th ed. Oxford, 1954.
2. Vernadsky, G., *Ancient Russia*. New Haven, 1946.
   Niederle, L., *Slovanské Starožitnosti*. Prague, 1925.
   Coon, C. S., *Races of Europe*. New York, 1939.
   Peisker, T., 'The Expansion of the Slavs', *Cambridge Modern History*, vol. II, chapter xiv.
   'The Asiatic Background', *op. cit.* vol. I, chapter xii.
   Allen, W., *The Ukraine — A History*. Cambridge, 1940.

3. Mongait, A., *Archaeology in the U.S.S.R.* Moscow, 1959.

Vernadsky, G., *The Origins of Russia*. Oxford, 1959.

Ancient Russia. Yale, 1943.

East, W. G., and Moodie, A. E., *The Changing World*, chapter xiv. London, 1957.

Fleming, R., 'Some Factors in the Development of Russia with Special Reference to European Russia', *Studies in Regional Consciousness and Environment*. Oxford, 1934.

4. Kendrick, T., *A History of the Vikings*. London, 1930.

5. Vernadsky, G., *Kievan Russia*. New Haven, 1948.

Macartney, C., 'The Pechenegs', *Slavonic Review*, vol. 8.

6. Vernadsky, G., *Russia and the Mongols*. New Haven, 1953.

7. Semyonov, Y., *The Conquest of Siberia*. London, 1944.

Treadgold, D., *The Great Siberian Migration*. Princeton, 1957.

Sumner, B. H., *A Survey of Russian History*. London, 1944.

Kennan, G., *Siberia and the Exile System*. New York, 1891.

Lantzeff, G., *Siberia in the Seventeenth Century*. Berkeley, 1943.

Fisher, R., *The Russian Fur Trade, 1550–1700*. Berkeley, 1943.

8. Sumner, B. H., *Peter the Great and the Emergence of Russia*. London, 1950.

9. The spread of the Russian lands at this period can be traced in vol. II of *Atlas Istorii SSSR*.

Allen, W., *op. cit.*

Konovalov, S., *Russo-Polish Relations — An Historical Survey*. London, 1945.

10. Skrine, F. H., *The Expansion of Russia, 1815–1900*, 2nd ed. Cambridge, 1915.

Skrine, F. H., and Ross, E. D., *The Heart of Asia*. London, 1899.

Kerner, R. J., *Russia's Urge to the Sea*. Cambridge, 1942.

Krausse, S., *Russia in Asia, 1558–1899*. London, 1899.

Baddeley, J., *The Russian Conquest of the Caucasus*. London, 1908.

Russia, Mongolia and China, 2 vols. London, 1919.

Sumner, B. H., 'Tsardom and Imperialism, 1880–1914', *Proc. Brit. Academy*, 1940.

11. Treadgold, D., *op. cit.*

Admiralty Handbook of Siberia. London, 1920.

12. Bowman, I., *The New World*. London, 1st ed. 1921, 4th ed. 1928.

Żółtówski, A., *The Border of Europe*. London, 1951.

13. Wilmot, C., *Struggle for Europe*. London, 1952.

Lloyd, T., 'The Soviet-Norwegian Frontier'. *Norsk Geografisk Tidsskrift*, 1955–1956.

Wagner, W., *The Origin and Genesis of the Oder-Neisse Frontier*. 2 vols. Frankfurt/M., 1957.

14. Carman, E. D., *Soviet Imperialism*. Washington D.C., 1950.

East, W. G., and Spate, O. H. K. (Ed.), *The Changing Map of Asia*, 3rd ed. London, 1956.

Dallin, D., *The Rise of Russia in Asia*. New Haven, 1949.

Beloff, M., *Soviet Policy in the Far East, 1944–1951*. Oxford, 1953.

East, W. G., and Moodie, A. E., *The Changing World*.

Lattimore, O., *Inner Asian Frontiers of China*, 2nd ed. New York, 1956.

Lattimore, O., *The Pivot of Asia*. Boston, Mass., 1950.

'New Political Geography of Inner Asia', *Geographical Journal*, vol. 119, 1953.

Tang, P., *Russian and Soviet Policy in Manchuria and Outer Mongolia*. Durham (N.C.), 1959.

15. Wilber, D., *Iran — Past and Present*. Princeton, 1958.
16. A useful modern discussion of the history of geographical exploration is Baker, J. N. L., *A History of Geographical Exploration*, London, 1945; also an atlas of exploration has been published in Moscow (1959), *Atlas Istorii Geograficheskikh Otkrytii i Issledovanii*.
17. Herodotus, *The Histories*. Ed. Selincourt. Penguin, 1961.
18. *The Geography of Claudius Ptolemy*. Edited and translated by E. L. Stevensen. London, 1932.
19. Beazley, C. R., 'The Text and Versions of John de Plano Carpini and William de Rubruquis', *Hakluyt Society*, 1903.
    Yule, H., 'The Book of Ser Marco Polo' (Ed. Cordier), 1903.
    Yule, H., 'Cathay and the Way Hither' (Ed. Cordier).
       Hakluyt (2nd series), vols. XXXIII, XXXVII, XXXVIII, XLI, 1915–1916.
    Morgan, E. D., and Cook, C. H., 'Early Voyages to Russia and Persia', *Hakluyt* (2nd series), vol. LXXII, 1902.
20. Schoy, C., 'The Geography of the Moslems of the Middle Ages', *Geographical Review*, vol. 14, 1924.
    Wright, J., *The Geographical Lore of the Time of the Crusades*. New York, 1925.
    Gibb, H., *Ibn Battuta — Travels in Asia and Africa*. London, 1929.
    'The Travels of Ibn Battuta', *Hakluyt* (2nd series), vol. CX, 1958.
21. Berg, L., *Die Geschichte der russischen geographischen Entdeckungen*. Leipzig, 1954.
       Chapter 1. 'Lomonossow und die erste russische Expedition zur Entdeckung einer nordöstlichen Durchfahrt.'
          3. 'Die ersten Russen in England.'
22. Berg, L., *op. cit.*
       Chapter 5. 'Die frühesten Nachrichten über den äussersten Norden von Sibirien'.
          4. 'Die ersten russischen Nachrichten über den Amerika.'
          6. 'Die Entdeckung der Beringstrasse durch Semen Deshnew.'
          7. 'S. P. Krascheninnikows Reisen durch Kamtschatka.'
          8. 'Die ersten russischen Weltreisenden.'
    Semyonov, Y., *op. cit.*
    Treadgold, D., *op. cit.*
    Andreyev, A., *Russian Discoveries in the Pacific and North America in 18th-19th Centuries*. Ann Arbor, 1952.
    Golder, F., 'Some Reasons for Doubting Dezhnev's Voyage', *Geographical Journal*, vol. 34, 1910.
    Mikhaylov, N., 'Iz Istorii Issledovaniya', *Sibir*. Moscow, 1956.
23. Berg, L., *op. cit.*
       Chapter 9. 'Die russischen Entdeckungen im Stillen Ozean.'
24. Berg, L., *op. cit.*
       Chapter 10. 'Die russischen Entdeckungen in der Antarktis.'
    Debenham, F., 'Captain Bellinghausen's Voyages', *Hakluyt* (2nd series), vol. XCI, 1945.
    Brown, R., *The Polar Regions*. London, 1927; contains summaries of exploration.
    Nordenskiöld, A., *The Voyage of the Vega*. London, 1881.

25. Freshfield, D. W., *Exploration of the Caucasus*. London, 1896.
26. Ravenstein, E., *Russians on the Amur*. London, 1861.
  Semyonov, Y., *op. cit.*
27. von Hellwald, F., *Russians in Central Asia*. Edited and translated by T. Wirgmann. London, 1894.
  Berg, L., *op. cit.*
      Chapter 15. 'Die ersten russischen Karten des Kaspischen Meeres und ihr Zusammenhang mit den Schwankungen des Meeresspiegels.'
          16. 'Iwan Kirilows Atlas des ganzen Russischen Reiches.'
          19. 'P. P. Semjonow-Tjanshanski als Geograph.'
28. Przhvalski, N. M., *From Kulja across the Tien Shan to Lop Nor*. London, 1879.
  Morgan, E. D., 'Prjevalsky's Journeys and Discoveries in Central Asia', *Geographical Journal*, vol. 9, 1887.
  Berg, L., *op. cit.*
      20. 'N.M. Prshevalski als Reisender.'
29. French, R., 'Geography and Geographers in the Soviet Union', *Geographical Journal*, vol. 127, 196.
  Examples of Soviet maps have been seen in those captured from the German wartime stores and in wall-charts published by the Moscow Pedagogical Institute and the Head Administration for Geodesy and Cartography in 1959 (*Topograficheskaya Karta 1: 50,000 and 1: 100,000*).
  Mellor, R., 'A New Soviet World Atlas', *Geographical Studies*, vol. 3, 1956.
      Another Major Soviet Atlas', *Scottish Geographical Magazine*, vol. 78, 1962.
30. Berg, L., *op. cit.*
      Chapter 24. 'Über die geographischen Forschungen an der Akademie der Wissenschaften der UdSSR.'
          25. 'Die geographischen Entdeckungen der Sowjetzeit.'
  Smolka, H., *Forty Thousand against the Arctic*. London, 1937.
  Schmidt, Y., *Voyage of the Chelyuskin*. London, 1935.
  Mellor, R., 'A Note on Soviet Arctic Research', *Geographical Studies*, vol. 3, 1956. Up to 1960 the Russians appear to have operated ten drifting polar stations.
Atlas Istorii Geograficheskikh Otkrytii, etc., plates show Soviet explorations.
Hooson, D., 'Some Recent Developments in the Content and Theory of Soviet Geography', *Annals Assoc. American Geog.*, vol. 49, 1959.
Armstrong, T., *The Northern Sea Route*. London, 1952.
  The Russians in the Arctic, 1937–1957. London, 1958.
Obruchev, S., 'Discovery of a Great New Range in North-East Siberia', *Geographical Journal*, vol. 70, 1927.

CHAPTER 4

Population Distribution and Ethnic Composition

COVERING one-sixth of the earth's land surface and comprising 209 million people (1959), one-fourteenth of world population, the Soviet Union has ninety times the area of the United Kingdom but its population is only four times greater. About three-quarters of the population live in European Russia, where around the big industrial towns and in the fertile farming lands of the south densities are comparable to Western Europe : elsewhere, with few exceptions, the country is sparsely inhabited. Slavonic Great Russians, Ukrainians, and Byelorussians form three-quarters of the population, but there are over a hundred other nationalities.

Long intercensal periods hinder the study of Soviet population as does only partial publication of demographic data, which are not always directly comparable between censuses.[1] Yet Russia was one of the first countries to have a true census, ordered in 1724 by Peter I, which enumerated in detail the population of European Russia but counted only Russians in the eastern provinces. Adjusted at various later dates, it remained the basis of population statistics until the 1897 census which listed all the population of the tsarist empire and also made an ethnic classification. The Soviet census of 1926 provided unrivalled detail on internal migration as well as on urbanization and national composition. A further census was held in 1939, but war prevented the complete publication of the results. After 1945, prolongation of wartime secrecy let few clues escape until 1956 brought the release of an official population estimate of 202·2 million, a figure lower than expected by Western demographers. The 1959 census gave a total of 208·8 million, an annual increase of over three million or 1·7 per cent.[2] It should be remembered that between 14 and 20 million people died in the 1941–1945 war and there was also a big loss of potential births, so that the population is probably about 30 million smaller than it otherwise would have been. Nevertheless, the rate of Soviet growth is above most of Western Europe though less than in many Afro-Asian countries.[3]

DOI: 10.4324/9781003172048-4

AGE AND SEX STRUCTURE

Published results of the 1959 census allow some examination of the national, but not regional, age and sex structure. The biological composition of Soviet population is generally better balanced than in Western Europe. A much larger proportion is formed by the younger age groups and the proportion of people over 60 is only half that of England and Wales. Three-quarters of the population have been born since the Revolution. War, nevertheless, changed the shape of the population pyramid. Loss

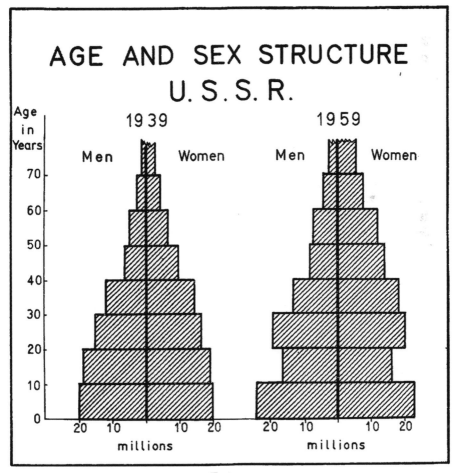

FIG. 20

Age and sex structure, 1939 and 1959. Compiled from Soviet census reports.

of births in the upset of war and the fall in the birth rate during and immediately after the war had reduced the proportion of people under 19 in 1959: for instance, the 16-19 age groups showed only a slight numerical increase and virtually no change in their share of total population, while the 10-15 age groups fell from 14·9 per cent in 1939 to only 8·2 per cent in 1959 and declined by over 11 million; even the youngest groups (9 years and below) fell from 22·8 per cent in 1939 to 22·2 per cent in 1959, though they numbered almost three million more. Better child care, it is claimed, besides a resurgence of the birth rate, will help to rectify these declines. The longer expectancy of life (from 32 years in 1897 to 68 years in 1959) has increased the proportion of older people: people over 50 comprised 13·4 per cent in 1939 and 18·6 per cent in 1959 and these age groups showed an increase greatly above the national average. Although the population in the economically productive age groups has risen from 102 million to 119·8 million in the past twenty years, they show an increase well below the national average. Possibly the heavy loss of manpower in the military age groups in the 1941–1945 war is reflected by an increase in the surplus of women from 7·2 million in 1939 to 20·8 million in 1959, though the male deficit has tended to rise since the 1897 census. Women formed 52 per cent of the population in the 1926 and 1939 censuses but 55 per cent in the 1959 count. The birth rate of 25 per 1,000, though lower than the 47 per 1,000 in 1897, is above Western European levels but comparable with the U.S.A. where there is also a comparatively young population. The Soviets claim the death rate of 7·2 per 1,000 is the world's lowest overall mortality rate and is 60 per cent lower than in 1940. Mortality in 1897 was 32 per 1,000! Rate of natural increase remains above the level of tsarist times. Occasional figures provided by Soviet handbooks show that in the countryside, natural increase is well above the national average but is low in the great towns. The countryside, with a much reduced death rate and a birth rate maintained at a high level, remains a focus of natural population growth, despite the strong outward migration since the Revolution which has reduced actual numbers of country dwellers. Natural increase, however, in the European countryside appears to be generally lower than in the eastern regions, where a youthful population of recent immigrants is no doubt a significant factor in the underlying causes of growth. The maintenance of a low death rate and a high birth rate remains a key of demographic policy, if the proportion of younger classes is to be held against increasing proportions in the upper age groups as the expectancy of life lengthens.[4]

POPULATION DISTRIBUTION

Despite a population of almost 209 million, the Soviet Union is thinly settled, with an average density of 24 persons per square mile, equivalent to sparse rural settlement in the British Isles. Over three-quarters of the country, densities are far below even this modest level and average about 5 persons per square mile. Most of the thinly-settled and almost uninhabited country is found in Asiatic Russia (Fig. 21).[5] Three out of four people live in European Russia, where a broad triangle of densities exceeding 25 persons per square mile lies with its base resting between the Baltic and the Black Sea and its apex in Western Siberia, forming a great wedge between the poorer northern coniferous forest and the dry steppe and semi-desert on the south: it covers almost entirely the deciduous forest belt with its lightly podzolized and brown forest soils suited to farming and the better-watered black earth of the wooded steppe, while it lies to the south and west of the main occurrences of *permafrost*. It represents the part of the Soviet Union with a reasonably reliable rainfall and relatively fertile soil suitable for sedentary agricultural colonization. Beyond it, a broad tract of country with much lower densities forms the modern pioneer fringe of Slavonic colonization. Southwards, densities of over 25 persons per square mile extend into the moister western grainlands of Northern Caucasia and along the northern footslope of the sparsely-settled Great Caucasus, which separates this belt from the distinctive native communities of Transcaucasia. South of the Turanian deserts, equally distinctive native cultures have remarkably high-density settlement in riverine and piedmont oases, with almost nilotic conditions, as well as in fertile mountain basins. In the northern tayga and the tundra, the main clusters of people occur only immediately around the few scattered mining and industrial communities, near northern ports, or in the agriculturally favoured middle Lena. A slender belt of settled country extends along the eastern sections of the Trans-Siberian railway, with population mostly found in

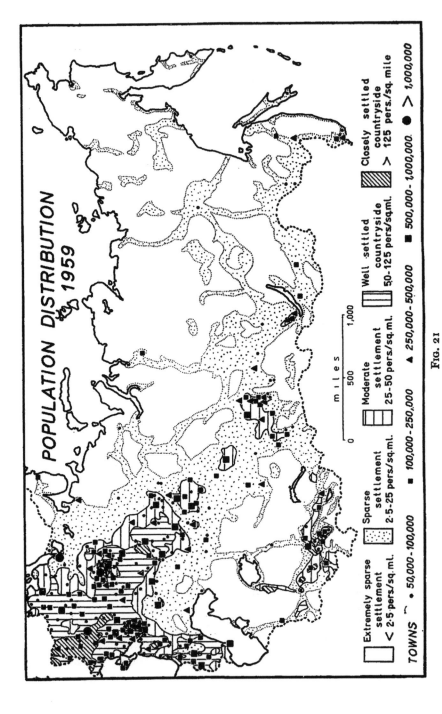

Fig. 21

Population distribution, 1959. *Atlas SSSR*, Moscow, 1962, pl. 98-99, and *Geograficheskiy Atlas*, Moscow, 1959, pl. 107.

the more fertile and attractive parts of the Amur, Zeya, and Bureya valleys, as well as in the Khanka plains and the mining and fishing communities around Vladivostok.

Within the settled triangle are considerable variations in the density and nature of settlement. The most heavily peopled areas, with about 100 persons per square mile (similar to mixed farming areas in the British Isles), lie west of the Volga and extend from the northern foot of the Carpathians across the better soils of the wooded steppe into the intensely developed black earth country, where there was chronic rural over-population in the nineteenth century and from which there has been strong outward migration. Such densities also occur around Moscow and eastwards towards the Volga. Locally, over 250 persons per square mile (high-density agricultural settlement in the British Isles) are found in the better-watered steppe, the deciduous forests of Podolia, the industrial Donbass, and the 'surburban' countryside around Moscow. This part of the country is rich in towns, with the largest concentration around Greater Moscow, where there are over seven million people. Others are the 'million' city of Kiev, Kharkov (930,000), Gorkiy (942,000), Kazan (643,000), Lvov (410,000), the several large Donbass towns (*e.g.*, Donetsk (Stalino) (710,000) and Makeyevka (358,000)), Dnepropetrovsk and Dneprodzerzhinsk and the port of Odessa. On the south, the lower Dnepr and Northern Crimea have a relatively thin population despite extensive cultivation, forming a countryside of large villages lining the gullies where water is available and open, sparsely-settled interfluves. The principal towns are the ports of Nikolayev (224,000) and Kherson (157,000) and the industrialized market town of Melitopol (95,000). Higher densities occur on the southern slope of the Crimean mountains where small holiday resorts like Yalta and Alushta lie amid vineyards. Other towns, including Simferopol (189,000), lie along the northern foot of the mountains, at whose south-western end the naval port of Sevastopol (148,000) stands by a fine natural harbour; on the east, Kerch (99,000) is an industrial town. North of the main belt of settlement, densities are mostly between 25 and 75 persons per square mile, stretching across Byelorussia, the Baltic republics, and North-west European Russia, with greater densities only locally near to big towns. Polesye is thinly settled in small villages on drier, sandy islands amid swamps. The population is generally greater along the drier morainic country of the Baltic-Byelorussian uplands, where there are many small and medium-sized villages and several important towns — Minsk (509,000), Smolensk (146,000), Vitebsk (148,000), and to the north, Vilnyus (148,000). Ports and fishing towns lie along the Baltic coast — notably Kaliningrad (202,000), Riga (605,000), and Tallin (280,000), while a group of small ports and resorts cluster at the head of the Finnish Gulf, dominated by Leningrad (3·3 million including suburbs).

The Volga region has few areas, except near Gorkiy and Kazan, where density exceeds 100 persons per square mile, though in parts of the forest and wooded steppe densities over 75 persons per square mile are found. Settlement is usually thicker on the high right bank not exposed to flooding, but below Kuybyshev and Saratov population density falls rapidly as the steppe is entered. Below Volgograd (Stalingrad), there is a sparse and partly nomadic population in the semi-desert, though the braided course of the Volga and Akhtuba provide water for cultivation to support a slightly heavier population. The river is lined by large towns marking trans-shipping points: in the upper reaches are Rybinsk (181,000), Gorkiy and Kazan at the confluence of the Oka and Kama; Ulyanovsk (205,000), Kuybyshev (806,000), and its nearby settlements in the middle reaches; Saratov (581,000), Volgograd (Stalingrad) (591,000) at the entrance to the Don-Volga Canal, and Astrakhan (294,000) at the mouth are passed by the traveller downstream.

Between the Volga and the Ural clusters of population are found in the new oilfields, including boom towns like Oktyabrskiy, whose population has rocketed to 65,000 since the early 1950's, Salavat in the Belaya Valley, and Almetyevsk, besides innumerable workers' settlements. Two-thirds of the 16·5 million people in the Ural live in towns and there is a relatively sparse rural population. Big towns include Sverdlovsk (777,000), Chelyabinsk (688,000), and Perm (628,000), and the new centres like Magnitogorsk, which has grown to 311,000 since the early thirties, and Novo-Troitsk (57,000), twenty-two times greater than in 1939. Most thickly settled is the low central Ural where there are over a hundred urban communities. With the exception on the western slope of industrial districts in the upper Kama around Perm and Berezniki and in the Belaya valley from Ufa southwards through the oilfields, densities are higher on the eastern slope where industrial towns line a belt of intense mineralization.

Within the mountains, settlement lies along the broad valleys, with only a sparse farming and trapping population in the forest-covered uplands.

Although more thickly settled than corresponding parts of Siberia, Northern European Russia has a scattered population of lumber workers and farmers or reindeer herders. Small concentrations of population are found in mining areas like the Ukhta oilfield or the Pechora coalfield near Vorkuta (55,000). In Karelia, industrial development is increasing population. The main centre is Petrozavodsk (135,000), while several mining towns (*e.g.*, Monchegorsk and Kirvosk) lie in the Kola Peninsula. Big ports include Arkhangelsk (256,000) and nearby Severodvinsk (79,000) and the ice-free Murmansk (226,000).

Siberia

Settlement in Siberia extends along the main routeways, particularly the railway lines, and where favoured industrial and mining areas have attracted settlers, or where conditions have been peculiarly suited to agricultural colonization as in the foothills of the Altay and the rich, rank meadows along some Western Siberian rivers. Agricultural colonization has also spread out from the wooded steppe into the northern forests as well as southwards into the drier steppe. The colonized lands along the railway are marked by large, industrialized market towns and river ports, such as Kurgan (145,000), Omsk (579,000), Tyumen (150,000), and the impressive Novosibirsk (887,000). In the Kuzbass coalfield, seven towns have more than 50,000 people.

The narrowing wedge of settlement extends eastwards along the Trans-Siberian railway to the shores of Lake Baykal, so that its apex includes such towns as Krasnoyarsk (409,000), Irkutsk (365,000), the coalfield centre of Cheremkhovo (123,000), and Angarsk, a boom town which has grown to 134,000 people since the late forties. East of Lake Baykal, a thin belt of settlement follows the railway, with isolated concentrations of people around Ulan Ude and the Chita mining districts. In the Amur basin, farming colonies have grown in the damp but fertile Zeya-Bureya valleys and at the Amur-Ussuri confluence near Khabarovsk (322,000). Small lumbering towns line the railway to Vladivostok, near which settlement in mining communities and around the fertile shores of Lake Khanka raises densities.

The Siberian tayga and tundra form an immense, sparsely-settled country into which the pioneer colonization by Russian trappers and even peasant farmers has slowly pushed, advancing along the routeways of the great rivers, leaving the vast inter-fluvial forests and swamps to a thinly-scattered, nomadic, native population. Beyond the agriculturally colonized areas, settlers are usually gathered in small mining, trading, or administrative settlements. In some of the most remote and least hospitable northern regions, transported prisoners have formed a part of the population. Although exploitation has led to growth amid the almost limitless forested and swampy wilderness, densities seldom exceed two or three persons per square mile outside the immediate vicinity of towns such as Norilsk (108,000) and its outport Dudinka, or Magadan (62,000) at the gateway to the Kolyma mining region. Russians are found even along the Arctic coast and on the offshore islands, amid a scattered native population in tiny isolated colonies associated with the operation of the Northern Sea Route.

Caucasia

The comparatively well-watered black earth steppes around Krasnodar and Stavropol are moderately-populated farming country with large villages. Townspeople dominate only in the lower Don around the industrial town of Rostov (597,000) and in the nearby coal-mining towns of the eastern Donbass. In the northern foothills of the Great Caucasus, petroleum resources have led to the growth of towns, and the better availability of water has produced greater localized densities than in the steppe farmlands. Towns include Krasnodar (312,000), Groznyy (240,000), Ordzhonikidze (164,000), and Stavropol (140,000). Densities fall eastwards as the country becomes progressively more arid, with exceptionally sparse population in the Kalmyk steppes. The high, mountainous Great Caucasus, sparsely-peopled by tribes living mostly in nucleated villages, separates the northern steppelands with their dominantly Russian population from the rich, warm lands of Transcaucasia with their own distinctive native settlement patterns. The humid Black Sea littoral, with its tea and fruit gardens and holiday resorts, as well as the drier parts of the Kolkhiz lowlands, support over 125 persons per square mile. Other concentrations are found around the industrial Kutaisi (128,000) and Zestafoni, while Tbilisi, commanding both north-south and east-west routes, has 694,000 inhabitants. The influence of relief on population distribution in Georgia is shown in the table below. Densities are lower in the drier, eastern parts of Transcaucasia, notably in the steppe of the Kura lowlands, and in the lower parts of the Armenian Plateau, though they nevertheless support usually more than 50 persons per square mile; but there are densely-settled, irrigated farming valleys around Yerevan (509,000) and Leninakan (108,000) and along the rivers Razdan and Araks. East and south of Lake Sevan, the higher parts of the Armenian Plateau have a sparse,

TABLE 3

POPULATION DISTRIBUTION BY ALTITUDE IN GEORGIA

Altitude (ft.)	Per cent of Territory	Per cent of Population	Average Density *
Below 1,600	23·6	56·9	210
1,600-3,300	22·6	27·6	108
3,300-4,900	16·6	9·1	49
4,900-6,500	17·4	5·7	28
Above 6,500	19·8	0·7	2·6

* Persons per square mile.

Based on Cherdantsev, G. N., *Ekonomicheskaya Geografiya SSSR*, Moscow, 1955.

nomadic population but drop westwards to the warm, sheltered valleys, with their vineyards, orchards, and grain growing, of the thickly-settled Nagorno-Karabakh region. Thickly-settled country is found in the oilfields of the Apsheron Peninsula around Baku (968,000 including suburbs) and Sumgait (52,000).

Central Asia

Soviet Central Asia shows great contrasts in population density between the fertile, crowded oases and the empty, inhospitable desert or lonely, high mountains, for water is the key to settlement. The poor steppe and semi-desert remain the home of nomadic Turkic peoples and densities are mostly between three to ten persons per square mile. Most sparsely settled are the poor country of the waterless Ust-Urt Plateau, the Kara Kum and Kyzyl Kum around the south and east of the Aral Sea, and the infamous Bet-Pak-Dala. Even in inhospitable poor steppe and desert, mineral wealth has brought the growth of mining towns like Karaganda (398,000), Balkhash (53,000), and the boom town Temir Tau, whose population has risen from 5,000 to 54,000 in twenty years.

Along the rich loessic foot of the Central Asian mountains is a countryside of large nucleated villages, like thickly-settled oases, wherever water and fertile silt are available from mountain streams or irrigation canals. Densities rise from 75 to well over 250 persons per square mile and some of the richer dry farming areas have 25-50 persons per square mile. Most outstanding is the Fergana valley around whose edge, on the delta fans of mountain streams, there are over 1,250 oasis cultivators per square mile, reaching 2,000 people per square mile around Andizhan. Similar concentrations occur on irrigated lands of the Zeravshan, where Samarkand has a population of 195,000; the Chirchik, with industrial Tashkent (911,000); around the desert delta of the Murgab near Mary (48,000) and along the foot of the Kopet Dag, with Ashkhabad (170,000), and near Alma Ata (455,000) and Frunze (217,000). The nilotic conditions of the riverine oases of the Syr and Amur Darya support densities of 75-125 persons per square mile, notably in the ancient delta around Khorezm. Population density falls rapidly with altitude, though crowded villages are found in well-watered and sheltered mountain basins. Elsewhere in the mountains, a sparse nomadic population moves up and down the slopes with the seasons.

THE GENESIS OF POPULATION DISTRIBUTION

The pattern of contemporary population distribution, a stage in a long, historical evolution, is the product of the mobility of the Eastern Slavs in their spread outwards from a primary dispersal across the European Russian Plain and into Siberia, the Caucasian isthmus, and even into Central Asia. The development of this vast *lebensraum*, extending from the Black Sea and the Baltic to the Pacific, has been one of the greatest and yet least known of the European colonizations. It has carried settlers northwards and eastwards into the coniferous forests as well as south and south-eastwards into the grasslands long commanded by Turkic and Mongol nomads. This colonization, mostly the search for land suitable for sedentary agriculture, brought the occupancy of a broad triangle of the better farming land whose base lies between the Baltic and the Black Sea and the apex in Western Siberia, which stands out in the modern population map as a belt of more densely settled country. Motivated by factors such as rural overpopulation, social and economic oppression in the homeland as well as, in later years, expansionist political aims, the Slavs have pushed back the native peoples into the more inhospitable grasslands and forests, or have swept around them, infiltrating their settlement areas and frequently absorbing them into the Russian cultural pattern. At some points, Slavs have impinged on, and absorbed into the Russian state territory, settled communities with their own high degree of social, economic, and political organization, notably in the Baltic littoral, the Caucasian isthmus and Central Asia. In contrast to the great eastern colonization, Russian expansion in the western march-lands has been restricted and expressed principally in political-territorial terms. Momentum of expansion has not been constant: nomadic invasions across the steppes have sometimes halted or pushed back the outposts of Slav settlement and have left a permanent reminder in complicated ethnic patterns; at other times, economic and social stagnation has slowed colonization but there have also been periods of intensified movement, of which one of the most significant has been the mining and industrial colonization sponsored by the Soviet regime. The expanding frontier of pioneer settlement, a key to Russian history, continues to advance, for there are still large, thinly-occupied territories to colonize within the state boundaries.

TABLE 4

DISTRIBUTION OF SOVIET POPULATION IN 1959

Density*	Area		Population						Urban %	Average Density*
	000 sq. miles	%	Total		Urban		Rural			
			mill.	%	mill.	%	mill.	%		
Under 26	6,695	78·0	30·0	14·3	14·9	15·0	15·0	13·8	50	4·4
26–124	1,579	15·4	109·5	52·4	47·8	48·4	61·8	56·6	44	95·0
125–249	278	3·2	48·9	23·5	22·5	22·6	26·4	24·2	46	176·0
Over 250	45	0·5	20·4	9·8	14·6	14·6	5·8	5·4	71	450·0
Total	8,608	100·0	208·8	100·0	99·8	100·0	109·0	100·0	48	24·0

* Density in persons per square mile.

Source: Calculated from table in 'Izvestiya Akademii Nauk SSSR', *Seriya Geografiya*, No. 6, 1959.

Too little information exists about the primary dispersal of *homo sapiens* in late Tertiary and Quaternary times to paint an adequate picture of the earliest colonization of the Russian lands.[6] The Slavs appeared from their cradle in the inaccessible and inhospitable forests and swamps of Polesye between the second and fifth centuries A.D., and began a strong outward movement. The eastern elements have formed the Great Russians, Ukrainians, and Byelorussians. Along the Baltic littoral, earlier peoples held their ground before the advancing Slavs; but penetration south to the Black Sea littoral and south-eastwards towards the Volga steppe was also stemmed between the fourth and fourteenth centuries by successive waves of nomads who commanded the grasslands. While only adventurers settled in the steppe fringe along the thinning southern edge of the forest constantly exposed to nomad raids, colonization north-eastwards against the weaker and probably less numerous Finno-Ugrians progressed steadily, particularly with the commercial growth of Novgorod, though the country was less suitable agriculturally. By the fifteenth century, however, nomad power was declining : in 1480 Moscow ceased to pay tribute to the Tatars, and a century later had opened the Volga basin to Russians who had already entered Siberia by the circuitous route round the northern Ural.

Adventurers and traders spread across Siberia to reach the Pacific by the middle of the seventeenth century. Even with the institution of exile to Siberia in 1582, the Russian population remained small, for peasant serfs were unable of their own free will to settle there, though some colonization of the Ural for iron smelting had begun.[7] By the early eighteenth century there were no more than 400,000 Russians in Siberia. During the sixteenth century, as Russian power spread south-eastwards against the weakening and thinning ring of nomads, the Cossack frontiersmen appeared, protecting the spreading sedentary agricultural population against nomad incursions. They played an important part in breaking Turkish power in Southern European Russia, which was opened to colonists from the north who turned it into a vast grainland, and they finally secured the earlier incorporation of the Don and Donets basins as far as the bank of the Volga. Freed from the danger of hostile intrusions, the black earth lands in the eighteenth and early nineteenth centuries became a land of colonization and opportunity. At the same time, territorial aggrandizement in the west at the expense of

Poland and Sweden incorporated several new national groups as well as the westernmost Eastern Slavs.

The nineteenth-century changes in population distribution came when the mobility of the Russians was probably less than for the previous three centuries. Territorially, it saw the spread of tsarist power in the west to Finland, Moldavia, and Congress Poland; but in the east, there were the conquests of Trans-caucasia and of the Turanian plains and Central Asian piedmont oases. All these were lands already well settled, offering little opportunity for large-scale Russian colonies, while the more thinly occupied country was too arid for agriculture.

By the mid-nineteenth century, the best black earth land in Central European Russia and the Ukraine was occupied, so that rural overpopulation began to appear and mounting social pressure culminated in the reforms and serf emancipation of 1861, releasing a flood of migrants to Siberia and to the growing industrial towns around Moscow, Leningrad, and in the Donbass. While the filling of poorer marginal lands in European Russia, mostly in the north, continued, the main colonizing effort moved into Western Siberia into the fertile wedge between the northern swampy forests and the southern steppe. A great wave of settle-ment followed the agricultural depression in European Russia in the 1870's and continued until the First World War, aided by the building of the Trans-Siberian railway after 1894. As the western lands of the Trans-Uralian peneplain filled, peasants moved further east along the often dangerous Sibirskiy Trakt, the gateway to the rich and well-watered grain-growing Altay foothills, where some also found employment in mining. Others settled on the rich dairying meadows of the Tobol, Iset, Tura, Ishim, and Irtysh. Some ventured into the steppe fringe near Omsk, Petropavlovsk, and even into the Kulunda steppe where the settler could feel safe now the nomads were subdued. Between 1885 and 1890, almost half the migrants went to the Altay, rising to 69 per cent in 1892, thereafter declining as the railway brought people who quickly settled the surplus land, after which more began to settle in the steppe fringe.

In the Amur and Ussuri valleys, taken in 1858, settlement needed, however, special encouragement, particularly before the completion of through railway communication. Some settlers were carried from Odessa by boat, but most came from within Siberia, with experience of the land's hard nature. There were also Cossacks, who had contributed to its seizure, and Ukrainian peasants, particularly Old Believers who migrated mostly to the Ussuri valley.

During the nineteenth century, over five million people settled in Siberia, so that the population rose from 575,000 to over eight million. In 1897, of 1·4 million European-born Siberians, over half had come from the black earth belt, where natural increase remained so high that its resources were taxed no matter how quickly peasants emigrated. By 1914 Siberia was a Russian country, with the best land already settled, so that a spread into the Kazakh steppe and Turkestan began.

TABLE 5

PERCENTAGE OF SLAVONIC RUSSIANS IN SIBERIAN POPULATION

	Census 1897	Estimate 1911
Siberia	82·0	86·5
Far East	63·3	74·0
Steppe	19·0	40·0
Turkestan	3·7	6·3

Source: Quoted in Treadgold, D., *The Great Siberian Migration*, Princeton, 1957.

The broad, settled triangle across the wooded steppe and the richer, deciduous forest was already apparent in the population distribution revealed by the 1897 census (Fig. 22).[8] In the south, the steppe of New Russia was well filled, though the southern industrial concentration had not sufficiently developed to stand out clearly on the map. North of the Leningrad-Perm railway, densities fell sharply, while east of the Volga the population thinned markedly, except in the Kama basin where there was forest exploitation, and in the central Ural with its growing industrial villages. Most striking was the settled belt already lying across Siberia between the forests and the steppe, lapping around the foothills of the Altay and Sayan, and extending into Transbaykalia. Densities were lower than in the old, settled lands of European Russia, but this was a country into which migrants were pouring by the new railway. A small, favoured area around Vladivostok and the Khanka Plain was filled mostly by military colonists. Tsarist conquest had also incorporated the

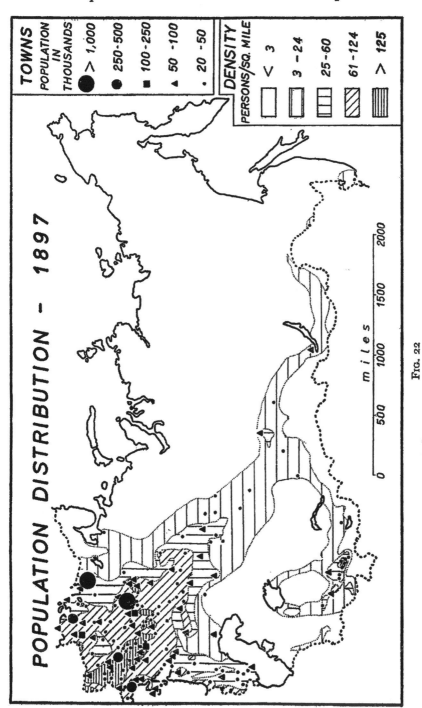

FIG. 22

Population distribution, 1897. Compiled from relevant sections of Supan's noted 'Die Bevölkerung der Erde' in *Petermanns Mitteilungen, Ergänzungshefte,* 27/28 (1898—1900) and 29/30 (1900—1903).

well-settled oases of Central Asia and other native agricultural communities in Transbaykalia, besides sparsely-peopled steppe, desert, and high mountains in these regions.

THE SOVIET PERIOD

The 1926, 1939, and 1959 censuses, while not directly comparable, provide a composite picture of population development.[9] Notable changes were the substantial growth shown in new and established centres of commerce and industry; the progressive thickening of densities along main lines of communication (notably

TABLE 6

URBAN AND RURAL POPULATION DISTRIBUTION, 1913–1959

Year	Total Population, mill.	Urban, mill.	Rural, mill.	Percentage	
				Urban	Rural
1913 *	159·2	28·1	131·1	18	82
1926 †	147·0	26·2	120·7	18	82
1939 †	170·6	56·1	114·5	33	67
1959 †	208·8	99·8	109·0	48	52
1960 † E.	216·0	108·0	108·0	50	50

* Within present boundaries.   † Within boundaries of period.
E. Estimate.

railways); and settlement in new areas. These came in response to the stimuli of the planned economy, re-creating some of the old mobility, but also leading in some areas to stagnation or decline from inability to adjust to new strategic and economic needs. The changes were attained principally by migration, characteristically from European Russia to new mining and industrial towns in the eastern regions rather than by continuation of tsarist agricultural colonization. Russia has ceased to be a dominantly agricultural country, for town-dwellers have risen from less than a fifth of the population in 1926 to a half in 1960: an increase of 81·8 million contrasted to a decline in rural population of 12·7 million. Movement has been not only from the countryside into nearby towns but also from the European countryside into towns in far-distant eastern regions.

The Revolution and its consequences brought unprecedented movements of population, so that a measure of stability did not return until the middle 1920's, when formulation of plans to intensify regional economic development and industrialization set the future population trends. A greatly increased demand for workers in established and new industrial centres started a strong migration to the towns from the countryside, where problems of rural overpopulation were accentuated by collectivization and the introduction of machinery. The new industrial workers and colonists came principally from the Slavonic settlement areas, so that Slavs now began to penetrate

TABLE 7

POPULATION DISTRIBUTION IN RUSSIA, 1897–1959

	1897		1926 ‡		1939 ‡		1959 ‡	
	mill.	%	mill.	%	mill.	%	mill.	%
European Russia	97·9 *	83·6	116·9	79·5	129·2	75·9	152·8	73·2
Transcaucasia	5·9	5·0	5·9	4·0	8·1	4·6	9·5	4·6
Siberia and Far East	5·7	4·9	10·5	7·2	16·7	9·8	23·6	11·3
Central Asia	7·6†	6·5	13·7	9·3	16·6	9·7	22·9	10·9
Total	117·2		147·0		170·6		208·8	

* Excluding Finland, Poland in their boundaries of period.
† Excluding Khiva and Bokhara.
‡ In the boundaries of the period.

where they had so far only been present in small numbers. Non-Slavonic groups, apart from prisoners, do not seem to have supplied much industrial labour until the late 1930's.

After the late 1920's, the flow of migrants generally exceeded the best pre-revolutionary years, for now the government had an unopposed hand. Not all migration was permanent: there were strong seasonal movements (as in tsarist times) and counter-flows, reflecting the heavy labour turnover, greatest in the arduous pioneering conditions of the Far East. Movement was, however, regulated: for instance, eastern regions were allocated areas of the European countryside from which to recruit, and for various reasons, movement to and from certain districts was prohibited for varying periods. Workers willing to live in remote places were offered incentives of housing, extra pay, travel expenses for families, and other benefits. Appeals were made to the patriotism of young Communists, and unemployed in European Russia were found jobs in the eastern regions. Undesirables were deported or directed to work in the least pleasant parts of Siberia; once there, they found it hard to return when their term was finished. Women were encouraged to migrate to areas where a predominantly male community was developing in order to build a balanced population.

There was a certain natural tendency for migration to follow set lines. European steppe-dwellers moved to the Siberian and Kazakh steppes, notably to Karaganda, where mining and industry were developing. Industrial workers from Moscow and Leningrad went to West Siberian and Ural industry, replaced at home by people from the adjacent countryside. Migrants on constructional work at new plants stayed to operate them, as at Stalinsk (now Novokuznetsk) and Magnitogorsk. The building of new 'combines' and towns absorbed many forced labourers, but they were used chiefly in the Arctic, where mining settlements grew in the Kola Peninsula (Kirovsk and Monchegorsk) and in the Taymyr-Yenisey area (Norilsk, Dudinka, Igarka), as well as in settlements connected with the working of the Northern Sea Route and in North-eastern Siberia, where *Dalstroy* built Magadan and worked the Kolyma ore fields. As Japanese power in Manchuria grew, strategic and economic policy dictated increased settlement in the Far East. Four thousand young Communists built Komsomolsk on the Amur, but an attempt to establish a Jewish national district failed.

Internal migration in European Russia was characterized by people leaving the villages because they did not like collectivization or because it and the new farm machinery made them redundant, while the towns began to offer a much better reward than farming. Despite the large natural increase in most country districts, there were 6·2 million less people there in 1939 than in 1926, when numbers had already been reduced by war and the great famines of the early twenties. The greatest loss was from the central black earth belt, the grainlands of the Ukraine and Northern Caucasia, and from the Volga steppes. From these districts came migrants to the industry of the Donbass and the Dnepr bend as well as to Moscow and its industrial satellites or the growing towns on the adjacent lignite field. Immediately before the war, interest began to shift to the east bank of the Volga, where discoveries of rich petroleum deposits heralded great expansion. There was also a movement from the marginal agricultural lands of the European North to new industrial settlements and to the Leningrad conurbation. Even in the eastern regions there was only a small increase in rural population since, for example, people left the countryside to work in developing industry in the Ural and Western Siberia. In Kazakhstan, rural population fell because of the repression of opposition to collectivization. In the oasis lands of Central Asia, where greater stability and improved farming techniques with rising living standards had followed Russian rule, rural population increased slightly as more irrigated land was brought into use and industries developed.

Development of population between 1939 and 1959 was complicated by the Second World War, as a result of which annexation of territory is thought to have accounted for 22·2 million out of the total increase of 38·3 million.[10] A lack of detailed Soviet statistics makes it difficult to assess regionally the proportion of growth due to natural increase and to migration. It is clear, however, in the eastern regions that growth has been considerably in excess of what might be expected from a 'normal' surplus of births over deaths and the inference is that the balance has come from immigration. In their latest plans for Siberia, the Russians recognize that 500,000 settlers annually for five years will be needed, while there is ample evidence to show that migration has been the means of peopling the virgin lands of Kazakhstan.

Rural decline was more widespread than before the war, since besides the drift from the European countryside, there was now a small loss throughout the east, including the pioneer lands of Eastern Siberia and the Far East; but in Kazakhstan, as a contrast, rural population had risen after the interwar depredations and with the new ploughing campaign. A further increase had also taken place in Central Asia, where more irrigation and a rising demand for its industrial crops raised the attractiveness of farming. The loss of the countryside was the gain of the towns, again the centre of growth, though on a smaller scale than interwar. In 1959, they contained almost half the Soviet population, and urban population rose by 65 per cent between 1939 and 1959, compared to only a 9·5 per cent increase in total population. The increase of almost 40 million was made up by nearly 25 million migrants from the country, about 8 million by natural increase, and 7 million by reclassification of rural as urban settlements; there was also a small addition from boundary adjustments to towns. The largest towns had, however, grown more slowly than before the war, while the big programme o town building in developing economic regions and in creating satellites to giant towns was reflected in an increase in the proportion of people living in the smallest towns. Wartime damage was an important factor in slowing town growth in European Russia, where Leningrad (which lost a million by starvation in the siege), Smolensk, Kerch, Kremenchug, Novorossiysk, Konstantinovka, Berdichev, and Vitebsk were still smaller than in 1939. Deceleration of town growth was marked by a fall in the number of boom towns from forty-eight to twenty-five, particularly noticeable in the Central Industrial region and the Ukraine, where there were only two boom towns compared to twenty-two pre-war. Three-quarters of the boom towns were now in the eastern districts, where no town appears to have risen by less than 20 per cent, though in European Russia eight towns had declined and thirteen had grown less than 10 per cent.

ETHNIC COMPOSITION AND DISTRIBUTION

Not all Soviet citizens are Russians: they belong to over 100 different national groups, defined on a composite basis of language, culture, history, physical type, religion, and in some instances simply because a group has formed a distinct community for a reasonably long time. Some nationalities are very large, like the 114·6 million Great Russians, and nineteen groups each exceed a million persons; but mostly they are small, and like the Yukagirs and Aleuts may number as few as 400 people each. The groups also vary widely in their material and cultural achievements, from some of the world's most advanced to others who are only just leaving the primitive hunting stages.[11]

The nationalities may be arranged into about a dozen major ethnic groups (Fig. 23). The Slavonic group forms over three-quarters of the population, spread into all parts of the country. It is dominated by the Great Russians, who form about 55 per cent of Soviet population, and includes the second largest national

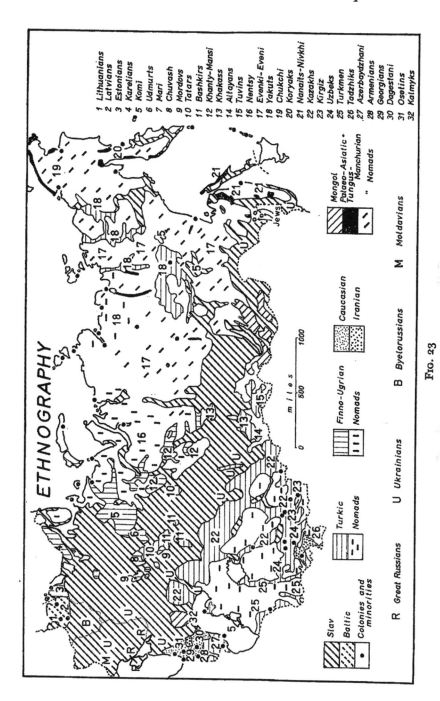

FIG. 23

Ethnography. *Atlas SSSR*, Moscow, 1962, pl. 96-97, checked against a coloured map in Tokarev, S., *Etnografiya Narodov SSSR*, Moscow, 1958.

1 Lithuanians
2 Latvians
3 Estonians
4 Karelians
5 Komi
6 Udmurts
7 Mari
8 Chuvash
9 Mordov
10 Tatars
11 Bashkirs
12 Khanty-Mansi
13 Khakass
14 Altayans
15 Tuvins
16 Nentsy
17 Evenki-Eveni
18 Yakuts
19 Chukchi
20 Koryaks
21 Nanaits-Nivkhi
22 Kazakhs
23 Kirgiz
24 Uzbeks
25 Turkmen
26 Tadzhiks
27 Azerbaydzhani
28 Armenians
29 Georgians
30 Dagestani
31 Osetins
32 Kalmyks

group, the Ukrainians, comprising over 17 per cent of total population. In comparison, the second largest group, the Turkic peoples, forms only 10 per cent of Soviet population, while the remaining major ethnic groups in no instance exceed 4 per cent. Despite its multi-national character, the U.S.S.R. is a predominantly Slavonic state — a Slavonic wolf masquerading in a sheepskin of multi-nationalism.

The Slavs are extremely diverse and there is certainly no unifying anthropological factor.[12] Apart from a common Indo-European origin and a strong linguistic connection, little identifies the relationships between some of them, but different historical experience and mixture with non-Slavs allow three main divisions to be distinguished within the U.S.S.R. The Slavs appear late in history, culturally associated with the 'burgwall' village early in the present era. Their original home appears to have been in the forests and swamps of the Pripyat basin, where they could maintain their identity amid a terrain difficult to approach. For reasons imperfectly understood, they began to grow rapidly in numbers during the second to fifth centuries A.D., spreading out from their marshland home. A temporary westwards expansion towards the Elbe was forced back by the eighth-century eastwards movement of the Germans; but their eastwards spread was more successful so that, despite setbacks, it may be considered to continue still in the contemporary colonization of Russia's Asiatic territories. In the process, Slav groups mixed among themselves as well as with peoples who came into contact with them, such as the Finno-Ugrians of the north, the Tatar peoples of the Volga, and even Mongols and Palaeo-Asiatics in Siberia.

The Great Russians (114·6 million) have their traditional settlement area in the forests of Central European Russia, in the swampy lands of the Oka and upper Volga, with the centre around Moscow. From here they have spread northwards to the Baltic and the White Sea, southwards into the black earth lands and the lower Volga, as well as eastwards across the Ural and into Siberia, forced outwards from a poor and harsh environment by cruel social conditions, historically a mobile element in the vanguard of migration. Outlasting Turkic and Mongol invasions from the east, they absorbed characteristics from these nomads which have set them apart from other Slavonic groups. Great Russian is rich in Turki-Tatar loan words. Their abilities to absorb and to intermarry have been important factors in the peaceful assimilation of territory in Siberia and even in Central Asia and Caucasia. Over 17 million Great Russians live outside their titular republic, the R.S.F.S.R., where they form 85 per cent of the population. In the other republics, Great Russians are usually the largest minority; but in the Kazakh S.S.R. they actually outnumber the Kazakhs (Fig. 24). The political leadership of the Eastern Slavs held by the Muscovite state and Great Russian numerical superiority has brought Great Russian to be the official state language and lingua franca and has introduced the Cyrillic alphabet into non-Slavonic languages within the country, besides the force it has imparted to the cultural influence of the Orthodox church under the Patriarch of Moscow.

The second largest group is formed by the 36·9 million Ukrainians (Little Russians). Physically diverse, they are usually distinguished by their language, which is more akin to the western Slavonic dialects. The Ukrainians developed in the wooded steppe and steppe, where they bore the brunt of many

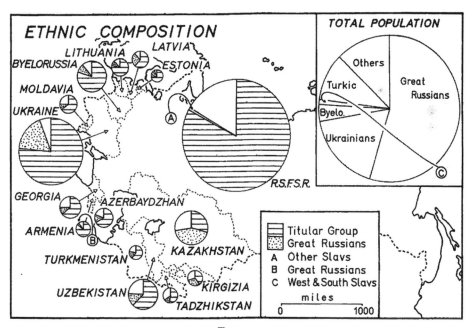

FIG. 24

Ethnic composition of union republics, 1959. Compiled from 1959 census reports. Circles are in proportion to size of population in each republic. The importance of Slavonic minorities in most republics of Central Asia and Transcaucasia should be noted.

nomadic invasions. In this frontier in the sixteenth century, though doubtlessly of earlier origin, there appeared the socio-military group of the Cossacks formed of outlaws and adventurers recruited in many different regions. Divided ultimately into twelve 'armies', as allies of the tsars they contributed substantially to Russian exploration, conquest, and settlement. For a long period, the Western Ukraine was under Polish influence, while the Carpatho-Ukraine (Ruthenia) was in contact with Magyar culture. In the west, the Ukrainians are mostly Uniates or even Roman Catholics, but in the east they are Orthodox. Their cultural centre is Kiev and relations with Great Russia have been coloured by a strong but complex nationalism.

The Byelorussians (7·8 million) are the smallest Slavonic group, developing in the mediaeval marchland between the Great Russians and the Poles, in the forests north of the Pripyat marshes and to the south of the Baltic morainic belt. They may represent a slavicized Baltic group, since they have some physical traits common to older Lithuanian stock. Long contact with Poland and Lithuania has left a cultural impression. Before 1939, half the Byelorussians, mostly Roman Catholics, lived in Poland, while those in the U.S.S.R. were Orthodox.

Small groups of Western and Southern Slavs are also found in the country, of which Poles (1·4 million) are the largest. Half the Poles live in Lithuania and Byelorussia, while another group is in Kazakhstan. Related to the Slavs are the Baltic Lithuanians and Latvians (Letts) and their subdivisions, such as the Latgals, numbering together 3·7 million. Although markedly influenced by Polish, German, and Russian culture, the groups have resisted submergence in their homeland of forest and swamp north of the great Baltic terminal moraines where they have preserved their archaic languages. The Lithuanians are Roman Catholics, one of many legacies of Polish influence, and the Baltic Germans introduced Lutheran faith to the Latvians. The German minority (1·6 million) has made a notable contribution to Russian language and culture.[13] Early German contact came through the Hansa at Novgorod and the Teutonic Knights in the Baltic lands. Many Germans became courtiers, advisers, and artisans to the tsars, and under Ivan the Terrible there arose the German suburb in Moscow. Peter the Great and Catherine II (herself German) encouraged Germans to settle in the recently acquired lands of the Volga and Black Sea littoral. Later, others settled in Volhynia and even in Transcaucasia, where there were twenty-three colonies near Tbilisi. By 1914, there were over 100,000 Germans in the Ural, Siberia, and even Central Asia. The mediaeval German colonization of the Baltic littoral left a well-established society of landowners and entrepreneurs in the towns, which remained dominantly German even late in the last century. After the Revolution and the First World War, the number of Germans declined; but German colonies around Saratov were given autonomous status and in 1939 had a population of over half a million. Immediately before the Second World War, Germans in the Baltic states were repatriated to the Reich.[14] Remaining German groups inside the Soviet Union were 'liquidated' to Siberia and Kazakhstan after 1941, but some western groups escaped to Germany.

The 2·2 million Moldavians, living mostly in their own republic in the Dnestr basin, are closely related to the Romanians, though they use the Cyrillic alphabet and are Orthodox. Important as a group are the 2·26 million Jews, who were limited in tsarist times to settlement in the 'Pale' of the Western Ukraine and Byelorussia, but there are also ancient Jewish communities in the Crimea and in Transcaucasia and Central Asia. In 1934, the Jews were given an autonomous district in the Amur basin, but the experiment was not successful; nowadays they are scattered throughout the country. Old trading settlements in the Black Sea littoral are represented by most of the 310,000 Greeks in the U.S.S.R. Scattered throughout Russia but mostly in the European lands are 132,000 Gypsies.

The second largest ethnic group is formed by the Turkic peoples, who number 21 million, about 10 per cent of the total population. Over half of them live in Central Asia and the southern steppes (11·7 million), and there are 5·9 million in the Volga basin, while 440 thousand live strewn over the vastness of Siberia. The Volga Turks have been strongly influenced by the Russians among whom they live: many are now sedentary industrial workers, particularly the Tatars (4·9 million), who came into the region as an advance guard of the devastating Mongol invasions. The Bashkirs (980,000), a nomadic people settled west of the Ural by Genghis Khan, are also becoming increasingly sedentary with the spread of mining and industry. Both groups have the status of autonomous republics, though they do not seem to form a majority of the population and live scattered among others, notably Great Russians. The Chuvash (1·47 million), a complex group, appear to have a mixed Turkic and pre-Turkic tradition, with strong Finno-Ugrian influence. The history of the Volga peoples is complicated by considerable cultural intermixing. Several Volga groups of indefinite ethnic character recorded in historical documents have disappeared, though archaeological exploration may clarify the present uncertainty.

Of the Central Asian Turkic groups, the largest and most important is the Uzbek (6·0 million), the vanguard of modern industrialization. Mostly sedentary agriculturalists, they have a markedly Persian culture. The Kazakhs (3·6 million) are a nomadic people with many Mongol traits, who fought hard against encroachment by Russia. Through resistance to collectivization, their numbers declined by a quarter between 1926 and 1939, partly by migration from Russian territory. The million Turkmen live in the oases at the foot of the Kopet Dag, but they are also nomads in the Kara Kum. The 970,000 Kirgiz are mountain nomads with strong Mongol influences. Around the southern shore of the Aral Sea live the 173,000 Kara-Kalpaks, a small sedentary tribe. There are also several smaller groups, such as the Uygurs and Dungans.[15]

The highly diverse people of Transcaucasia also include some three million Turkic peoples. The Azers (Azerbaydzhanis) form the bulk (2·9 million) and live in the steppes of the lower Kura, to which they migrated from Central Asia, around the southern shore of the Caspian Sea, so absorbing many Persian influences. Other groups include the Balkar and Karachay, both of whom were accused of wartime collaboration with the Germans and deprived of their national status until reinstated in 1957. The Turkic groups of Siberia number only 440,000, of whom 236,000 are Yakuts, a people who were pushed into the middle Lena basin by Tungus and Mongol peoples from the south. They adapted themselves to a sub-arctic environment and remain horse- and cattle-breeders, as well as reindeer-herders in the far north, where they are found among the related Dolgans. Their language has become the lingua franca of the Siberian forests. In the upland basins and surrounding mountains of the Sayan live 100,000 Tuvinians, incorporated into the Soviet Union in 1944. In the Altay are found the Altaians, Khakass, and Shorsi. The Turkic peoples are predominantly Moslem, although belonging to different sects, while a few, like the Yakuts, are Orthodox.

The Caucasians comprise 7·3 million people, mainly the Japhetic populations of Georgia and Armenia, who form together well over half the total, plus some forty sub-groups living in the mountains, particularly in the remarkably diverse Dagestan where there are no less than thirty different nationalities, which also include Turkish and Iranian peoples. Most are Moslems, though the Georgians are Orthodox, except around Batumi. The Armenians follow a Gregorian rite of their own. These peoples are sedentary by tradition but some pastoralists are found in the mountains of Dagestan and in Armenia on the high plateau east of Lake Sevan.

The Finno-Ugrians (4·2 million), widely scattered and extremely diverse nationalities, appear to have originated in the Altay country and to have been forced north-westwards by Central Asian nomads, later to be driven further north or absorbed by the expanding Eastern Slavs. They now live chiefly in the Ural-Volga region and include the Mordovs (1·28 million), Udmurts (623,000), and the Mari (504,000), while in the north live the Komi (431,000) and related tribes. The North-west European group includes the 969,000 Estonians, Protestants influenced by German culture, and the Ingrian Finns. In the poor, glaciated country north of Leningrad live 169,000 Karelians and some 93,000 Finns. In 1940, the Karelians were accorded their own republic, but the industrial development of the country brought in large numbers of Slav settlers, who now comprise 70 per cent of the population. There are also scattered remnants of tribes such as the Veps, Vods, and Ishori in the upper Volga basin who were not submerged by Russian settlers. In the Khibiny Peninsula, the reindeer-herding Saami (Lopari) are related to the Lapps. In Northern Siberia, reindeer-herding Finno-Ugrians include the Selkups, Nentsy, and Khanty-Mansi, though among some their origins are obscured by contact with other peoples. They are mostly of Orthodox faith with remains of primitive shamanism, and all have been influenced by contact with Russians.

In Soviet Central Asia and in Transcaucasia, there are about two million people with Iranian affinities. The 1·4 million Tadzhiks, mountain nomads of the Pamir and associated ranges, form the largest group. They are Sunni-Moslems, but the small mountain tribe of Yagnobts are Ismailian Moslems, recognizing the Aga

Khan as their head. Both predate surrounding Turkic peoples. The largest Caucasian group is the Osetin (410,000), though there are also 59,000 Kurds, besides smaller groups like the Tats and Talysh. The Mongol tribes of Baykalia, although comparative late-comers, for long formed an obstacle to Russian spread. The most important are the 253,000 Buryats, the purest Mongol stock, but there are also several smaller groups related to them. In the steppes of the lower Volga, the Buddhist Kalmyks (106,000) migrated to the region in the early eighteenth century. In 1941, they were accused of collaboration with the Germans, and nothing further was heard of them until they were 'reinstated' in 1957.[16] The Mongols are mostly Lamaist Buddhists, except for the small Sart-Kalmyk Moslem people in the Altay. The vast wastes of Central and Northern Siberia are sparsely settled by nomadic tribes predating the Turkic and Mongol expansions. The Tungus-Manchurians live scattered from the Yenisey to the Pacific coast. The two main tribes, Eveni and Evenki, number together 33,000, and there are several very small groups. The main concentration is in the Central Siberian Plateau. In the most inaccessible poor forests and tundra of North-eastern Siberia, the offshore islands and the marsh-protected delta of the Amur, live earlier inhabitants, the Palaeo-Asiatics, scattered among later tribes. Already several groups are extinct and others have succumbed to extensive russification. They are usually distinguished by their economy, depending on fishing for coastal dwellers or on hunting and herding for tribes in the interior. They are related to indigenous North American tribes and do not exceed 20,000 in number. The largest group is the Chukchi (12,000) and there are also 6,300 Koryaks and 1,100 Eskimos. To this group may be ascribed the 'hairy' Ainu of the Kuril Islands and Sakhalin, whose actual anthropological affinities are enigmatical.

Long and complicated racial history has produced great physical diversity between and even within national groups, though the relationships may become clearer with more detailed investigation of blood-group structures. Criteria of physical anthropology may only be applied on a generalized scale and far more cognizance has been taken of cultural, historical, and linguistic traits in formulating 'nationality' than of physical attributes. Essentially a linguistic concept, the Slavs show great physical diversity, tending to medium stature and broad-headedness, with lighter skin and hair colourings in the north-western European districts. On the west, they have the features of mixed nordic-alpine Europeans, while towards the east and south-east these traits are augmented by intermixture with dark-skinned and narrow-eyed Asiatic elements. The Finno-Ugrians have white to yellow-brown skins, black to dark-brown hair and high cheek bones, though they have also absorbed 'nordic' characters in the west, besides mixing with the diverse Eastern Slavs. Among some of the more primitive northern groups there are traces of mongoloid features. East of the Volga, across Siberia and into Central Asia, people have straight hair, mostly black, with skins ranging from whitish-yellow to yellowish-red. Medium to broad heads, apart from the Eskimos, are common, like broad faces and narrow eyes. The Mongols have coarse, black hair and yellow to yellow-brown skins, with broad heads, flat faces and noses, and the characteristic eyefold. Traces of mongoloid features can be found among many groups in Siberia and Central Asia, notably the related Tungus. On the west, the Mongoloids seem to have formed an intermixture with primitive Europids, so that Central Asian groups such as the Kazakh-Kirgiz and Turkmen have dark colourings but long faces and frequently prominent noses. Europid features are found among the enigmatical Ainus of the Far East. Throughout these vast territories, the population tends to be of medium stature. The numerous nationalities of the Caucasian isthmus belong generally to the Indo-European peoples and are a branch of the alpine stock. They are brown-skinned with wavy hair and many show semitic facial features. Some mountain groups have a tall stature, though like Mediterraneans they are broader headed.[17]

NOTES ON CHAPTER 4

1. Lorimer, F., *The Population of the Soviet Union: History and Prospects* (Geneva, 1946) forms a most useful statistical and historical summary. Also useful are the relevant chapters in Kirk, D., *Europe's Population in the Interwar Years*, Geneva, 1946.
2. On 1 January 1961, estimated at 216 million; mid-1963 at 225 million.
3. *Demographic Yearbook of the United Nations* (New York, annually) is a useful source of comparative statistics.
4. Results of the 1959 census were published piecemeal in the Soviet press, e.g., *Pravda*, 10.5.59 and 4.2.60.

5. Konstantinov, O., 'Some Conclusions about the Geography of Towns and Urban Population in U.S.S.R. based on 1959 Census', *Trans. Soviet Geography*, vol. I, 1960. A detailed population map has appeared in *Geografischeskiy Atlas dlya Uchiteley Sredney Shkoly* (1960) and in *Atlas Selskogo Khozyaystva SSSR*, Moscow, 1960.

6. Hrdlička, A., *The Peoples of the Soviet Union*. Washington, 1942.
   Mirov, N., *A Geography of Russia*. New York, 1951.

7. Treadgold, D., *The Great Siberian Migration*. Princeton, 1957.

8. *Petermanns Mitteilungen, Ergänzungshefte*, 27/28 (1898–1900) and 29/30 (1900–1903) carry the relevant sections of Supan's famous *Die Bevölkerung der Erde*.

9. Lorimer, F., *op. cit.* and Frumkin, G., *Population Changes in Europe since 1939*. Geneva, 1951.

10. Roof, M., and Leedy, F., 'Population Redistribution in the Soviet Union, 1939–1956', *Geographical Review*, vol. 79, 1959.
    Roof, M., 'Soviet Population Enigma Reconsidered', and Selegen, G., 'First Report on the Recent Population Census in the Soviet Union', both in *Population Studies*, July 1960.
    Pokshishevskiy, V., *Geografiya Naseleniya Vostochnoy Sibiri*. Moscow, 1962.

11. Lorimer, F., *op. cit.*
    Podyachikh, P., *Naseleniye SSSR*. Moscow, 1961.
    Pravda, 4.2.60.
    Tokarev, S., *Etnografiya Narodov SSSR*. Moscow, 1958.
    Coon, C., *Races of Europe*. New York, 1939.
    Kolarz, W., *Russia and her Colonies*. London, 1952.
       The People of the Soviet Far East. London, 1954.
    Wheeler, G., *Racial Problems in Soviet Muslim Asia*. Oxford, 1960.

12. Niederle, L., *Slovanské Starožitnosti*. Prague, 1925.

13. Schleunig, J., *Die deutschen Siedlungsgebiete in Russland*. Göttingen, 1955.
    Stump, K., 'Wo leben noch Deutsche?', *Digest des Ostens*, September 1959.

14. Conquest, R., *The Soviet Deportation of Nationalities*. London, 1960.

15. Caroe, O., *Soviet Empire*. London, 1953.

16. Conquest, R., *op. cit.*

17. Hrdlička, A., *op. cit.*

CHAPTER 5

Town and Village and Territorial Administration in the Soviet Union

ONE of the striking results of Soviet policy has been the urbanization of the country, although historically the town in Russia has played an important rôle in the growth of nationhood and in the integration of incorporated territory into the Russian state. Before the Revolution less than a fifth of the population lived in towns; by 1939, a third was town-dwelling and by 1961, one Soviet citizen in two lived in a town. The four and a half thousand urban settlements vary not only in size but also in form, function, and origin. The growth of towns in Central Asia, Transcaucasia, the Baltic lands, and the extreme western European territories which lay until recently beyond Russia's control, was associated with non-Russian cultures. Even the truly Russian towns have been influenced from outside: Byzantine culture contributed through the church, while German ideas have penetrated via Poland and the Baltic German colonies, and there have been Persian, Turkish, and even Chinese ideas in Transcaucasia and Central Asia. Foreign artisans were used to build towns, of which Leningrad bears a vivid stamp. Contrary to some Soviet claims, the Eastern Slavs, though they lived in stockaded encampments, appear to have first received the concept of a town from Scandinavian and Byzantine contacts.[1]

While factors of the geographical environment have been important in moulding the Soviet towns, it is perhaps best to classify them by their historical genesis, though naturally most contain elements of several historical periods.[2] The most ancient towns were developed by oasis civilizations in Central Asia: Mary, though not always on the same site, is among the oldest cities in the world, while Samarkand (as Maracanda and Afrosiab) dates from the third millennium B.C. There are also the ancient Greek colonies of the Black Sea littoral — in Classical times thriving commercial towns, though few have remained in continuous

L

 DOI: 10.4324/9781003172048-5

occupation. Still early in relation to European development are the Transcaucasian towns with their marked Persian influences. Apart from an ephemeral influence of the Scandinavian Varangians in the European lands, the Germans made an important contribution. Many towns of the Baltic littoral were founded by Hansa merchants and Teutonic Knights and had considerable influence on the form of the Russian town. German civic law spread both through German colonists and through Polish-Lithuanian influence, turning the stockaded, wooden villages of the Russians into towns in a legal sense. A common feature in towns was the close association in their early stages of the defensive-administrative function and commerce, arising from the common interest of site. The defensive-administrative walled enclosure, the *gorod*, was frequently closely followed by the rise within its own walls or nearby of the civilian settlement, the *posad*, of merchants and artisans, while in northern Russia the cathedral or monastery often replaced the fortress (*kremlin* or *dyetinets*). Towns grew along river routeways which crossed the Russian plains, particularly on the portages between river systems where defence was militarily desirable and where markets arose at break of bulk points. The first towns appear to have developed by the tenth century, notably Kiev, Novgorod, Smolensk, and Murom; but the main period came in the twelfth and fifteenth centuries (Fig. 25). In the north, towns were founded during the great period of Russian monasticism between the thirteenth and fifteenth centuries, when the stability imposed by the *Pax Tatarica* opened trade routes and encouraged commercialism. The expansion of Muscovy against rotting Tatar power after the fifteenth century led to the establishment and spread of towns, with a primarily defensive function, in the steppe and forest fringe. The sixteenth and seventeenth centuries, the time of the rapid spread across Siberia, brought the foundation of fortified posts which quickly grew into wooden towns with the imposition of undisputed Russian power. Some became prison towns to which the banished were sent. The development of industry in the eighteenth century, appearing first in the villages of the forest belt, led to the growth of industrial villages consisting of huts for the serfs near to the works, and these were common in the upper Volga, in the central parts of Great Russia, and in the Ural. Out of them arose the planless, squalid, industrial towns of the nineteenth century, while similar unpleasant places grew

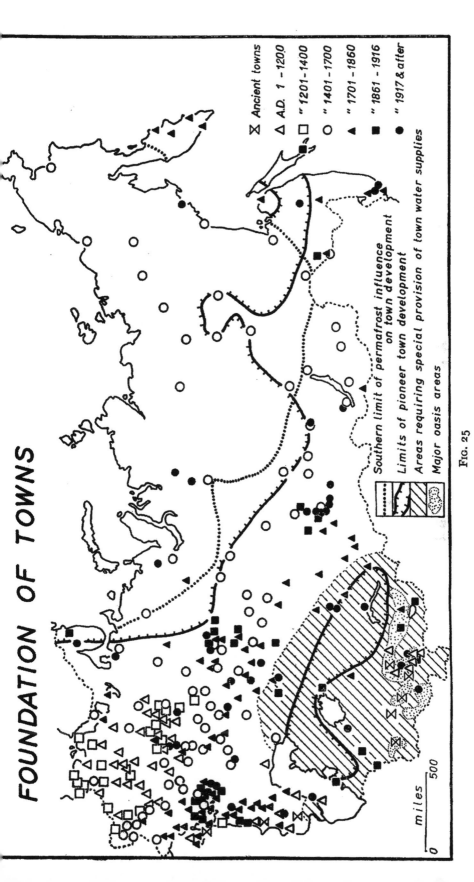

FOUNDATION OF TOWNS

⊠ Ancient towns
△ A.D. 1 – 1200
□ " 1201 – 1400
○ " 1401 – 1700
▲ " 1701 – 1860
■ " 1861 – 1916
● " 1917 & after

Southern limit of permafrost influence on town development

Limits of pioneer town development

Areas requiring special provision of town water supplies

Major oasis areas

miles
0   500

FIG. 25

Foundation of towns. Compiled from *Atlas Istorii SSSR*, Moscow, 1954–1960, three volumes, and *Voprosy Geografii 38*, Moscow, 1956, p. 229. Only the largest towns founded in the Soviet period are shown, while many of the Siberian 'towns' have hardly risen above large villages.

on the Donbass coalfield. The late eighteenth century also brought the foundation of market towns in the Ukrainian steppe, the Black Sea littoral, and Northern Caucasia, while in the nineteenth century territorial spread incorporated native towns in Transcaucasia and Central Asia, alongside which grew new Russian settlements. These colonial towns with their gridiron pattern and liberal use of ground, contrasted with the old, crowded, native walled towns.

Finally, the Soviet period brought new town foundations based on Communist ideas, for the dissemination of whose ideologies they form a 'proletarian base'. The important rôle of towns in the expanding share of the economy taken by industry has brought prolific new foundation as well as feverish growth in many older towns. The definition of urban communities has become an economic concept, although the criteria vary slightly between republics and from time to time. A town usually requires a minimum population of 1,000 adults, more than 75 per cent industrially employed, whereas a 'settlement of town type' has less adult population and a smaller share working in industry. A workers' settlement, for instance, needs 400 adults and at least 65 per cent engaged in industry. Once a community has become a town, its further progress is marked by the organ of local government to which it is subordinated as it rises up the municipal ladder, first under the *rayon*, then the *oblast*, and finally the *republic*. If it does not prosper, it may revert in status to a village. A town may have less than 3,000 people and a 'settlement of town type' at least up to 20,000 inhabitants.

In 1959, 82·6 million people lived in 1,694 towns and another 17·2 million in 2,922 'settlements of town type', whose geographical distribution, as might be expected, showed a close relationship to overall population distribution already described. Almost half the town-dwellers lived in 147 towns with over 100,000 people, which accounted for half the increase in urban population between 1939 and 1959, despite a slower rate of growth than in the period 1926–1939. Altogether, towns with over 50,000 people contain sixty per cent of the urban population (Fig. 21) and lie mainly in the broadest part of the settled triangle in European Russia which contains the three 'million' cities. Only three of twenty-two towns with between 500,000–1 million inhabitants are in Siberia (Novosibirsk and Omsk) and Central Asia (Tashkent). Of the towns with 100,000–500,000 people, twenty-four are in Siberia and fifteen in Central Asia contrasted to eighty-four in European Russia (including the Ural and Transcaucasia), which contains 224 of the 299 towns with more than 50,000 people.

Soviet policy of developing mining and industry under intense physical difficulties in the sub-arctic and semi-desert has been reflected in the growth of urban communities in such regions. Several of the towns have grown to considerable size [*e.g.*, Norilsk (108,000) and Murmansk (226,000), both north of the Arctic Circle, and Arkhangelsk (256,000), Magadan (62,000), and Vorkuta (55,000) or Karaganda (398,000), Temir Tau (54,000), and Balkhash (53,000) in the semi-desert]; but there are also many smaller towns. Such communities have been the principal means of settling country beyond the pioneer fringe of the sedentary areas, and the pattern is reflected in the very high proportions of urban population in such regions. For example, townspeople comprise nine-tenths of the population in Murmansk *oblast* and the 190,000 town-dwellers (62,000 in Magadan City alone) form 81 per cent of the population in Magadan *oblast*. In arid Karaganda *oblast*, over 800,000 townspeople form three-quarters of the population and half of them live in the adjacent towns of Karaganda and Temir Tau. In contrast, those districts where 30 per cent or less of the population is urban are the well-settled agricultural districts of Central European Russia, the Ukraine, or the oasis lands of Central Asia, where townspeople may form less than a fifth of the total.

The clusters of towns which appear on the distribution map do not usually form the closely-knit, urban agglomerations of Western Europe, for the towns stand apart at considerable distances, often separated by tracts of almost virgin country, as in the Ural or Western Siberia. Moreover, government policy, though not outstandingly successful, frowns upon the formation of conurbations and even giant towns. The largest and most compact group lies around Moscow, where eight towns with over 50,000 people stand within 15 miles of the centre of the capital. Probably another million people live in smaller communities within the same radius. These satellites, lying along more than a dozen railways into the capital, supply the bulk of the 500,000 daily commuters into the city. An extension of the Moscow boundaries in 1960 makes it now the largest city in the world in area, though the satellites are growing more rapidly than the metropolitan nucleus.

A more scattered group of towns is found on the Donbass coalfield, where underlying structure exercises some influence on town location. Within an area 160 miles long by 100 miles broad, roughly six million people live in towns, of which fourteen have populations exceeding 50,000 people. The largest towns, mostly mining coking coal or with large iron and steel plants, lie at the western end of the field, like Donetsk (Stalino) and Makeyevka (together 1·06 million), some seven miles apart, and Gorlovka (293,000), while Lugansk on the north has 274,000 people. The northern anthracite mining towns are smaller and those on the eastern, less-developed part of the field are also small, apart from Shakhty (196,000) and Novoshakhtinsk (104,000). Zhdanov (284,000) on the Azov littoral is closely associated economically with the coalfield, like Rostov on the Don delta (597,000). The mining towns are frequently composed of several settlements around mines and industrial plants, linked to a central core.

The Ural region has over 300 towns and workers' settlements spread over an area the size of England and Wales, where almost six million people live in towns with over 50,000 people and another four million in smaller places. The towns lie in the basins of the upper Kama and the Belaya on the west and along the belt of intense mineralization on the eastern slope. On the eastern slope, a small group of towns lies around Krasnoturinsk (62,000), Serov (98,000), and Nizhniy Tagil (338,000). Two groups are seen in the low central sections of the range : around the railway junction of Sverdlovsk (777,000) and near Chelyabinsk (688,000), where there are several lignite mining towns, and Kamensk-Uralskiy (141,000). In the south-eastern Ural is the large modern iron and steel town of Magnitogorsk (311,000). On the western slope, in the upper Kama, Perm (628,000) is the largest town and to the north lie Berezniki (106,000), Chusovoy (60,000), and Kizel (60,000). To the south, in the Belaya valley are Ufa and Chernikovsk (together 546,000) and the oil boom towns, including Salavat (60,000) and Sterlitamak (111,000). A rapidly growing group of towns lies in the Ural valley, centred around the metallurgical towns of Novo-Troitsk and Orsk (together 233,000), as well as the small Blyava-Mednogrosk and Kuvandyk.

The Kuzbass coalfield has seven towns with over 50,000 people each, though its area is little less than the Donbass. The largest is Novokuznetsk (Stalinsk) with 377,000, and Prokopyevsk and Kiselevsk, less than ten miles apart, have together 412,000 people. The total urban population of the coalfield and nearby ore field is about two million. Leningrad, together with its suburbs, has over three million people, while the cluster of oil towns near Baku on the Apsheron Peninsula number a million, excluding the nearby industrial Sumgait (52,000).

Small clusters are also found around the oases of Central Asia, on the Black Sea littoral between Adler and Sochi, in the Dnepr bend as well as on the Moscow lignite field near Tula. There are also many examples, both within the clusters and outside, of close spatial association of two or three towns along a railway or on opposite banks of a river; for example, Ufa-Chernikovsk, Bryansk-Bezhitsa or Barnaul-Chesnokovka, and at Gorkiy.

SOME EXAMPLES OF RUSSIAN TOWNS

Moscow, capital of the U.S.S.R., with a population of over five million is probably fifth of the world's greatest cities.[3] The Russian town best known to foreigners, it lies in a central part of the Russian Plain amid the head waters of several river routeways and protected by marshes and forests, particularly on the east in the great swamps of the Oka and Klyazma, though its southern flank was more exposed to raids from the wooded steppe. Its nodal position and its easily-defended location were considerable factors in its rise to leadership among the Russian princedoms.

A settlement appears to have existed before the first reference to it in 1147, when there was a wooden fort on the present site of the Kremlin, the 130-foot-high hill overlooking the broad, swampy meander of the Moscow river and protected on either flank by the marshy valleys of the Yauza and Neglinka. In the fourteenth century, permanent walls were built and later strengthened. Separated from the Kremlin by the large Red Square, the commercial districts of the *Kitai Gorod* (Tatar for Fortified Town) developed and were surrounded by a wall built in 1534. From the fourteenth century, the *Kitai Gorod* was the seat of a powerful merchant class. By 1520, Moscow is reputed to have had 45,000 houses and to have been larger than London. As the town grew, the White Town, the home of courtiers and richer citizens, spread around the central nucleus, and was enclosed within a wall in

1586. Further colonization led to expansion beyond these walls, where the Earth Town, named after its earthen ramparts, the home of many of the tsars' servants, spread outwards and extended across the Moscow river. In the 'suburbs' beyond the Earth Town were quarters for foreigners (*e.g.* the German Quarter of the north-east) though they also lived in the central parts of the town. Many of Moscow's street names reflect the functions of their districts at this period. In the seventeenth century, the Kremlin lost its rôle as a fortress and took its present elaborated form. In the eighteenth century, the suburbs were surrounded by an outermost rampart around which were fourteen customs barriers (*zastavy*). The more exposed flanks on the south and south-east were marked by a line of fortified monasteries, which at this time were also elaborated. Despite the brick Kremlin and the many fine churches and monasteries, the appearance of the town remained village-like, with roughly-paved streets, gardens, and wooden houses with vast courtyards and outhouses, as well as squalid hovels. Around the outskirts lay the wooden palaces of the rich, some disguised under a stuccoed exterior to give the appearance of stone, the older ones in traditional style but the newer ones copying Western forms.

In the fire of 1812, when Moscow fell to Napoleon, three-quarters of the houses were destroyed, but the town was quickly rebuilt. Despite the removal of the capital to S. Peterburg, the prosperity of Moscow steadily increased for new industries brought commercial wealth. The old walls were pulled down and replaced by wide streets, while the walls of the Earth Town were turned into gardens (the Sadovaya), leaving the characteristic rings seen in the street pattern of modern Moscow. The outer parts of the town became factory sites, notably along the banks of the Yauza in the south-east, in Baumann and Nogatino, and across the river in Zamoskovorechye. Railway development brought a ring of terminal stations around the town, roughly along the line of the old barriers. Brick and stone began to be used more freely, though wooden houses remained still very common even to the present day. After emancipation, country people flooded into Moscow, the population rising from 600,000 in 1871 to 1·6 million in 1912.

New importance came after the Revolution, when Moscow once again became capital. Industrialization increased its national significance for it contributed the most important manufactured articles, and the superstructure of Soviet bureaucracy attracted much labour. Its population had risen to two million by 1926 and to 4·1 million in 1939. A large development plan was started in the early 1930's, removing the last of the old walls and widening streets, so that much of the centre of the city was recast. New squares were created, parks opened, and large new public buildings and hotels erected. Public transport was improved by the construction of the *metro*, whose monumental stations are now tourist attractions. Completion of the Moscow-Volga Canal added to the value of the river and brought port and industrial development on the south at Nogatino and on the north at Khimki. Housing was given comparatively low priority, though blocks of drab flats appeared. Since 1945, accelerated development has brought large new residential quarters on the south in Cheremushki and the Lenin Hills as well as on the north in Ostankino and Khimki. Along broadened thoroughfares (some with fly-under junctions), imposing blocks of flats have been built in the gardens of squalid wooden or brick tenements, which remain, however, a common sight in Moscow's side streets.

The town is a strange mixture of architectural styles, reflecting its varied history. There are the fine old buildings of the Kremlin (with its Italian influences), St. Basil's Cathedral, and the old fortified monasteries; the simple baroque of the palaces, frequently executed in wood and stuccoed; and less pleasant nineteenth-century buildings in brick and in wood. In the early years of the Soviet regime, a few buildings in glass, steel, and concrete were erected, but most new buildings are in the heavier, bizarre style of the later Stalinist period, reflected in five post-war skyscrapers. Since Stalin's death there has been a reversion to simpler forms, more Western in appearance, seen in the new housing quarters.

Leningrad (formerly S. Peterburg and Petrograd), a remarkable contrast to Moscow in its elegance and lightness, was created by the Western-minded Peter I, who made it a 'window on the west', the most westerly point in Russia then accessible to open water, through which European ideas might enter.[4] Founded in 1703, the site is on low, marshy islands in the Neva delta, liable to autumn flood when prolonged south-west winds pond back water flowing from the river. Lying on a narrow isthmus between Lake Ladoga and the Finnish Gulf, it was protected by fortresses at Schlüsselburg (Shlisselburg) and the island of Kronstadt (Kronshtadt). Some 40,000 peasants were assembled to build the town, though they were frequently decimated by plague. Italian and other foreign architects and artisans were also employed. Large buildings had to be built on deep piles driven into soft, estuarine materials which restricted their height, imparting the characteristic low skyline broken by high, thin spires. The spaciousness of the town is given by its wide, radiating street plan and the broad arms of the Neva and the drainage canals. The central part grew around the first foundation of the Peter-Paul Fortress, but the old parts of the town are now surrounded by more suburbs of a character found in all Russian towns. Along the shore of the Gulf of Finland lie pleasant resorts, and nearby are communities around former palaces, like Petrodvorets, Pushkin, and Pavlovsk. The Leninport, a complex of wharves and docks at the mouth of the Bolshaya Neva on reclaimed ground and well-placed to serve the export of north-west Russian produce, with easy contact with the Volga basin, is one of the largest Soviet ports despite winter closure by ice. Leningrad's industries grew traditionally on the manufacture of imported raw materials for government requirements and still form one of the main producers of high-quality chemicals and specialized engineering goods.

Kiev, the Ukrainian capital, stands on the Dnepr in the marchlands between the northern forests and the southern steppe, where contrasting economies meet at an important north to south routeway.[5] The earliest settlement (perhaps pre-ninth century) stood on a section of the high right bank, about 300 feet above the river, isolated by deeply-incised valleys and easily defended. This site contains today several of the most important historical monuments. To the south, on the right bank, lay the Pechersk Monastery near to the Prince's village of Berestovo, in the seventh century admired by Latin writers as being larger and more splendid than any other centre in Christendom.

Weakness stirred by internal strife contributed to Kiev's capture by the steppe nomads about 1230 and until the eighteenth century its fortunes vacillated. With the spread of Russian power to the Black Sea coast, the town again began to grow rapidly. Buildings spread along the high bank between the deeply-incised ravines, while at Podol at the confluence of the Pochaina with the Dnepr, on a low river terrace, new industrial suburbs and the main river port grew out of an old quay and artisans' quarter. The great industrial expansion of the 1930's led to growth of factories both on the right bank and across the river at Darnitsa.

A typical site of the older Russian towns is a high bank or river terrace overlooking a defendable river, usually at a routeway intersection or the end of a portage. The nucleus has frequently been an enclosed *gorod* containing a fort or even a cathedral or monastery.[6] Both Kiev and Moscow have these characteristics, found also in Smolensk, Chernigov, and Pereyaslavl. Novgorod, whose commerce led it to greatness despite the agricultural poverty of its surrounding countryside, was founded in the ninth century on an important routeway leading from the Baltic via the Neva and Volkhov to the Black Sea along the Lovat and the Dnepr or across the Valday Hills to the Volga basin. On a low bluff on the west bank lay the kremlin within whose walls was also the cathedral. Around this nucleus and separated from it by a broad square was the administrative and ecclesiastical settlement (the Sofiyskaya Storona), surrounded by a moat. At an early date a bridge across the Volkhov linked it to the merchants' and artisans' quarter (the Torgovaya Storona) — also a walled community, whose trade centred on a large square. Here were colonies of Visby and Hansa merchants. Novgorod was noted for its fine wooden buildings, perhaps one reason why it was so thoroughly destroyed by Ivan the Terrible, who is said to have slaughtered 60,000 people. Another old town and once trade rival of Novgorod is Pskov, a tenth-century foundation, whose kremlin lies on an easily-defended tongue of land at the confluence of the Velikaya and Pskova, while nearby the fourteenth-century town lies surrounded by the remains of a wall and a moat, beyond which there is the rectilinear plan of the modern town. Similar sites are found at Suzdal, Rostov-Yaroslavskiy, Yaroslavl, Tver (Kalinin), Smolensk, and Vladimir (known even by the twelfth century for its stone architecture), and at Tula, where the town has been much modified by later development. Nizhniy Novgorod (Gorkiy), founded in 1221, stands at the confluence of the Oka and Volga, on a dissected tongue of land 300 feet above the river, surmounted by the kremlin. The modern town has developed on lower surfaces around the kremlin, though deeply-incised gullies interrupt the rectilinear street pattern, and steep roads lead to the newest parts on the low terrace besides the river. The low terrace on the opposite bank was the site of one of the largest Russian trade fairs. The establishment of the town was primarily as a defence against the inroads of nomadic Volga tribes; but the fair seems to have been founded in rivalry to the one held by the Tatar khans of Kazan.

The old towns in the Baltic provinces, despite many Russian influences, were founded by German colonists on the sites of earlier Baltic communities. With their strong Hansa connections and the support of the Teutonic Knights, they bore features common among northern German towns. The largest is Riga, lying on a sandy plain on the banks of the wide Western Dvina, where it contains several small islands providing bridging points, and easily defended by several small lakes. The old town, with its castle, cathedral, and many guild foundations, has the crooked, mediaeval street plan and retained its walls and moat until 1857. Riga held Lübeck Law until it was organized on the model of Russian towns in 1877. The rapid growth in the nineteenth century produced a gridiron pattern of outer suburbs in a mixture of Russian and German styles. The Germans remained dominant in the town's population until after the inflow of Russian and Latvian workers for the new industries established by German initiative in the late 1880's. In the Hansa period, Riga traded along the easy route provided by the Western Dvina with the interior. In the nineteenth century, timber from the river's 5,000 square miles of drainage basin was floated down to the port, which became the largest timber exporter in Russia, while railways brought grain from the Ukraine. During the interwar years, Riga was cut off from its hinterland by the international frontier and the poor relations existing between Latvia and the U.S.S.R. Since reincorporation into the Soviet Union, however, Riga has again become one of the largest Russian ports and a major manufacturing town. Tallin (Revel) stands on a low, raised beach around which the Glint forms a broad amphitheatre. The Danes captured the site from the Estonians in 1219 and sold it to the Teutonic Order in 1346. It took Hansa membership and Lübeck Law, though the town was divided between the aristocracy, who held the cathedral and castle on a 140-foot-high hill, a remnant of the Glint, and the merchants of the 'lower town' on the raised beach. The older buildings have a distinct German Baltic style in contrast to the newer Russian buildings. It serves as a winter port for the Gulf of Finland when Leningrad is closed. Narva, on the west bank of the Narva river, is another town reputedly founded by Danes and later inhabited by Germans, passing to the Teutonic Order in 1347. On the eastern bank, the Russians founded the fortress of Ivangorod in 1492, which became the working-class suburbs for the large, nineteenth-century Krengolm textile mills.

The older Siberian towns, dating from the seventeenth and eighteenth centuries, were originally fortified wooden trading posts. Their wooden walls and towers, preserved in Yakutsk until early in the present century, surrounded low, wooden houses lying in large courtyards with high fences, wooden churches (often brightly painted), and wooden public buildings, which lined streets paved with wood. In Tobolsk and Omsk and other south-west Siberian towns, the original settlements were around fortresses sited on right-bank terraces. Most towns, irrespective of date of foundation, have rectilinear street patterns. They lie typically at route intersections between rivers and portages or tracks inland. The building of the east-west railway in the south produced the growth of large market and industrial towns at the intersections with rivers, of which the best example is the impressive Novosibirsk whose population since foundation in 1894 has risen to over 800,000.

In Central Asia, there are many ancient towns built by indigenous oasis civilizations and influenced by Persian and Indian culture, though existing towns suffered heavily in the Mongol devastations and some were completely destroyed. Samarkand dates the original foundation from the third millennium B.C. and Khodzhent (Leninabad) is contemporaneous. Though not on the original site, Mary (Merv) traces its origin to Antiocha Margiana of Classical times. Several towns were founded by the Alexandrian conquest. The towns were absorbed into Russia in the late nineteenth century.

The Russian quarters, with broad tree-lined streets and European-style buildings and churches, contrast with the native walled towns with narrow, crowded alleys, bazaars, caravanserais and mosques (some of great if decayed splendour). The Russian and native quarters usually lie adjacent to each other, with a Russian citadel dominating the old town. In the Russian quarter, wooden houses with shuttered windows looking out on to the tree-lined streets stand amid their own gardens and contrast with the windowless outer walls of the courtyard houses lining narrow alleys in the native town. Since the Revolution, crowded and insanitary native towns have been partly cleared and improved, while buildings and wide avenues in the Soviet style have been built, but these only emphasize the shabbier outskirts. Industrialization has affected towns like Tashkent and Samarkand more than the smaller places, so that Bukhara, for instance, retains much of its native character. The native towns of Transcaucasia, with strong Persian and some Turkish influence, date from early in the Christian era. Tbilisi was founded in the fifth century after discovery on the site of warm, health-giving springs. Commanding north-south and east-west routes, it became the Georgian capital. The old town lies in the narrow part of the Kura gorge, dominated by a fort and the royal palace above the winding passages of the bazaars, while the broader part of the gorge upstream is occupied by the gridiron street pattern of the Russian town around the railway station. Baku was originally sited on a commanding hill at the southern side of a broad bay on the Caspian shore, possibly replacing the ancient town of Baila submerged by a change in the sea level. It is commanded by a fortress within whose walls lies most of the native town; but the modern town, with its refineries and nearby oil wells, has grown in a dreary gridiron pattern in a wide amphitheatre of the hills.

During the nineteenth century, industry haphazardly turned villages into straggling, untidy, unplanned towns, with poor housing and vacant lots among which lay factories and railway yards. Since the Revolution, the Soviet authorities have tried to replan these towns. Modern if shoddy blocks of flats have replaced wooden hovels, vacant lots have been filled by parks, and streets planted with trees. The centres have the imposing local government buildings of modern Russian towns. In contrast are the new 'towns of socialist realism', spaciously laid out with the streets radiating from the main works' entrance and lined by blocks of flats arranged in neighbourhood units provided with the necessary services.[7] Special quarters exist for the 'creative intelligentsia'. Examples are Karaganda, founded 1928, formed by a group of coal-mines around a central administrative centre, and Magnitogorsk (1929) built adjacent to its steelworks in the shadow of Magnitnaya Gora from which the ore comes and alongside a reservoir on the Ural river to supply water in this dry region. Even Norilsk above the Arctic Circle has the same imposing centre as towns further south. Of the smaller towns, Volzhskiy, near the Volgograd (Stalingrad) hydro-electric barrage, became a town in 1954, when it contained 150 two- and three-storey houses (flats?) with central heating, five schools, four club-houses, nine kindergarten and crèches, a park with summer theatre, cinemas, and a stadium.[8]

MORPHOLOGICAL CHARACTER OF THE CONTEMPORARY RUSSIAN TOWN

Few detailed studies of Soviet towns have been published and large-scale maps and plans are not available. Fig. 26 represents the typical elements in the contemporary town drawn from experience in small and medium-sized provincial towns. The largest towns have features generally not unlike Western and Central Europe; but the small towns are quite alien and often appear village-like in their buildings, layout, and the great amount of open space. Towns annexed since 1945, such as Lvov, Uzhgorod, and Mukachevo, still bear a deep impression of their former owners. Of the truly Russian towns, perhaps the most lasting visual impression is their similarity, so that it is difficult to pick out features lending individuality to particular communities, which doubtless arises from the widespread standardization throughout Russian life.

Despite forty years of Soviet rule and the upheavals of the period, Baedeker's description of the Russian town published in 1914 remains remarkably true.[9] 'The towns of Great Russia generally cover a great deal of ground, and are all laid out on the same pattern. The centre is occupied by a spacious square, from which radiate broad and badly-paved streets crossing each other at right angles. In the central part of the town the houses are built of stone and painted white, yellow or pink. The public buildings seldom possess any architectural merit. In the suburbs the houses are of wood and stand at considerable distances from one another. The only effective architectural features are the large churches, which generally stand in open spaces at the intersection of streets. Their gilded, silvered or brightly-coloured cupolas are very conspicuous. . . . A large proportion of the population wear uniforms. . . .'

Today the towns contain extensive greenery in formal parks,

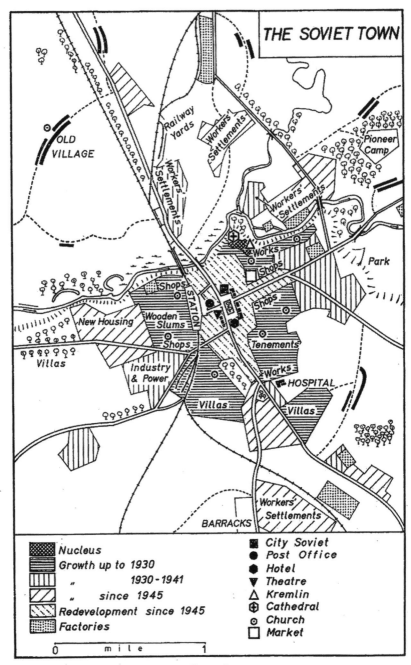

THE SOVIET TOWN

OLD VILLAGE

Railway Yards

Workers' Settlements

Workers' Settlements

Pioneer Camp

Workers' Settlements

Works

Shops

Park

Shops

New Housing

Wooden Slums

STATION

Shops

Shops

Tenements

Villas

Industry & Power

Shops

Works

Villas

HOSPITAL

Villas

Workers' Settlements

BARRACKS

Legend:

- Nucleus
- Growth up to 1930
- „ 1930-1941
- „ since 1945
- Redevelopment since 1945
- Factories

- City Soviet
- Post Office
- Hotel
- Theatre
- Kremlin
- Cathedral
- Church
- Market

0   mile   1

FIG. 26

Morphology of the Soviet town. This is an attempt to construct an idealized diagram of the morphology of the typical Soviet town about 1960. Data were collected in towns in European Russia.

tree-lined streets, and gardens, so that the small towns look like masses of buildings scattered in a forest when approached from the surrounding farmland. To maintain such greenery, even in the arid south, water is used liberally — sprayed from water carts or by hoses and hydrants. In Kharkov, the parks have been greatly extended since the war, as in many other towns. Nearly every town has its large Park of Rest and Culture.

Buildings range from the traditional styles of the churches, monasteries, or fortresses, which have been carefully preserved, like the Cathedral of the Assumption and the town walls in Smolensk, or the fourteenth-century fortress (*dyetinets*) in war-damaged Novgorod. The numerous wooden buildings vary from stately palaces in imitation-Western styles, like the Sheremetyev palace at Ostankino, Moscow, to trim, suburban 'private' villas with their high fenced gardens or the squalid wooden slums of most towns. The great mass of buildings are undistinguished brick or neglected, colour-washed stucco. Modern public buildings are mostly in the monotonously flamboyant Stalinist style, though occasionally more austere buildings of simpler glass and concrete erected in the early nineteen-thirties have been rebuilt in the same style after war-time damage (notably in Minsk and Kharkov). There has been a reversion to simpler styles, though shoddy and untidy workmanship adds an unnecessarily dilapidated appearance. There is a great contrast between the lightness, spaciousness, and dignity of Leningrad and the barbarity and oppressiveness of Moscow's skyscrapers or Kiev's ornate *Kreshchatik*, the main thoroughfare.

Driving along broad, well-paved tree-lined streets, the transition from country to town is rapid and without marked ribbon development. The town centre focuses on a large square, with flower-beds, trees, and a huge statue, bordered by modern public buildings containing local government, industrial, and party organizations, a theatre, and an hotel. In the ground floors are the state *Univermag* department store (sometimes a building for itself), the well-stocked *Knigtorg* bookshop, and the rather disappointing food store labelled *Gastronom*. Along the main roads leading from the centre stand large public buildings and blocks of 'luxury' flats, frequently boxlike buildings with an imposing, colonnaded portico of uncertain classical order. A square is commonly found before the impressively large railway station, whose grandiose exterior conceals the paucity of its open platforms.

Away from the centre, standards drop quickly and most property is shoddy and neglected. Tenements in brick, stucco, and wood line poorly cobbled or even unpaved streets, while there are also less imposing public buildings. The dreary appearance of the tenements standing around dusty courtyards is broken only by the gay window-box flowers in summer. The ground floors of many buildings contain shops and there are kiosks selling cigarettes, newspapers, and food in the streets. Both large and small factories are found in this part of town, while the ground floors of the tenements sometimes contain small workshops. The churches, of which few remain in use, look particularly neglected and are used as stores or garages. Propaganda slogans replace the billposters of the West and neon signs are generally only found on buildings in the centre of the towns. Street lighting is poor. Some squares contain gardens, though one usually serves as the bus station for country and inter-town services, while trams and trolley buses run along the main urban thoroughfares. Taxis also form a considerable part of the relatively light traffic which consists mainly of lorries. Another bare patch of ground is the collective farmers' 'free-market', which is one of the busiest places in town; to judge by the crowds it is preferable to buy here than in the state shops.

The suburbs contain the remains of old villages swallowed by the spreading town. In the black earth lands, suburban development has often been influenced by the occurrence of erosion gullies. Most suburban housing consists of small wooden houses, which in some quarters form privately-owned villas set in large, fenced gardens and lining dusty, unpaved roads. They draw their water from street standpipes, also to be seen in the central parts of towns. Some wooden housing forms slums, though here it is common to see new flats rising in the gardens of the hovels they are to replace. The largest modern factories, usually with extensive railway facilities and tips, are found in this part of the town. Nearby stand the settlements for their workers, whose street pattern sometimes focuses upon the works' entrance. The more modern settlements are designed as neighbourhood units of large blocks of flats, which in the largest towns may be six or more floors high and complete with schools, shops, cafés, and nearby playing-fields, sometimes incorporating the ideas seen in the new 'towns of socialist realism'. Here are also the summer camps for children and the large Park of Rest and Culture, besides the inevitable barracks to provide sentries who guard most bridges in the Soviet Union.

Slowly but with gathering momentum the towns are spreading into the countryside. On the lignite field south of Moscow, already mines and factories are spreading amid fields and forests and the industrial Shchekino creeping nearer to the idyllic setting of Tolstoy's home, *Yasnaya Polyana*. Kharkov is growing outwards as new neighbourhood units are built, though they are separated from each other by broad but relatively steep-sided erosion gullies. At Kiev, whole new suburbs are being built on the low east bank of the Dnepr, raised above inundation level by sand pumped from the river bed.

THE VILLAGE

The social and economic order of the village has moulded Russian history. Even today, with half the population living in towns, the village remains an important element in the landscape and retains its significance in the settlement pattern of the countryside. It is not unexpected in a country with such physical and ethnic diversity that the form and distribution of the village varies considerably between regions.

The village is the common unit of rural settlement throughout the Slavonic lands. In northern European Russia, villages are usually small, comprising less than a dozen houses, and widely scattered, lying on patches of drier ground, notably sandy terraces above the flood plains of rivers and avoiding the wet and not uncommonly swampy interfluves. There are also scattered outbuildings in the fields among the forest and the swamp. The villages are laid out in a simple line along a routeway or watercourse, consisting of small wooden houses, often richly carved on the gables and window frames. In the central Great Russian districts and the North-west, where the ground is drier but there is nevertheless a high water table, villages are larger and lie on the higher and drier watersheds or on small knolls and moraines. The wooden houses are larger and a village may contain a few dozen houses, again along a straggling street, though there are loosely-nucleated clusters on drier mounds amid wet ground, as in Polesye or the Meshchera. Several villages may be grouped to form one collective farm.

The steppe villages cluster along the banks of rivers or in the wet bottoms of gullies, for water here attracts rather than repels settlement as in the north. Each village has a large pond to collect water for the dry season. A valley location gives protection from bitter winter winds sweeping across the open treeless steppe, which only contains a few scattered huts; yet in some places it is impossible to see any houses from the grainland interfluves between the incised valleys and *ovragi*. The valley villages, containing perhaps a few hundred houses, stretch for miles along streams, but there may be loose nucleations around ponds or where springs are found at valley heads. Some villages have the 'chess board' pattern of planned settlements, as for example in the larger former Cossack villages (*stanitsy*) or newly-laid-out farm villages, some of which are to develop as agricultrual towns (*agrogoroda*).

Building materials vary. In the north, the wooden or log house is most common; but in the south, the general style is the frame house covered by bricks, wattle and daub, stone or clay, with a colour wash (frequently blue) applied in some districts. Roofing materials are nowadays mostly metal sheeting, though old houses and outhouses have thatch. Throughout the country, the mass-produced, standardized, wooden or asbestos sheeting houses are tending to replace traditional forms, though farmers often add such traditional features as a veranda or carved window frames. Every peasant house stands in a fenced garden patch shaded by trees and with small outbuildings, where fowls, bees, and pigs are kept and vegetables grown.

The nineteenth-century land reforms loosened village structure and progressive peasants left to set up separate farms in the fields, adding an element of dispersal which varied from district to district according to the success of the movement. Nowadays, the patterns of the state and collective farms set the norms of rural settlement, so that the old distinction between villages with and without churches has gone and the important villages are now the farm centres, with village *soviets*, machine and tractor stations (often using an old church), grain elevators, schools and clubs. Here are new barns and sheds for livestock. The smaller villages, hamlets, and scattered huts on the farm are the homes of various 'brigades'. In the new wheatlands, villages established as the centres of state farms have become virtually agricultural towns.

Other nationalities have adopted modified forms of the Russian village and house. For instance, unlike the Finns who live in scattered farms, the Karelians adopted the loose-street village and wooden house, though often in a two-storied form with an unusually wide barn at the rear, also found in the Volga basin. Chuvash settlements form loose-street villages, generally away from main routeways, while Tatars, who are not inclined to dig wells, site their villages in a confused mass along streams or round springs. The Bashkirs in the nineteenth century gave up their conical tents for a settled life in villages modelled on the Russian form but with more primitive housing. The Moldavians live in large villages in the hilly *kodry* country and have a house similar to the Ukrainian frame house.

Isolation from Great Russia and the distinctive social conditions enjoyed by the Baltic provinces produced a pattern of hamlets of twenty or so houses or scattered single farmsteads, where the tendency to disperse was strengthened by nineteenth-century and inter-war land reforms. Collectivization consequent on reincorporation into the Soviet Union has renewed nucleation. Houses have been wooden though larger than in Russia, but these are being replaced by standard Soviet forms.

Lowland settlements in Caucasia have been influenced by Russian ideas; but the mountain villages are compact masses of houses built on valley benches which offer a defensive site and water supplies. The houses are flat-roofed, partly subterranean, and built of stone. Watch towers are common, particularly in Dagestan, Georgia, and Svanetiya. In the valuable irrigated oasis lands of Central Asia, less influenced by Russian ideas, houses are pressed into the smallest area possible. Villages resemble towns and have crowded, flat-roofed courtyard houses and narrow alleys. The flat roofline of these commonly walled communities is broken by towers and minarets of mosques. The Russian house is, however, spreading among the nomadic tribes as they take to more sedentary life, though 'brigades' working away from the principal settlements still use the traditional Central Asian *yurt*. The *yurt* in the eastern form with a low, rounded top in contrast to the western cupola roof, is widespread in Southern Siberia, notably among the Mongolized tribes. But there are also wooden *yurts* and even conical and pyramidal tents of felt, bark, or logs. Subterranean huts are typical of the winter dwellings in Northern and North-eastern Siberia, though tents

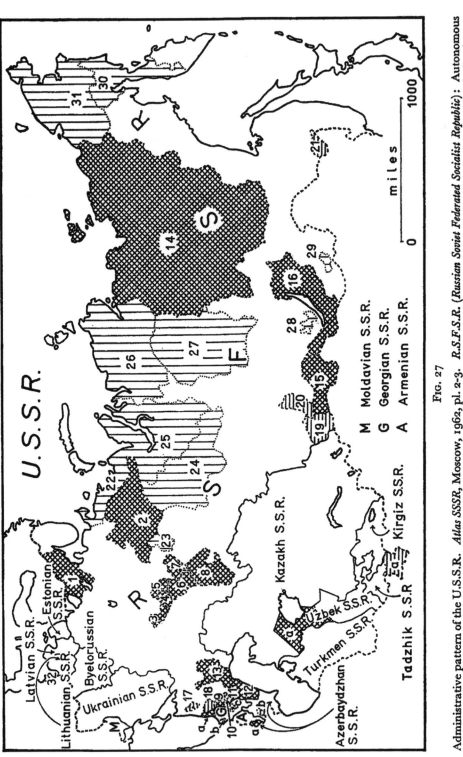

FIG. 27

Administrative pattern of the U.S.S.R. *Atlas SSSR*, Moscow, 1962, pl. 2-3. *R.S.F.S.R. (Russian Soviet Federated Socialist Republic)*: Autonomous republics: 1 Karelian, 2 Komi, 3 Mordov, 4 Chuvash, 5 Mari, 6 Tatar, 7 Udmurt, 8 Bashkir, 9 Kabardino-Balkar, 10 North Osetin, 11 Chechen-Ingush, 12 Dagestan, 13 Kalmyk, 14 Yakut, 15 Tuvan, 16 Buryat. Autonomous oblasts: 17 Adegey, 18 Karachev-Cherkess, 19 Gorno-Altay, 20 Khakass, 21 Jewish. National okrugs: 22 Nenets, 23 Komi-Permyak, 24 Khanty-Mansi, 25 Yamal-Nenets, 26 Dolgan-Nenets, 27 Evenki, 28 Ust-Orda Buryat, 29 Aga Buryat, 30 Koryak, 31 Chukchi. Kalingrad oblast (32) is detached but subordinate to the

166

and huts, sometimes raised on stilts, are used in summer. Settlements are scattered along rivers or in open patches in forests. Nevertheless, again the Russian *izba* is spreading.[10]

THE ADMINISTRATIVE-TERRITORIAL GEOGRAPHY OF RUSSIA

Old Russia was governed by a multiplicity of princedoms, slowly absorbed by the growth of Muscovite power, the remains of which were swept away in 1708 and replaced by large 'governments', whose number was later increased as Russian territory expanded. The modifications in their boundaries were, however, seldom fast enough to mirror the broadening territorial, social, and economic horizons. In 1917, the country was divided into over seventy governments and just under twenty provinces, which were subdivided into a large number of districts and rural parishes created by the reforms of 1861. The units had an exclusively administrative-territorial function and had remained in some instances unchanged since the Empress Catherine's time, so that the changes in population distribution and the growth of industry in the nineteenth century were not reflected. Generally, the governments in the more thickly-settled parts of the country were smaller than in the remoter regions, but their boundaries sometimes cut new industrial districts and no marked considerations were given to relate the boundaries to the distribution of ethnic minorities. They were administered from small or medium-sized towns, mostly places historically important or older commercial centres. New industrial towns were seldom selected: Ivanovo-Voznesensk, a major textile town of almost 60,000 people in 1897, was subordinated administratively to Shuya, a town of less than 5,000 people, while the adjoining 'villages' of Orekhovo (Vladimir Government) and Zuyevo (Moscow Government), together employing the largest labour force after Moscow and Leningrad, remained administratively separate.[11]

The Revolution recast the single centralized empire as a Union of Soviet Socialist Republics, though very strong centralized functions remained but were not stressed in public. In the new Constitution, the union republics were given wide autonomy and the right to leave the union if they wished. Because of this right, only those national groups whose territories had a common frontier with an outside power, a consequently peripheral location, could qualify for such status, providing they had reached the necessarily high standard of national and social achievement. National groups surrounded entirely by other Soviet territory or whose cultural and social levels were less advanced were given the status of autonomous republics. Still lower levels might attain rank as an autonomous *oblast* or a national *okrug* or even a national *rayon* or village *soviet*. At each stage down the ladder, economic and political autonomy was reduced. The lower levels of national autonomy were incorporated into the larger union republic framework.[12]

The government of the U.S.S.R., apart from the highly centralized functions wielded by the *Politburo*, is exercised through two chambers. Deputies to the Supreme Soviet of the Union are elected on the basis of one to each 300,000 of the population irrespective of national affiliation; but the deputies of the Supreme Soviet of Nationalities are sent on a basis of 25 deputies from each union republic, 11 deputies from each autonomous republic, 5 from each autonomous *oblast*, and 1 from each national *okrug*. Both chambers have an equal right to initiate legislation, which is considered adopted if passed by a simple majority in each chamber.

The subdivision of territory is based on the recognition of the 'national principle', so that boundaries are drawn to reflect the limits of ethnic groups. These are only changed if the ethnic distribution changes. For example, in 1940, after incorporation of the Baltic republics, small eastern frontier areas of Estonia and Latvia with a Russian majority were transferred to the R.S.F.S.R., while Lithuania received its majority region around Vilnyus. During the inter-war years, when Romania held Bessarabia, a small Moldavian republic was created on the east bank of the Dnestr and received some Ukrainian settlement areas to make it a viable unit. In 1945, when Bessarabia was incorporated into the Moldavian S.S.R., these areas returned to the Ukraine; and likewise, in the ethnically complex Fergana valley whose physical unity was broken to satisfy national aspirations, though local industrial trusts are administered supranationally. Some national areas have been created artificially, like the Jewish autonomous district in the Far East, though it did not attract many Soviet or other Jews. National areas have also been swept away: during the war, the Crimean Tatar A.S.S.R. and the Volga German A.S.S.R. were dispersed. Several Caucasian groups and the Buddhist Kalmyks of the lower Volga suffered the same fate but most were 'reinstated' in 1957. While progress has usually been up the ladder of autonomy, increased Russian immigration coupled to changing political conditions brought the reduction of the Karelo-Finnish S.S.R. to an autonomous republic in 1956. These national areas do, of course, often contain large minorities of Great Russians, Ukrainians, and others.

Areas of the larger union republics not occupied by small, national districts are divided into territorial-administrative units based on planning concepts evolved in the late 1920's, when functional territorial units were developed on which the framework of development programmes might be pinned. These basic units are the *oblast* (or *kray* in less developed areas) and the subservient *rayon*. Once the boundaries of national areas were delineated, certain common principles were used to define all other levels of administrative-territorial structure and these applied also to the administration of the national areas. Each unit is designed to serve as the hinterland, supplying local needs as far as possible to a 'proletarian centre', usually an industrial town or workers' settlement, generally a route focus from which ideological ideas can be disseminated to the countryside. In each district, a diverse ('complex') economy is to be created, though (perhaps paradoxically) local specialization should not be overlooked. This is explained as an historical legacy arising from regional specialization in major branches of the tsarist economy when particularly favoured areas were intensively developed. These have been retained in Soviet times as a beneficial territorial division of labour, but they have been overshadowed by heterogenous economic development in the regions, which helps to make them locally self-supporting and reduces the waste in crosshauls by rail and river. Achievement of completely independent economies is impossible because of the geographical inequality in the distribution of natural resources and conditions, which tend to select certain areas for particular developments (*e.g.*, cotton growing in the nilotic conditions of Central Asia). All regional development has, however, to be conceived on a long-term basis at both a local and an 'all-union' plane, so the units must be flexible to adapt to changes, and frequent realignments of the *oblast* boundaries are common.

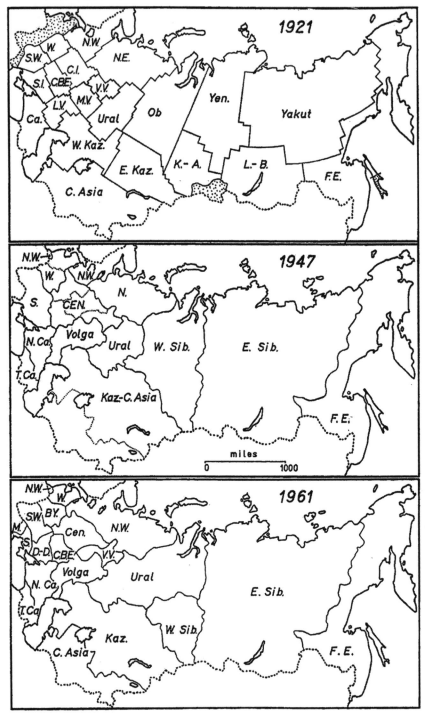

FIG. 28

Major economic regions. The three dates chosen represent important stages in the evolution of these units. The 1921 units were ultimately rejected, but the later divisions have been used. Many other arrangements have been suggested. The recent division of the country into *sovnarkhozy* and the newest management regions may tend to overshadow these regions. Usual compass point abbreviations. BY.: Byelorussia; Ca.: Caucasia; C.B.E.: Central Black Earth; Cen.: Centre; C. Asia: Central Asia; C.I.: Central Industrial; D.-D.: Dnepr-Donets; F.E.: Far East; Kaz.: Kazakhstan; K.-A.: Kuznetsk-Altay; L.-B: Lena-Baykal; L.V.: Lower Volga; M.: Moldavia; M.V.: Middle Volga; N.Ca.: Northern Caucasia; S.I.: Southern Industrial; Sib.: Siberia; T.Ca.: Transcaucasia; V.V.: Volga-Vyatka; Yen.: Yenisey.

In the less developed parts of the country, the districts are large, though small districts are characteristic of the more intensively developed parts of the Soviet Union. In Siberia, intra-district development tends to be eccentric, with the centres lying in the south along the Trans-Siberian railway where it crosses north-flowing rivers, in the most intensively developed areas, and characteristic of heterogeneity created by association of an industrially developed area with undeveloped forest or steppe. Subdivision of large districts as the economy expands tends, however, to homogeneity, already seen in small *oblasts* in Central European Russia; around Moscow, *oblasts* are dominantly industrial, but in the central black earth belt there is a clear agricultural character. It is common to split a homogenous region : for example, the Donbass coalfield is divided on ethnic grounds between the R.S.F.S.R. and the Ukraine. In the latter, it is further divided between Donetsk (Stalino) and Lugansk *oblasts*, both containing large areas beyond the coalfield; in the former, the eastern coalfield forms part of the composite Rostov *oblast* stretching far into the grain-growing steppe. Recently, Soviet theory has increasingly favoured specialism and homogeneity. Five *oblasts* in Northern Kazakhstan, exceeding in area Britain and West Germany together, have been grouped into the Virgin Territory (*Tselinny Kray*), centred on Akmolinsk (Tselinograd), containing 90 per cent of the Kazakh grainlands. Despite a predominantly grain-growing economy, the new territory also possesses varied, actively-exploited mineral wealth.

A series of major economic regions, varying in number from time to time with fluctuations in Soviet thought, are used for broader economic planning, statistical, and fiscal purposes, but they are without any administrative governmental function. They originated in the Kalinin Commission (1921–1922) which sought to formulate electricity generating regions for GOELRO. Using the principles of regional definition outlined above, the Commission produced twenty-one regions, but they did not sufficiently regard the integrity of national units and were rejected by the Party Congress. For instance, the Ukrainian and Russian sectors of the Donbass were combined in a large southern region ; the then small Byelorussian republic was put into the western region of the R.S.F.S.R. so that its national development was overshadowed by the larger neighbour ; an amalgamation of the Transcaucasian republics with the Northern Caucasus was also not accepted. Prototypes were set up in the industrial Ural and in the agricultural Northern Caucasia; and the administrative-territorial pattern which appeared with the New Economic Policy bore a clear imprint of the *Gosplan* ideas.

The interwar *gigantomania* in Russian life introduced in pro-posals for the third five-year plan a concept of *makrorayons*, regions larger and fewer in number than formerly conceived, to aim at maximum self-sufficiency, designed to reduce the mounting burden on transport. Their weakness has been described as a failure to regard 'geographical realities': no single region con-tains all the 1,500 or more substances needed to feed modern industry, for natural resources are unequally distributed, so that even the Ural region coming nearest to the ideal is deficient in fuel resources, and the potentially rich Siberian regions lack suitable conditions for production of sufficient foodstuffs. The Central Asian republics are needed to concentrate on industrial crops not otherwise possible in the U.S.S.R. at the expense of their own food output. Nevertheless the idea contained attractive strategic considerations.

War interrupted implementation of the idea, after which a new division into thirteen regions appeared in 1947. Though the criteria used in definition were not revealed, apart from the principles of the 'complex' economy and 'specialization', they appear to be affected by the *makrorayon* theory. The grouping of *oblasts* and other units into these major economic regions was attacked because clear affinities of districts to particular centres appeared to have been ignored. Criticism was also levelled against the small number of regions, later increased to fifteen and still later to seventeen, though a division into over twenty regions has been proposed. Opinion among Soviet geographers on the most satisfactory division appears to be deeply divided. Large regions are supposed to hide the distinctive character of particular areas and to conceal intra-regional differences; many 'objectively existing or planned production projects' were not represented accurately, while self-sufficiency led only to duplica-tion of effort and burdened transport. Even so, these large regions are commonly used as a basis of regional description in Soviet textbooks.

The decentralization begun in 1957 created regional economic commissions (*sovnarkhozy*), with wide local autonomy to promote industry and constructional work, whereas formerly local control was limited to agriculture and small local industries. Major industrial undertakings regarded as nationally important were controlled by their respective ministry. Under the new system, single *oblasts* or groups of *oblasts* have their own planning and budgetary responsibilities, claimed to give greater efficiency at both local and national levels, particularly in development of close spatial association of inter-related industries not possible under the compartmentalized ministerial system. The new system is said to reduce waste and duplication and to make possible reductions in high-cost, low-efficiency units developed under a too rigorous application of regional autarchy. The existing principles of regional definition apply in the form of the new councils : their centres are industrial towns, their boundaries follow existing *oblast* boundaries, and they respect the integrity of national autonomy. Supposedly they reflect local feeling and loyalty, though some reports suggest that such feelings have brought claims for units not workable in content and size. Nevertheless, they seem to be flexible enough to allow future amendments as the economy changes, and already the number has varied from 103 to 105. Each autonomous republic forms one unit, but the larger union republics may be subdivided — the Ukraine now (1963) consists of fourteen commissions, Uzbekistan has five, and there were nine *sovnarkhozy* in Kazakhstan until recent reorganization. Kiev *sovnarkhoz* has eight million people, but the Kamchatka commission has only 200,000. The Yakut republic, immense in size, forms one *sovnarkhoz* as does the small Kabardino-Balkar A.S.S.R., while economic structure varies from heavy industry in Donetsk (Stalino) *sovnarkhoz* to the swamp and forest economy of Tyumen *sovnarkhoz*.

NOTES ON CHAPTER 5

1. Tikhomirov, M., *Towns of Ancient Russia*. Moscow, 1959.
   Mongait, A., *Archaeology in the U.S.S.R.* Moscow, 1958.
2. 'Geografiya Gorodov', *Voprosy Geografii*, no. 38. Moscow, 1956.
   Saushkin, Y. G., *Economic Geography of the Soviet Union*. Oslo, 1959 — chapter on 'Towns in the Soviet Union'.
      'The Study of a System of Cities of the Soviet Union', Trans. *Soviet Geography*, vol. I, 1960.
   A series of articles on Soviet towns appeared in *Soviet Geography*, vol. III, 1962. In November 1962, eight towns were claimed to have over a million inhabitants (Moscow, Leningrad, Kiev, Baku, Gorkiy, Tashkent, Novosibirsk, and Kharkov).

3. *Moskva — Putevoditel.* Moscow, 1959.
   Moskva — Planirovka i Zastroyka Goroda 1945–1957. Moscow, 1958.
   Still useful is Cornish, V., *The Great Capitals.* London, 1923.
4. *Leningrad — Putevoditel.* Moscow, 1959.
5. *Kiev — Plan Shkema.* Kiev, 1958.
   Kiiv — Putivnik-Dobidnik. Kiev, 1956.
6. Guides are also available to Kharkov, Minsk, Novgorod, Odessa, Riga, Vilnyus, and Zhitomir. Historical background can still be obtained from *Baedeker's Guide to Russia.* Leipzig, 1914. See also Tikhomirov, M., *op. cit.,* and *Cities of Central Asia.* Central Asian Research Centre. London, 1961.
7. Schwarz, G., *Allgemeine Siedlungsgeographie.* Berlin, 1959.
8. Kovalev, S., 'Die Entwicklung des Netzes der städtischen Siedlungen', *Petermanns Mitteilungen,* vol. 99, 1955.
9. Baedeker, K., *A Guide to Russia.* Leipzig, 1914.
   Mellor, R., 'The Soviet Town', *Town and Country Planning,* February 1963.
10. Voeikov, A., 'Le Groupement de la population rurale en Russie', *Annales de Géographie,* vol. 18, 1909.
    Tokarev, S., *op. cit.*
    Saushkin, Y. G., *op. cit.* Chapter on 'Settlement'.
    Information has also been gathered in the field in Russia.
11. Feigin, J., *Standortverteilung der Produktion im Kapitalismus und Sozialismus.* Berlin, 1956.
12. Alampyev, P., 'Tendencies in the Development of Major Economic Regions', *Soviet Geography,* vol. I, 1960.
    'Ekonomicheskiye-adminstrativenyye Rayony SSSR'. *Izvestiya Akademii Nauk,* 1957.
    Chambre, H., *L'Aménagement du territoire en l'U.R.S.S.* Paris, 1959.
    Constitution of the Union of Soviet Socialist Republics. Moscow.
    Komar, I., 'Major Economic Geographic Regions of the U.S.S.R.', *Soviet Geography,* vol. I, 1960.
    Lutskiy, S., 'O Generalnom Ekonomicheskom Rayonirovanyye SSSR', *Izvestiya Akademii Nauk,* 1957.
    Mellor, R., 'Trouble with the Regions — Planning Problems in Russia', *Scottish Geographical Magazine,* vol. 75, 1959.
    Shabad, T., 'Soviet Economic Regionalization', *Geographical Review,* vol. 43, 1953.
    Shimkin, D., 'Soviet Economic Regionalization', *Geographical Review,* vol. 42, 1952.
    Taubert, H., 'Die ökonomisch-administrativen Rayons der Sowjetunion', *Petermanns Mitteilungen,* vol. 102, 1958.
    Tokarev, S., and Alampyev, P., 'Problems of Improving Territorial Organization of the National Economy and Economic Regionalization', *Soviet Geography,* vol. II, 1962.
    Voprosy Geografii, no. 47, 1959, is entirely devoted to the question of economic regionalization.

CHAPTER 6

Geography of Agriculture

ABOUT a fifth of the land area of the U.S.S.R. is used agriculturally, compared to over a half in the U.S.A. Immense areas of Siberia are suited only to forestry or herding with small patches of arable or meadow. In the poor steppe and semi-desert and the mountains of Soviet Central Asia, as well as in mountainous Caucasia and Southern Siberia, activity is limited to low density pastoralism. Despite large plains, winter cold and summer drought make successful crop raising hazardous, while low yield can be expected from great areas of soils of mediocre fertility in Northern European Russia and parts of Asiatic Russia. The Russian farmers' problems in a difficult physical environment have not been lessened by the tasks assigned them by the Soviet authorities.

TABLE 8

AGRICULTURAL EMPLOYMENT 1939–1959

	1939	1959
Total population } millions	170·6 *	208·8
Rural population }	114·5 *	108·8
Percentage of gainfully employed:		
Industry	30·1	36·9
Agriculture	50·1	38·8

* Boundaries at 17th September 1939.

Source: *Selskoye Khozyaystvo SSSR*, Moscow, 1960.

The attempt to be as self-sufficient as possible has put the burden of feeding the people, an increasing proportion of whom live in towns, almost completely on home production, further taxed by rising living standards and the need to grow industrial crops. Mining and industrial development beyond the bounds of established food-producing areas, in sub-arctic Siberia and arid

 DOI: 10.4324/9781003172048-6

Central Asia, has made it desirable to grow food locally despite the unfavourable environment. Better strains of crops and livestock, mechanization, and improved husbandry have been sought to increase output as rural population and the share of agriculture in the employed population have fallen.

PHYSICAL FACTORS AFFECTING AGRICULTURE

In a country composed essentially of lowlands, climate and soil rather than slope, aspect, or altitude are the principal factors which restrict expansion of the farmed area and hamper current farming. The great distances from the sea of much of the country mean a generally low precipitation and an important relationship between it and evaporation. In the north, the combination of low temperature with a moderate precipitation (15-20 inches) leads to water-logging of the ground, while in the south, hot summers and precipitation of 12-16 inches falling in spring and early summer create the necessity for careful moisture conservation, accentuated by the unreliability of the rainfall (Fig. 29). Indeed, from north to south, as the thermal regime becomes more favourable, precipitation decreases or occurs at seasons when its effectiveness is reduced by intense evaporation. Eastwards, increasing continentality shortens the growing season and the frost-free period, while the rapid change between seasons reduces the planting and harvesting times. Long summer days in high latitudes help to offset some of these effects, but over half the country has a growing season of a hundred days or less, not all frost-free, and a widespread occurrence of *permafrost*. In the great winter cold, much of Asiatic Russia has only a thin snow cover and early autumn and late spring frosts make it risky to sow autumn or winter wheat and rye. In the southern and eastern steppeland, aridity arising from the low and variable annual precipitation is made worse by hot, dry winds scorching the ground in summer, with high loss of moisture by evaporation, and soil erosion or dust-bowl formation an ever-present menace. Grain yield is reduced but labour requirements increase from a necessity to dry farm. In some areas, cultivation is only possible with irrigation, which can introduce problems of salt-pan formation unless planned with caution.

About a third of the country may be discounted agriculturally

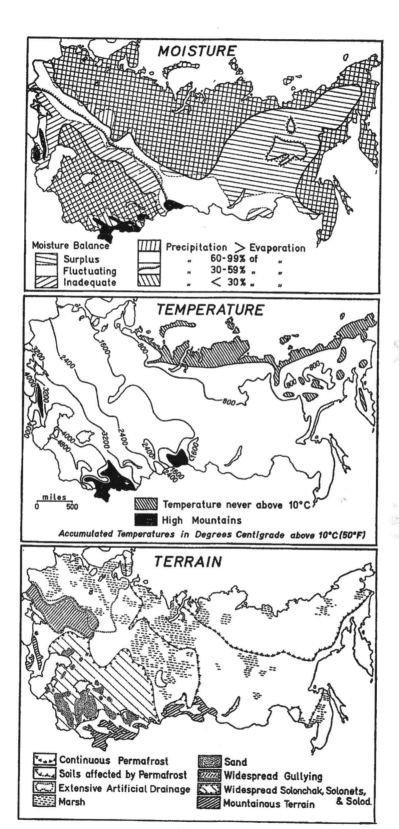

MOISTURE

Moisture Balance
Surplus
Fluctuating
Inadequate

Precipitation > Evaporation
„ 60-99% of „
„ 30-59% „ „
„ < 30% „ „

TEMPERATURE

miles
0   500

Temperature never above 10°C
High Mountains

Accumulated Temperatures in Degrees Centigrade above 10°C (50°F)

TERRAIN

Continuous Permafrost
Soils affected by Permafrost
Extensive Artificial Drainage
Marsh

Sand
Widespread Gullying
Widespread Solonchak, Solonets,
Mountainous Terrain   & Solod.

FIG. 29

Physical problems of agriculture. Compiled from Soviet atlases.

because of climate, composed principally of about 15 per cent in the tundra and 10 per cent in the desert, while only a small part of the semi-desert may be used for grazing. Almost half the area is covered by great northern forests, with unfavourable climate and soils, much of which lie above *permafrost*. Under a tenth of the forest belt is used for arable and meadows.

The widespread poor and mediocre soils hamper farming, even when climatic conditions are otherwise suitable. The best is the black earth, but even including much degraded types it covers not more than 12 per cent of Soviet territory, two-thirds located in Southern European Russia. The Siberian black earth, smaller in area and poorer in quality, has about a quarter of its area not well suited to agriculture because of salt-pan formation. Siberia, including the adjacent districts of Kazakhstan, however, contains the greatest part of the black earth yet unused. This soil, which retains its remarkable fertility despite heavy cropping and limited application of fertilizer, is highly friable and careless farming has already lost much good land to sheet and gully erosion. The soil gathers and stores winter moisture for a rapid and luxuriant growth in spring, but the crop yield is influenced by availability of summer moisture when the dry *sukhovey* blows. Frequency of rain failure is the principal hazard of the black earth belt.

The forest podzol, cold, acid, and poor in humus, is the most widespread soil, covering almost half the area. With suitable fertilizer application it can be used agriculturally; but the commonly high water table makes artificial drainage usually necessary. The quality is best in the southern forest belt and wooded steppe, while in drier Eastern Siberia the soils are markedly less acid. This belt may offer the greatest scope for extending the cultivated area.

Undesirable salts in the upper layers of chestnut, brown, and grey soils in the dry steppe and semi-desert hamper farming in South-east European Russia and in Asiatic Russia. When wet they become sticky and unworkable and when dry they are iron-hard clods. Capillary movement of moisture caused by surface evaporation draws the harmful salts to the surface, so that irrigation poses many potential difficulties. *Solonchak* is not suited to agriculture because of alkalinity; *solonets* may be reclaimed by replacing sodium with calcium through application of lime and gypsum; but *solod* is almost unreclaimable because it lacks humus and salts. The thick loessic soils and the silts and muds brought down by rivers along the southern fringe of the desert belt can be extremely fertile when irrigated and do not usually accumulate harmful salts.[1]

HISTORICAL FACTORS IN AGRICULTURAL DEVELOPMENT

In the Kievan period, the peasants, largely free, supplemented shifting agriculture by hunting and gathering beeswax and honey in the forests. During the Mongol-Tatar invasions, allodial landlords (*boyars*) were replaced by *pomeshchiki*, holding land conditional on their military service, notably during the sixteenth century in the central, southern, and western parts of the country. At this period, the natural, largely self-sufficient economy was replaced by commercial capitalism, and shifting cultivation replaced by three-field rotations. Many free or lightly-dependent peasants were enserfed, and serfdom finally made absolute in 1649. Peasants who had settled in the marginal lands along the steppe frontier were enserfed by the spread of Russian power to these areas in the eighteenth century. Russian serfdom, unlike mediaeval Europe, produced for the market and not only to the capacity of its own and landlords' stomachs.

Peasants in Estonia, Kurland, and Livland had been freed between 1816 and 1819, though the landlords, mostly of German descent, still owned the land. French and Prussian influence brought a partial land reform in Poland, and there were many free peasants in Bessarabia. In Russia proper, at emancipation in 1861, about half the peasants were serfs, who paid either a tax (*obrok*) to their landlord or worked for him on his land (*barshchina*), fixed at three days a week though frequently exceeded. In some districts, duties were combined. *Obrok* was common in the North and Centre (Fig. 30) where landlords were mostly absentee and little use for serfs' labour could be found on the poor land, much of which was held by monasteries and the Crown. In the West and the black earth lands, where grain, hemp, and flax were grown, *barshchina* provided landlords with cheap labour to cultivate market crops. *Obrok* was, however, growing more common, since it gave landlords a living and paid-up peasants a chance to seek more lucrative work in towns. Few free peasants, apart from

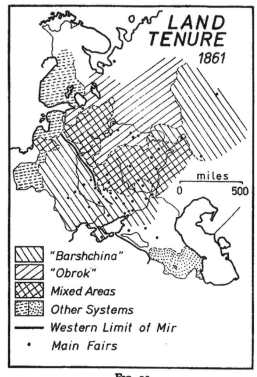

FIG. 30

Land tenure, 1861. *Atlas Istorii SSSR*, Moscow, 1955, Vol. II, pl. 15.

German colonists and descendants of ancient freemen (*odno-dvortsy*), owned their land. During the eighteenth century, the *mir* (commune), which controlled redistribution of strips and crop rotations, began to dominate peasant society: a convenient unit through which the State might administer the countryside, while it relieved landlords of administration without giving independence to the peasants.

Emancipation left peasants without capital and guidance to improve their techniques and output, although they began to leave overcrowded, rural districts to move into towns or into empty lands in Siberia. Holdings acquired at emancipation varied considerably in size between districts. Large numbers of peasants accepted 'pauper holdings' of one-quarter the normal size to avoid paying redemption dues. In the black earth lands, holdings were generally smaller than before emancipation, as landlords were more interested in their full share of the valuable land. Allotments per male peasant on State lands averaged 23 acres; on estates of the Imperial family 15 acres; and on private estates, about 9 acres. Redemption values were greatest where labour was most valuable, since the charges (payable over forty years at 6 per cent) were based on both land and labour. Landlords frequently retained the best land which would have made balanced peasant holdings, like the river meadows of the North and North-west. Estates were left without capital to replace cheap labour by machinery, particularly in the North where poor, acid podzols did not give a fair return. As landlords sold out and retired to the towns, the richer and more progressive peasant (*kulak*) replaced them. Although people left the country-side, emancipation did not solve the land-hunger of a rural population increasing more rapidly than allotments of land were made to it. The average peasant holding in 1877 was 35·5 acres, but it did not support a family except in the rich black earth areas, since output was kept low by the old three-field rotations, the repartitions by the *mir*, and the inability of peasants to buy fertilizer, which meant keeping a larger area than necessary fallow each year. The repartitional system dominated in the North, Centre, North-east, and South-east; in the steppe, about two-thirds were so held and a quarter of the land in the West and South-west, but it was unknown in the Baltic lands.

Further reform in 1905 tried to end the land-hunger by abolishing the *mir*'s authority. Holdings were to be consolidated, though by 1916 only a fifth of peasant families had managed to consolidate the whole or part of their holdings. Fewer still had left the village, which remained the dominant settlement unit. Most peasants (65 per cent) were poor, depending on secondary occupations, and annually two million sought seasonal work in the grainlands. *Kulaks* comprised only 15 per cent. The country-side was not altogether self-supporting and the forest lands imported grain from the surplus of the South, which was small in bad years. Export of grain from the Ukraine through Odessa began after completion of the railways in the late 1860's, which also aided movement to the Baltic ports, S. Peterburg, and Moscow, replacing older movements up the Volga. The rapacious plundering of the grainlands was seldom shared by the peasants, who had hardly enough to support a family and were usually unable to augment their living by secondary occupations like their compatriots in the North.[2]

AGRICULTURE AND THE REVOLUTION

The Revolution brought a return to communal farming in collectivization, although different to the older, small-scale peasant farming, which would doubtless have been unable to produce a large surplus to feed a growing urban population. The development of the new system has not been smooth, for the State has followed a vacillating policy in its struggle for supremacy over the peasants and natural conditions have brought wide year-to-year variations in harvests.

The *mir* reappeared in the Revolution over much of the thickly settled black earth lands, forcing back into communal village tenure peasants who had managed to separate their holdings. Peasants seized estates, but had insufficient experience to run them, and many capable, richer peasants and estate factors who might have helped were killed or fled. Pillaging destroyed valuable animals and equipment, while fields went out of cultivation as peasants held more ground than they could till. Disruption in trade and transport brought a catastrophic fall in food deliveries, while confiscations and lack of manufactured goods to buy discouraged peasants from selling in markets, and by early 1918 the towns were starving. The countryside reverted to subsistence farming as hiring labour was forbidden. A hot, dry summer in 1921 caused a disastrous crop failure in the Ukraine, the Volga, and the Ural, in which ten million people reputedly perished.

Between 1916 and 1922, the cultivated area fell by 38 per cent, and in Kazakhstan it declined by over a half and the number of animals from four million to half a million. In the Kuban and Black Sea provinces, about half the horses and cattle and over 80 per cent of the sheep and pigs were lost. Cats and dogs disappeared. The remedial New Economic Policy introduced the Rural Code, recognizing peasant ownership of stock, equipment, and produce. Forced redistribution of land stopped and compulsory produce deliveries were fixed and no longer based on available surplus. There was a rapid increase in the sown area, though still below 1913; but in 1927 another poor harvest and insufficient grain deliveries brought fresh accusations against the peasants, foreshadowing a ruthlessly ambitious experiment in agricultural economics to set the pattern of Soviet farming.

With the *mir* still active and strip cultivation of large open fields used over much of the country, it was comparatively easy to introduce new collective farms employing machinery and advanced crop rotations on larger fields, while surplus labour was to be released. Collectivization brought a new wave of slaughtering, because peasants did not like to pool their animals but also hoped that the State would provide better ones, which affected meat and milk output into the late thirties. Crop yield fell because of the lack of animal manure and a still inadequate supply of artificial fertilizer, most marked on poor podzols. Disinterested and frequently unco-operative peasants were not won to collectivization because many farm presidents were good party men but poor farmers. A good harvest in 1930 allowed heavy export of grain to pay for imports of machinery; but the 1931 harvest was a partial failure in a hot, dry summer followed by a heavy growth of weeds in 1932. Circumstances forced peasants into collectives, now starting to receive machinery from foreign and Russian factories, which helped to make the 1933 harvest the best since the Revolution and to save the poorer one in 1934. The pace of collectivization varied regionally: the grainlands of the Ukraine and Northern Caucasia were the first to be reorganized, because initially they possessed more machinery and larger fields, besides their great importance to the welfare of the country. Resistance was strong among Cossacks on the Terek and Kuban, where many state farms were started. In the North, in Transcaucasia and in Central Asia, where peasant economy was more mixed, longer was needed to develop collectives; while a long and bitter struggle was waged against Kirgiz and Kazakh nomads.

Collectivization progressed steadily, claiming over nine-tenths of all peasant households by 1940 and spreading into territories incorporated by Russia. It was halted by the German invasion which in the first year spread over half the arable land, producing 38 per cent of the cereals and 84 per cent of the sugar beet, on which lived 88 million people. The comparable loss of food-producing area to population went far to avert famine in unoccupied Russia. War-time shifts in agriculture were less than in industry, since the cultivated area could not be readily moved eastwards. War-time demands on money and equipment prevented work on new irrigation and drainage schemes, which in any case would have needed several years to mature. The existing shortages of machinery and fertilizers were increased.

In the immediate post-war period, dry summers in 1946 and 1947 gave poor harvests; but by 1948 reconstruction in the western lands was claimed to be complete, though effects of war in the countryside could be seen in the late 1950's. The collective principle has not been relaxed; but individual farms have been allowed greater freedom in management or crop rotations and machinery. Strenuous efforts have been made to raise yield on already cultivated land, while attention has been given to increasing the cultivated area by drainage in glaciated areas of European Russia and by ploughing 'virgin' or long-fallow land in Siberia and Kazakhstan. New irrigation and ameliorative measures have been started in Central Asia and Western Siberia. Natural conditions still remain important: the good harvests of 1956 (a record), 1958 (extraordinarily good), and 1962 have been offset by the disappointments of the 1955, 1957, and 1959 harvests.[3]

SOVIET FARMS

Farming in the Soviet Union is conducted either by collective farms (*kolkhozy*), which control about two-thirds of the sown area, or by state farms (*sovkhozy*), which are usually found where the organization of collectives is not possible or desirable or where new techniques have to be developed. There are very few districts where state farms hold most land (Fig. 31). Both have been adapted to meet special conditions such as grain-growing, animal-raising, or even fur-farming and fishing. In 1959 there were 31·7 million collective farm members and 6·6 million workers on state farms, while a further 10 million people, members of farmers', industrial, and office workers' families, were engaged in 'individual subsidiary husbandry'. There remained only 92,000 individual peasant farms. On the collectives, women formed 57 per cent of the labour force and 41 per cent on state farms. Altogether, the 65·5 million collective farmers and their dependants, living in 18·5 million households, formed 31·4 per cent of total Soviet population.[4]

Ownership of land, equipment, and stock is vested in the collective, but theoretically the individual is free to leave, though he may not take land with him even if he previously owned it. The farm produce is sold to state marketing agencies at agreed

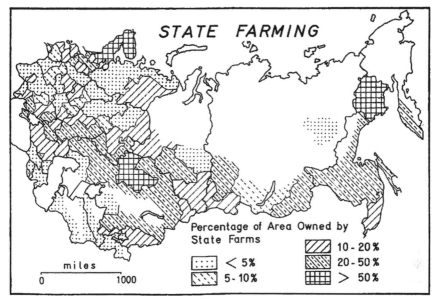

FIG. 31

State farming. Based on tables in *Selskoye Khozyaystvo SSSR*, Moscow, 1960, and *Posevnyye Ploshchadi SSSR*, Moscow, 1957.

prices and the income divided among the members after paying obligations. Payments to members are based on 'labour days'; for example, one labour day may be equivalent to reaping a given area or attending to a certain number of animals. The farm may retain a part of its crop to distribute or to sell at agreed prices to members. Each collective farm family is entitled to a small private plot and some livestock whose produce may be sold on the 'free' market, making a considerable contribution to Soviet food output, but this is tolerated rather than encouraged by the State. Collectives formerly hired machinery from machine and tractor stations, but since 1958 they have had direct control of their equipment, sold to them by the stations, now reduced to repair depôts, though farms always owned some of their own carts and lorries.

The most important settlements in the countryside are now the central villages of collectives, which may have schools, shops, or veterinary units, as well as new farm buildings and silos. The main inadequacies are poor, unpaved, rural roads, lack of piped water and housing, though modern and mostly wooden houses are replacing the old huts.

The fall in the number of collectives has been the result of amalgamation or conversion into state farms, though their total sown area has risen. The average sown area per farm rose between 1950 and 1959 from 2,384 acres to 5,720 acres, closer to the suggested optimum of 6,175-7,410 acres, although some amalgamations have produced vast farms. The smallest *kolkhozy* are usually devoted to special cultures, while the biggest are frequently grain-producing or animal-raising. In the glaciated lands of Northern and North-western European Russia where the terrain is rough and interspersed with lakes and bogs, *kolkhozy* are smaller than in the rolling, open black earth plains, and the

TABLE 9

AGRICULTURAL *KOLKHOZY* AND *SOVKHOZY* 1918-1959

	1918	1928	1932	1940 *	1950	1958	1959
Collectives: thousands	—	33·3	210·6	235·5	121·4	67·7	53·4
Sown area: mill. acres	—	3·4	—	280·7	298·8	314·8	321·8
Households: millions	—	—	—	18·7	20·4	18·3	18·5
Percentage collectives:							
Households	0·1	1·7	61·5	84·3 *	96·7	99·7	99·8
Area	—	2·3	77·7	89·4 *	98·5	99·9	99·9
State farms: number	—	1,407	4,337	4,159	4,988	6,002	6,496
Sown area: mill. acres	—	4·3	—	28·6	31·8	129·6	133·1
Machine-tractor stations	—	6	2,446	7,069	8,414	345	34

* Within present boundaries. Before territorial expansion, 96·6 per cent of the households and 99·9 per cent of the sown area were collectivized.

Source: *Selskoye Khozyaystvo SSSR*, Moscow, 1960.

smaller fields are scattered between patches of forest, so that it is more difficult to use machinery. The largest collectives lie in the immense open steppe of the southern Ural and Kazakhstan, where sown area averages over 17,000 acres, and large farms are found in Siberia.

Described as typical, a *kolkhoz* south of Moscow, a union of four pre-war farms, each developed around a village nucleus, has an area of about 5,900 acres, of which 3,900 acres are annually under the plough for crops such as sugar beet, potatoes, and fodder. Cereals occupy another 1,500 acres. The remainder consists of pasture along a small river, the plots of *kolkhoz* families, and some woodland. A part of the farm is devoted to vegetable

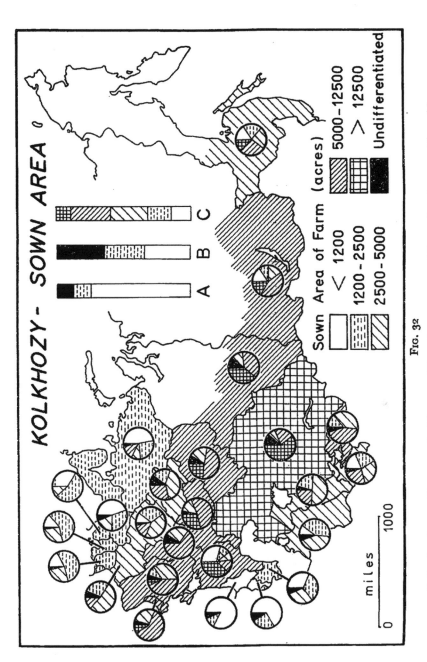

FIG. 32

Collective farms. Sources as in Fig. 31. The columns represent the national composition of farms by size: column A is for 1933, B for 1950, and C for 1959. The shading shows the average sown area of collectives in the republics and economic regions, while the circles show the division of the sown area among farms according to size in these administrative units.

gardening for the Moscow market. Management is by a committee of 8 workers and a non-working president. There are 6 administrative staff. Work is divided between 120 workers in a cereals brigade, 125 workers in a market-garden brigade, 2 brigades of 90 workers each to attend to fodder crops and meadows, while a fifth brigade of 50 workers cares for the orchards, and the sixth brigade of 60 workers (more in winter) looks after livestock which includes 200 dairy cattle. Personnel in the brigades does not change. About eight tractors are usually at

TABLE 10

THE SHARE OF DIFFERENT TYPES OF FARMS IN AGRICULTURE
(Participation as percentage of the national total, 1959)

	State Farms *	Collective Farms	Collective Farmers
Total sown area	30·0	66·4	2·7
Grain crops area	32·0	66·8	1·0
Technical crops area	12·3	86·7	0·7
Potato and vegetable area	12·0	42·5	31·1
Fodder crops area	33·3	65·8	0·8
Number of animals :			
Horned cattle	16·6	49·8	23·3
Cows	12·0	37·6	34·4
Pigs	24·0	50·1	18·8
Sheep	22·5	56·3	16·1
Goats	3·2	15·8	46·5

* Includes other state agricultural enterprises.

Source: *Selskoye Khozyaystvo SSSR*, Moscow, 1960.

work and three-quarters of the field work is by machinery. In an area of high rural population density, the farm population is about 2,000.

A more specialized *kolkhoz*, in the Meshchera swamp of the Oka valley, comprises 1,600 acres, with a substantial part in orchards and soft fruit gardens, glasshouses, and frames. The produce nearly all goes to Moscow. A large labour force of 431 workers from 300 families is needed by the small but specialized cultivation. The problems of some farms are illustrated by a collective near Pinsk which has 15,200 acres, of which only 1,500 acres are arable distributed on eighty-two sandy islands among swamp and wet meadow. In Transcaucasia, *kolkhozy*, quite small in area, concentrate on tea, wine, citrus fruits, and even tung oil, while in Uzbekistan cotton is an important crop and sugar beet in parts of the Western Ukraine. Nowhere do these crops exclude the production of other crops and livestock as well.

The state farm is run more like a factory, with a manager, paid employees, and its own machinery. Holding under a third of the sown area, these farms serve as training and experimental farms or to develop latent resources in areas where collectives cannot be formed. They are usually larger than collectives, with

TABLE 11

LAND-HOLDING BY ENTERPRISES IN THE U.S.S.R. 1959

	Total Land Surface	Total Agricultural Area	Plough- land	Of which in use *
	←-------- Million hectares --------→			
State farms	253·8	169·7	64·7	61·3
Collective farms	744·1	310·2	153·6	146·4
Collective farm peasants	6·4	5·8	5·2	5·2
Other workers	1·8	1·7	1·3	1·3
State land fund †	216·8	76·5	1·2	0·5
State forest fund †	926·4	12·2	0·8	0·5
Total ground	2,233·3	608·9	233·5	220·8

* Excludes unsown fallow.
† Excludes land loaned long term to state and collective farms.

Source: *Selskoye Khozyaystvo SSSR*, Moscow, 1960.

an average sown area of 20,000 acres, and more highly mechanized. The desert-fringe karakul farms have an average holding of 489,000 acres (though only a small area is sown), while cotton-growing farms have a sown area averaging 12,800 acres worked by 1,981 employees. In the new grainlands, a state farm may hold 75,000-125,000 acres (not all sown) and have a staff of 1,500 including office workers, agronomists, and technicians. In contrast, a farm like *Sovkhoz Industriya* near Apatity in the Kola Peninsula carries out research into cultivation on poor soils under Arctic conditions. It produces milk, potatoes, and vegetables for nearby industrial towns and has 8,000 forcing frames and 10,000 square yards of hot houses. Locally produced fertilizer is used. At Igarka on the lower Yenisey, a similar farm lies on a sandy *permafrost*-free island in the middle of the four-mile-wide river, surrounded by large masses of water several degrees warmer than generally prevailing air temperature, and supplies fresh milk and vegetables to adjacent settlements. Production costs are high, but such farms eliminate more expensive transport of fresh foods from low-cost producers into such distant and inaccessible regions. Though regarded usually as grain producers, state farms engage in many different branches: of the 6,496 farms in 1959, 985 were grain-growers compared to 1,210 vegetable, fruit, and potato producers and 1,874 chiefly milk and meat producers. Since pre-war days, there has been an increase in fur-breeding farms from 6 to 63, while since the early fifties, horse-breeding farms have dropped from 171 to 72. Over three-quarters of the grain-growing farms are in Kazakhstan and almost two-thirds of the sugar beet farms in the Ukraine. Two-thirds of the farms growing cotton are in Uzbekistan and a similar proportion of the tea farms in Georgia.

SOME CONTEMPORARY PROBLEMS OF RUSSIAN AGRICULTURE

A growing population with an increasing proportion of urban dwellers and a rising standard of living make maintenance of self-sufficiency in food more difficult. To maintain it, it is not only necessary to raise yield of crops and livestock but also, if possible, to expand the cultivated area. Within already cultivated areas, notably in Europe, there are possibilities, by forest clearance or reclamation of marsh or gullied land, to extend the sown areas, which may also be increased by watering lands in the steppe fringe and in Central Asia; while in Siberia, extension of the cultivated area may be the greatest possibility of increasing production to avoid long-distance food transport.[5]

In the forest belt, about a third of pasture and hayland and over a tenth of arable requires artificial drainage. Soil may also be improved by liming or even treatment with waste ash from Estonian oil shales to reduce acidity and by using biologically active organic fertilizers, such as ploughing in green crops or dung-phosphatic compounds. Drier land may be cleared of trees and used directly, but on some morainic areas stones are costly and laborious to clear, though providing useful material for filling underground drains. Reclamation of water-logged ground is more difficult, but where peat is absent simple drainage is usually adequate, and if it is less than a foot thick it is usually removed and land afforested before cultivation. Large drainage schemes have been started in Polesye, around Lake Ilmen, and in the Oka valley, where emphasis has been on providing meadow rather than arable. New varieties of crops to mature in the short growing season and on poor soils, with considerable resistance to frost, have been developed, with particular attention devoted to wheat.

In contrast to the north, with its surfeit of moisture, the southwards advance of the cultivated area is affected by availability of moisture in areas of low and unreliable precipitation. Irrigation has been practised since ancient times in the piedmont oases of Central Asia and Transcaucasia. Many ancient canals, reservoirs, and underground galleries exist, while there are traces of large dams on the Murgab ruined during the Mongol invasions; and elaborate codes governing the use of water were a feature of the oasis cultures. In Central Asia, where over 10 million acres are irrigated, the rivers provide the foci of the systems, though hardly a small stream descends from the mountains without its water being used. Irrigated land, however, forms under a twentieth of the total area, although it contains two-thirds of the sown area. The best rivers carry much silt and mud in suspension, forming a valuable natural fertilizer, and are not too fast to be easily controlled. The greatest need is for dams to pond back flood water and to release it gradually during the dry season.

Water may also be obtained from wells, particularly important in semi-desert pastoral regions; and most common are tube wells, 60-75 feet deep, constructed chiefly since the early 1940's. Artesian wells, usually 350-450 feet deep, have been dug in the Ust-Urt Plateau. A modern well using wind-generated electricity can provide water for 4,000-5,000 sheep daily. Ancient water-wheels moved by camel or mule also raise water. In the Kopet Dag, water from underground galleries is fed into canals. In porous desert soils, particularly in piedmont loess, about a quarter of the water is lost to seepage, and canals and channels are now commonly concrete-lined. Experiments in extremely arid parts have been made in collecting copious dew, while at Krasnovodsk water is distilled from slightly saline Caspian water.

Irrigation has to be carefully controlled to prevent increased salinity, especially in soils in shallow depressions with a high water table. Formation of soil akali, disastrous to plant growth, may be prevented by treatment with calcium sulphate. Sulphur is also used to neutralize alkalinity by oxidation.

On the Syr Darya, the Fergana valley canals collect water from rivers flowing into the valley and distribute it to oases along the mountain footslope. The works, first completed in the inter-war years, have recently been extended and the irrigated area has increased by 350,000 acres since 1913. A large dam on the Kassansay near Namangan, when complete, will regulate water entering the North Fergana Canal, which already waters 86,000 acres. Syr Darya water is also used in other systems: one of the older systems, extended in Soviet times, is the Golodnaya Steppe west of Ursatyevskaya, with some 495,000 acres already watered, and work begun in 1958 will extend to 1,120,000 acres. A canal from the Arys to Turkestan waters the foot of the Kara Tau. Extensive irrigation canals around Tashkent date from before the Revolution. Near Kzyl Orda, the Tas Buget dam waters 295,000 acres for rice-growing and provides summer water for herds brought from arid, interfluvial deserts.

The great Chu Canal, begun in 1940 but not yet complete, will ultimately water 173,000 acres around Frunze. Water is also drawn from the Chu river, which supports a third of the population of the Kirgiz S.S.R., while on its upper reaches the Orto-Tokoy dam regulates spring and winter flood water.

So much water is drawn from the Zeravshan around Samarkand and Bukhara that only in exceptional years does it reach the Amu Darya. Near Bukhara, over eighty main canals water 1,030,000 acres of rich piedmont loess. The larger system around Samarkand is fed from the Pervomaysk dam, while the artificial Uzbek Sea at Katta Kurgan regulates the irregular flow of the Zeravshan. The use of the Kashka Darya waters is also highly developed.

The annual run-off of the Amu Darya is two or three times greater than the Syr Darya; but together they supply water for half the irrigated land in Central Asia. The greatest works on the Amu Darya are in reaches above Kerki, on the Pyandzh, Vakhsh, and Surkhan Darya. Below Kerki no tributaries enter the river, so that it gradually loses volume from loss of water by seepage and evaporation as well as to irrigation. The river carries a load of silt said to exceed any other in the world, which makes sedimentation a serious consideration in planning dams and canals. The Vakhsh system comprises over 1,000 miles of canals which water about 150,000 acres. There is also a highly-developed system in the Gissar valley around Dushanbe, while a further 250,000 acres are watered by the Kyzyl-Su and Yakh-Su. Amu Darya water is now carried to the Murgab oasis by a canal along the rough line of the Kelif Uzboy across the Kara Kum desert. Another canal has been started across the Karshi Steppe towards the lower Zeravshan. Important works also exist near Khorezm and Chimbay. Use of the upper waters of the Amu Darya is complicated because a quarter is drawn from tributaries on the Afghan bank. A remarkable project to carry water from the Aral Sea to the Caspian along a 600-mile canal in the old Uzboy depression appears to have been shelved in 1953.

Completion of the Kakhovka barrage on the lower Dnepr provides water for the Black Sea steppes and Northern Crimea and the Tsimlyansk reservoir supplies the Salsk Steppe. Plans also envisage use of Volga water to irrigate east bank lands. In Northern Caucasia, several schemes exist on the Kuban, Kuma, and Terek and there is an incomplete system on the Manych. The dry steppes of the lower Kura and Araks are also watered by canals which will be extended to use water from the Mingechaur reservoir. The irrigation system of the Razdan in Armenia is also being extended, using an increased outflow from Lake Sevan.

Without respect to natural hazards and possible long-term effects of cultivation, the Russians have tried to develop grain-growing in a broad belt of virgin and long-fallow soils, mostly black earth and associated forms, extending from Northern Caucasia into Western Siberia in the wooded steppe and steppe, where almost 100 million acres were ploughed between 1954 and 1960. This ploughing campaign has followed the earlier ones (1928–1932 and 1940–1944), both of which had helped to increase the cultivated area of Siberia. The main limitation is a low and annually variable precipitation, varying from 12 to 15 inches in the north to 10 inches in the south, and at least one year in five drought may ruin the crop. Moisture is also lost by wind blowing away the light winter snow cover, by run-off in heavy spring showers, and by intense evaporation in dry summer air. At harvest time, strong, dry winds are an added hazard. Such conditions create dangerous conditions for soil erosion and dust-bowl formation.

The last free land in the European black earth belt appears

to have been occupied in the early thirties. More recently it has become possible to use poorer, degraded black earths for arable, including some *solonets* areas (with expectedly greater risk) and large reserves of chestnut and brown soils, chiefly in Kazakhstan, though they contain big patches of *solonets* and alkali soils and uncertain precipitation. Under suitable conditions, these soils are said to produce superior wheat to the black earths. Ultimate success depends on adequate fallowing (not always appreciated by Siberian farmers), suitable ploughing, and correct moisture conservation, besides improved crop rotation.

Rehabilitation of eroded land and prevention of further erosion also helps to extend agriculture. Widespread gullying throughout southern Russia is a testimony to past erosion as a result of overcultivation by inadequate techniques. Large numbers of ponds have been built to retain moisture from winter snow and to preserve the water table. Shelter belts also help to preserve the water table and hold wind-blown topsoil. A grandiose plan for shelter belts published in 1954 has been only partly fulfilled, though a number have been planted in Northern Caucasia, in the Volga basin, and in the Central Black Earth region. Afforestation has been undertaken along watersheds and erosion gullies as well as on shifting sands. Success depends on the correct choice of trees (mostly deciduous) and adequate labour in the early years of growth : overgrowing of young trees

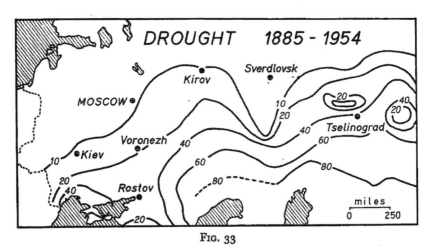

FIG. 33

Drought, 1885–1954. Percentage frequency calculated from Malygin, E., *Zasukhi v SSSR*, Moscow, 1958.

by steppe vegetation is a problem, while *kolkhozniki* frequently cut young trees or graze animals among them.

Output has also been raised by introduction of crop rotations using perennial grasses, claimed to restore the crumb structure to the soil, and legumes and beans to restore soil nitrogen and to provide fodder for stock, which in turn gives valuable manure. Such plants are difficult, however, to raise successfully in a dry, continental climate, while the system reduces the acreage under bread grains. Fertilizers have also been important in raising yield. In the southern lands, in the absence of animal manure, artificial materials are the main source. Russia does not lack natural resources to produce such fertilizers, though the most important deposits lie far away from the principal consuming areas.

DISTRIBUTION OF THE PRINCIPAL TYPES OF FARMING

Regional variations in farming, despite attempts to diversify the pattern and to overcome adverse local conditions, still represent an adjustment to prevailing natural conditions, notably soil, climate, and location, though some differences may be attributed to contrasting farming traditions among Russian and non-Russian peoples. Mixed farming, which has spread at the expense of other types, predominates in the main settled areas, though cropping patterns change with passage from the severe, northern winters and poor podzols to the milder winters and fertile black earths in the south, while east of the Volga farming reflects the growing continentality (Fig. 34). On lands where arable farming is not generally possible, as in the steppes, semi-desert, the high mountains, or the tayga and tundra, livestock-keeping is characteristic. In the northern forests, only patches of agriculture occur amid lumbering, breeding and hunting valuable fur animals, and some herding. In some richer mountain valleys and in the desert oases, 'garden' (*bakhchi*) farming is common. Near big towns, a market-gardening and dairying 'suburban' economy is typical. Everywhere are found suitably adapted collective and state farms.[6]

The mixed farming region is associated with the principal settled areas of European Russia and Siberia. Its northern limit

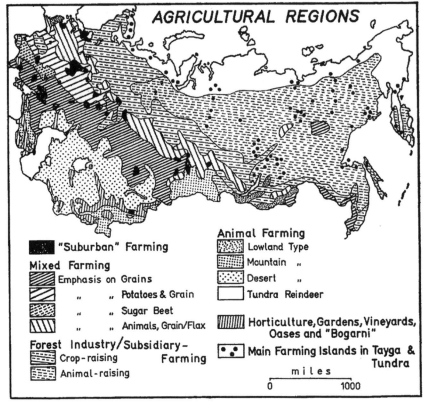

FIG. 34

Agricultural regions. Compiled from recent Soviet atlases, but the pattern has been simplified to bring out the major contrasts.

lies in the southern fringe of the northern coniferous forests, roughly coincident with the northern limits of wheat-growing, and the southern limit is governed by increasing aridity and by altitude in the high mountains of Caucasia and Southern Siberia. Isolated patches occur in Transbaykalia, the Far East, and even in the anomalous soil and climatic conditions of the middle Lena.

The north-western part, in Byelorussia, the Baltic republics, and around Leningrad, is characterized by flax and dairying. The damp climate and heavy summer showers are good for flax but make the ripening of wheat difficult, though the acreage is increasing. The rotation of flax, which makes heavy demands on the poor soil, with fodder grasses leads to an association with dairying. Rye is also grown, since it stands the damper climate better than wheat, while oats flourish. Rye is the main winter crop; spring crops are barley, oats, and wheat. Potatoes are widely grown, particularly around Leningrad where they occupy, along with vegetables, about a third of the sown area and support a thriving pig industry. Livestock-raising is being developed with the growing of green maize for silage. The Baltic republics are important for dairying, while the estates (now state farms) in former German East Prussia concentrate on grain and dairying (chiefly Friesian cattle). It is a countryside of small fields interspersed by forest and marsh and there are rank meadows in most valleys.

The Central European region grows potatoes, spring wheat, and other grains. South and east of Moscow, acid podzol gives way to brown, woodland earths, and further south lies the rich black earth. In the wooded steppe, the more fertile soils combine with warmer, drier summers and milder winters to allow a wider range of crops. On the south, winter rye and spring and winter wheat, potatoes, and sugar beet form the main crops, while sunflower is also grown, and more recently maize has been introduced. Throughout, good pasture found along the sluggish rivers is used for animals, also fed on silage and waste from agricultural industries. Fodder is provided by the complex rotations developed since 1917. This is one of the most intensively farmed parts of the mixed farming region, but the overpopulation of tsarist times has been relieved, while the dependence on cereals has been reduced by introducing new industrial crops and stall-reared cattle, for which breeds like the Red Tambov have been developed. Special crops include *makhorka*, green tobacco, and hops, notably near Bryansk. Apples and pears are widely grown, particularly around Michurinsk. Much eroded land has been reclaimed and further loss prevented in this open countryside whose large fields are broken by occasional patches of forest.

The swampy Oka lowlands form a sub-region, producing soft fruit and vegetables. Around Murom and Arzamas, seed-raising is important and also cucumber-growing, and there is horticulture on the rich, humid soils of the Vladimir Opolye. Throughout this part of Russia and elsewhere, poultry and bee-keeping form important secondary sources of income, though honey from the Mari districts is most famous.

On the black and brown earths of the wooded steppe of the Western Ukraine and Moldavia, mixed farming shows an emphasis on sugar beet, which is suited to the long, warm summers and an annual precipitation of between 20 and 24 inches. The crop has introduced rotations including winter wheat, grasses, and coarse grains. Waste from the refineries, which dot the landscape, feeds cattle and pigs and has founded a meat and dairying industry. The southern part of the mixed farming belt stretches across the Ukraine and east of the Volga into Siberia. Associated with the best black earths and with chestnut and brown earths, this forms the Soviet granary. Occupied late in Russian history, the grain-growing economy expanded in the late nineteenth century with the building of railways, while since 1954 this has been the belt of strenuous efforts to plough and cultivate 'virgin' and long-fallow lands.

West of the Don winter wheat predominates, while sunflower, maize, and industrial crops are grown and perennial grasses have been introduced into rotations. Where too dry for maize and wheat, millet is grown. Although there is little natural pasture, waste grain, rotation grass, and silage provide for stall-fed animals. There is also cultivation of some rice in the flood-plains of the Dnepr, Dnestr, and Southern Bug, while the vine is found on gravel terraces in sunny, sheltered valleys in Moldavia and the Ukraine, but cotton-growing has declined. Horse-breeding, formerly important, has also declined. Grains are ideally suited to the grassland climate of long, warm summers and an annual precipitation between 15 and 20 inches, unfortunately unreliable in the south, though occasional hot, dry winds may damage the crop. Frequent strong winter winds also blow away the light snow cover and expose the friable soils to erosion. The frost-free period lasts from 140 days in the north to 200 days in the south. An assured water supply and use of machinery are keys to success. Villages line valleys or gullies where water is obtainable and the large open fields lie on the treeless interfluves. It is a bare countryside with a vast, empty horizon.

East of the Don annual precipitation is light (8-12 inches) and variable, with a maximum in early summer, and water has to be provided by wells and irrigation schemes. Apart from *solonchak* and *solonets*, the black and chestnut soils of the dry steppe are extremely fertile when correctly cultivated. Spring wheat is preferred, because there is a danger of frost, from which the light winter snow gives little protection. Two-fifths of the sown area in the lower Don and Northern Caucasus is occupied by spring wheat. This region supplies about a fifth of Soviet wheat, maize, and sunflower. Other crops include kenaf, castor seed, and rice (about 15 per cent of Soviet output), grown in flooded backwaters of the Kuban, Terek, and Don. The vine is also grown. The coastal plain makes possible two maize crops a year producing quality seed. Beef cattle, sheep, and poultry are raised, while animals are also brought for fattening from the Caucasian mountains and from the dry steppe. In the Stavropol and Kuban district, large state farms are frequently centred around old Cossack settlements.

In the lower Volga and southern Ural, grain-growing and animal-raising are in most respects similar to the Northern Caucasus and lower Don. Wool sheep become important, while millet is found in the driest parts. Near the rivers fruit, typically apples, cherries, melons, peaches, and apricots, and vegetables are grown, particularly in the Volga valley below Volgograd. Eastwards, growing of spring wheat, millet, and maize, but also sunflower, has moved into the newly-ploughed, drier steppe of Western Siberia and Kazakhstan.

In Western Siberia, along the line of the Trans-Siberian railway, grain-growing gives way to dairying as the emphasis in mixed farming. Excellent natural pastures, partly flooded in spring, exist in the valleys of the Western Siberian rivers. Developed in the late nineteenth century, dairying concentrates on articles easily transported to major consuming centres outside the region. Ameliorative measures, particularly prevention of spring flooding, and better transport, are keys to expansion in what is still pioneer country. Since the Second World War, mixed farming has spread further into the Altay foothills and the Minusinsk steppe as well as eastwards towards Lake Baykal. Scattered patches are found in Transbaykalia and in the Far East, where Russian colonists have abandoned traditional methods and use Chinese or Manchurian systems. For example, grain is sown in ridges to keep it above the water which stands long on the fields in spring. Many new varieties have been developed for cultivation and Chinese-Manchurian forms cultivated. Transbaykalia grows mostly grain, but the Amur-Ussuri basin is more varied. Soya beans and perilla are grown by Russians, Chinese raise millet, and Koreans grow rice. Sugar beet introduced about 1920 flourishes in the Khanka Lowland. In the Zeya and Bureya valleys, sugar beet, cereals, and sunflower are grown and pigs and cattle kept. Fruits are mostly Manchurian. The better soils (resembling black earths) and short, hot summer of the middle Lena allows grain-growing and even melons, but emphasis is on cattle-raising.

Pastoral farming occurs where natural conditions prevent arable farming or sometimes as a result of historical conditions. Animals are also widely kept in the mixed farming regions, since arable farming and livestock-raising are often complementary. Differing in the nature and purpose of their stock and in their seasonal activities, but with a low density of animals and a semi-nomadic economy to which the collective *artel* has been adapted, pastoral patterns may be divided between the southern lowlands and mountains of the arid belt and the northern forests and tundra. Pastoral farming has not only suffered from encroachment by arable farming but also from the great loss of animals during the bitter struggle to collectivize the nomads.

The Central Asian poor steppe and semi-desert are largely sheep-breeding, for wool, meat, and tallow. Cattle are kept only on the fringe of the cultivated lands and pigs are reared by Russian settlers. Horses have been important in the past and camels are still found on the eastern shore of the Caspian and along the southern frontier. In winter animals are taken from the mountains into the lowlands, while in summer they return and those remaining in the lowlands move from poor to richer steppe. The main effort has been to increase the density of stock by drilling large numbers of different types of wells to water pasture and provide drinking pools. Aircraft direct flocks to pastures and maintain contact between farms and herding groups, while improved weather forecasting allows flocks to be moved to safety from storms or drought, when in the past such disasters cost thousands of animal lives. In Kzyl Orda *oblast*, shelter belts have been planted to protect flocks wintered there. Winter pastures lie mostly in the river valleys of the south, while spring and summer pasture is most common in Karaganda and Aktyubinsk *oblasts*.

Pastoralism in the Central Asian mountains is characterized by seasonal movement of flocks and herds up and down the slopes. In Kirgizia, flocks are taken down to the foothills and piedmont zone in winter, where natural fodder is augmented by growing suitable crops, though some animals remain in high, sheltered valleys in winter. As the grass on the lower slopes withers in spring, the animals are driven back up the mountains. Horses are usually kept on the lower slopes, with cattle on the middle pastures, and sheep on the highest. Herdsmen still use the traditional felt *yurt*, pitched in little clusters near some brook.

The Caucasian pattern is more diverse, with much greater areas in the mountains used for crops, while there is also forest shelter, so that the density of animals is higher. Extensive movements take place from the Caucasian ranges and the Armenian plateau to winter pastures in the Kuma-Kura steppes and even the Terek and Iori valleys. Meat, wool, and cheese deliveries are handed to marketing agencies in the winter encampment. Flocks remain in the lowlands until late spring when, with renewed strength, they move to the summer pastures. There is also a similar movement to and from the limestone uplands of the Crimea.

In Southern Siberia, settled and semi-settled pastoralism is replacing the old migrationary habits. Cattle and sheep are the principal livestock; but in the poorer interior of the Altay and Sayan, herds of *maral* and reindeer or goats are found. Horse-breeding is important. Yaks are also kept along the Mongolian border. In Tuva, three-quarters of the animals are sheep and the district has the largest number of animals per head of population in the country. In the south-west Altay, cattle may be pastured on mountain slopes all the year: elsewhere there is seasonal migration. Stall-feeding increases with growing of fodder crops.

In the wooded tundra and tundra, reindeer-herding of the old native economy has been adapted to modern Soviet methods. In winter, herds find shelter and fodder in the wooded tundra and northern tayga fringe, living off lichens and mosses which they grub from below the thin snow cover. In summer, tundra plants and grasses provide nutrition. Modern methods prevent overgrazing of pasture, of which several hundred acres are needed for each animal, for these require ten years to regenerate. Fodder is imported for herds in winter in case heavy snow covers natural fodder. The long treks of the past from one pasture to another have been reduced by planned grazing, though frequent short movements occur. Each herder, almost exclusively from Siberian tribes, looks after 200-300 animals.

The tayga is little used, though around the fringes and along accessible waterways there is lumbering, and hunting is found in the interior. Around the southern edge and along the rivers, patches of cultivation are found as well as fur-breeding farms, and there is also meadow. Russians are active in farming, but natives are hunters or herders. Poor soils, short cool summers, and widespread *permafrost* hamper agriculture, though some northern farms have been established.

'Suburban' farming is found around towns in both the mixed farming and other districts. The largest area is around Moscow and its satellites, though it includes much forest and mixed farming. Farms concentrate on supplying fresh vegetables, milk, and meat to the large towns, as these do not stand movement well in the absence of suitable long- and medium-distance transport. Plots belonging to workers are also important here, forming over a tenth of the area sown to potatoes and vegetables in Russia, as well as containing many animals. Glasshouses and frames are found on most farms: in Moscow *oblast*, there are over a million frames and 200,000 square yards of glasshouse. Some farms use nightsoil, like the Lyubertsy farm which receives 200,000 tons annually from Moscow.

In the Caucasian isthmus, growing of crops like tea, citrus fruits, the vine, and even tung nuts and other fruits is carried on partly on irrigated land, while fruit gardens and vineyards are typical of the mountains of the Crimea and the reaches along the Volga-Akhtuba. Irrigation is important in Central Asia, where oases lie in sheltered valleys and along the piedmont loess belt as well as along the desert rivers. Conditions make possible in some parts raising two crops each year, and where precipitation increases, dry farming (*bogarni*) is practised. The emphasis has shifted from food crops to industrial crops, like cotton, which cannot be grown as satisfactorily elsewhere in Russia; but there is also growing of rice, notably by Korean settlers, and many native fruits, and the grape. In these areas, few animals are kept because of the lack of fodder and pasture, though trees are planted along irrigation ditches for shade and wood. In Central Asia and Transcaucasia, mulberry is grown for silkworms.

There are three types of oasis: those almost surrounded by desert, like Murgab and Tedzhen; oases of the mountain foot-slope, like Fergana and Tashkent; and oases in the higher foot-hills, such as the upper reaches of the Vakhsh and Kashkadarya. Oases in the plains and lower foothills concentrate on cotton, vines, and fruit, while those in the higher foothills produce fruit, vegetables, and even grain, often on the surrounding *bogarni*. An example of the foothill type is the Fergana valley, sheltered by

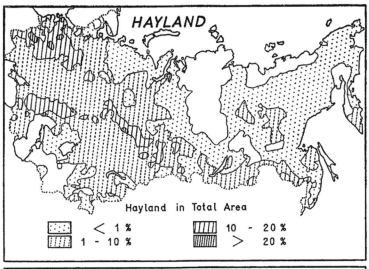

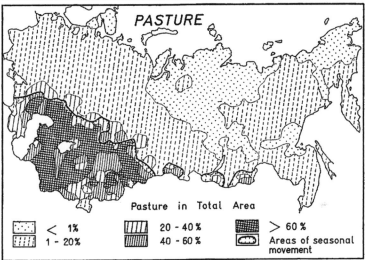

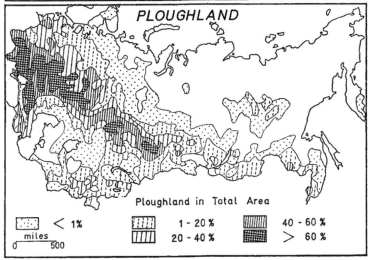

FIG. 35

Land use. *Atlas Selskogo Khozyaystva SSSR*, Moscow, 1960, pl. 92, 93, 94.

high mountains and watered by the Syr Darya and its tributaries, with hot, dry summers but a cool winter. The frost-free period lasts for 210-230 days a year, while there is a high accumulated temperature. Precipitation averages 7-8 inches rising to 12 inches in the foothills. Where water can be supplied on the piedmont loess, more than one crop a year is possible. Sands in the centre of the valley have been little used, but will be developed for cotton. The valley supplies about a quarter of Soviet cotton and it rears two-thirds of the silk in the Uzbek republic. *Bogarni* farming rises to 3,600 feet, while between 3,000 and 6,000 feet there are walnut forests and mountain pasture.

Sheltered and irrigated valleys in Dagestan and other parts of Caucasia produce a wide range of crops, but the main centre of sub-tropical crops is the Kolkhiz Lowland and the western slopes overlooking the Black Sea. Adzharia supplies three-quarters of Russian citrus fruits, while tea comes particularly from Georgia around Makharadze and terraces overlooking the Black Sea, and tung also comes from this area. Medicinal plants, ginger, and bamboo are also grown, and grapes form the basis for the well-developed wine industry. Arid eastern Transcaucasia is less developed, except in the Lenkoran Lowland and near Baku.

RELATIVE IMPORTANCE AND DISTRIBUTION OF PRINCIPAL CROPS

A changing share of the expanding acreage among the principal crops has been marked by a fall in the proportion sown to grains, which nevertheless still occupy almost two-thirds of the area. A rising share has been sown to hay and fodder crops. Grain-growing is characteristic of the steppe and wooded steppe, where long, warm, dry summers are suited to cereals and the European regions give the highest yield and a more stable return than Asiatic Russia where precipitation is less reliable.[7]

Wheat was formerly concentrated on the better black and brown earths, but policy has been to reduce specialization in any one area where crop failure might have serious national consequences. In the older wheat areas, new crops have been introduced; while a northwards advance has been achieved by developing quick-ripening and frost-resistant strains, which are claimed to grow as far as 60° N. Preparation of seed for germination before sowing has enabled wheat to be grown in Baykalia and the Far East. Winter wheat, which gives a better yield, is now the most important grain in the Ukraine, Northern Caucasia, and the Crimea : it has been introduced experimentally into Siberia where spring wheat is sown, since little protection is given from frost under a thin snow cover. To hold the snow to prevent frost from delaying spring sowing, stubble is left standing in autumn. Yield is also reduced by prevention of tillering through too rapid a rise in spring and early summer temperatures. A dry spring can be disastrous as it allows wind to blow away top soil. Heavy summer showers or wind may flatten a standing crop, making the use of large combine harvesters difficult. When

TABLE 12

PROPORTION OF SOWN ACREAGE UNDER PRINCIPAL
CROP TYPES

Crop	1940		1950		1959	
	Mill. Acres	%	Mill. Acres	%	Mill. Acres	%
Spring grains	179·0	48·1	164·0	45·3	207·0	42·7
Winter grains	94·1	25·4	90·1	25·0	88·5	18·3
Fodder and hay	44·4	12·0	51·2	14·2	129·0	26·8
Potatoes and vegetables	24·6	6·6	25·8	7·1	28·6	5·9
Industrial crops	29·1	7·9	32·0	8·4	31·3	6·3
Total sown area	371·2		363·1		484·4	

Source : *Selskoye Khozyaystvo SSSR*, Moscow, 1960.

mature, the crop is harvested by heavy machines whose weight can be carried by the hard, dry ground. Threshed wheat is fed directly into lorries which carry it to elevators. Concentration is on hard bread grains. A major export before 1914, the interwar level fell to about 600,000 tons, though it has again risen to about 6 million tons with exports mostly to satellite countries.

Rye, traditional forest belt cereal and making a wholesome but less nourishing bread than wheat, has fallen from 69 million acres in 1913 to 42 million in 1959 under pressure from wheat-growing. Almost entirely winter sown, it remains important in Northern European Russia because of its tolerance to climatic conditions ; though in the intensely hard winters of Northern and Eastern Siberia it is spring sown. Oats form a fodder crop principally in Central and Northern districts; while barley, sown further north, is cultivated in Karelia and in the lower reaches of the Ob and Yenisey, its lower moisture needs also allow its substitution for oats in drier areas. It is used chiefly as fodder with some for malting. The sown area has remained fairly constant, but spring barley has declined and is grown in the Black Sea littoral, Northern Caucasia, and in the Central Asian foothills where it can mature before the hot season.

Maize-growing has been greatly expanded for grain, green fodder, and silage, though the grain acreage has tended to rise in relation to the others, but the proportion for grain (3·5 per cent) is much below the U.S.A. (32 per cent). It has been introduced into the newly-ploughed regions and is grown in the south-eastern foothills of Kazakhstan, but the main centres are the Southern Ukraine and Northern Caucasia as well as Moldavia and the central black earth belt, where it is primarily a grain crop. In the north, where grain will not ripen, it is for fodder and silage. Its cultivation is a key to the policy of raising livestock output.

Millet, because it is resistant to drought though susceptible to frost, is found in the drier parts of the Ukraine and central black earth belt as well as in Central Asia, where it has been put in rotations in place of wheat. Buckwheat, from European Russia, Baykalia, and the Far East, is used as a feeding stuff and for porridge. Rice cultivation, mainly of lowland types where irrigation is available, occurs in Central Asia, around Tashkent, Samarkand, Fergana, and Kzyl Orda (where the Tas Buget dam increases the watered area) and a new area at Ush Tobe near Taldy Kurgan. Rice paddies are also found in the Amu Darya delta. Koreans have developed rice-growing in the Khanka lowlands near Vladivostok and have contributed to cultivation in Central Asia. Favoured valleys along the Black Sea littoral, such as drained backwaters of the Kuban near Krasnodar, have also grown rice.

Industrial crops are grown in both established and new agricultural areas, and several new crops have been introduced. Although cotton has been grown for a long time within the present boundaries of Russia, quality has been poor in the past. The first five-year plan (1928) began expansion of cotton-growing, which has encroached on the food-producing area of Central Asia, where the most favoured areas are sheltered valleys with abundant water, plentiful sunshine, and freedom from frost. The 'nilotic' conditions have been increased by completion of several major irrigation projects, though in the middle and lower Amu Darya and Syr Darya winters are too cold to be ideal. Uzbekistan, notably around Tashkent, Samarkand and Fergana, has over half the sown area in the U.S.S.R. Almost all the crop comes from collective farms. In Tadzhikia, the Gissar and Vakhsh valleys are known for long staples drawn from American and Egyptian varieties and policy now concentrates on quality rather than quantity, so that marginal producers in Northern Caucasia and the Ukraine have been reduced in area, but Azerbaydzhan is rising as a long staple grower.

Flax will grow on a wide range of soils and in cool, moist summers, but it rapidly exhausts the soil and is thus grown in rotations with fodder crops. Fibre and seed may not usually be drawn from the same plant, since the best varieties for seed are not those for fibre, while cultivation and climatic conditions require to be different. Northern flax is largely for fibre and bushy (southern) flax for seed. Russia is a major world producer. Hemp is grown in European Russia for both fibre and seed, while jute, and its substitute kenaf, and ramie (Chinese nettle) have also been grown, besides less known types like kendyr, a Central Asian fibre producing a tough, resistant cloth. Silk is an ancient product of Central Asia and Transcaucasia, but its expansion is unlikely in face of competition from synthetic fibres. Nowhere are natural conditions suited to rubber trees (*hevea brasiliensis*), but in the early thirties a group of small dandelion-like plants bearing a rubbery latex were discovered in the Tyan Shan and were widely introduced into Central Asia and even Southern European Russia, particularly *kok sagyz* and *tau sagyz*. Potato alcohol output is associated with synthetic rubber production.

Sunflower, the most important oil-bearing plant, is used widely in food, such as margarine, while the seed is chewed. It is adaptable to many conditions, providing there is a dry ripening season and suitable pollinating insects. Waste from oil extraction is a useful cattle food. It is widely grown in the black earth belt, the Ukraine, Northern Caucasia, and in Northern Kazakhstan. The soya bean, a recent introduction, comes almost entirely from the Far East, with half the sown area in the Amur district.

It cannot stand frost in the seedling stage and has little resistance to drought, while yield is reduced by low night temperatures. Tung, a small-scale cultivation in humid Transcaucasia, is used in paints and varnishes, but the waste is poisonous to cattle. Other oilseeds include rape, hemp, linseed, castor, and mustard, while cotton seed has been developed. Russia imports groundnuts, soya beans, and other vegetable oils.

Sugar is provided principally from beet, though small amounts of cane have been grown in Transcaucasia. Refined sugar output is between 10 and 12 per cent of beet tonnage. It is also an industrial crop, producing alcohol. Almost two-thirds of the sown area is in the Ukraine and most of the remainder lies in the black earth belt in Northern Caucasia and the middle Volga as well as Moldavia. In the Ukrainian wooded steppe about a tenth of the area is under beet. Although it is machine sown, mechanized harvesting is limited. Sugar making begins in October and lasts until April, though the harvest starts in September and lasts until the ground is too hard under frost. Refined sugar was imported until 1936 and begun again in 1954, though much is later re-exported. Honey was the traditional sweetener and bees are widely kept in the forest belt and elsewhere by collectives and individuals.

Of the area sown to vegetables and potatoes, collectives accounted for 48·1 per cent and state farms 7·3 per cent, but 44·6 per cent lay on the plots of collective farmers and allotments of industrial workers. Potato and vegetable growing are most important near the great centres of population. Away from these areas, potatoes are frequently used for pig swill, starch or alcohol production; some 47 state farms produce potable alcohol, while much vegetable production goes to tinning or pickling, particularly cucumbers and cabbage. Vegetable-growing has been developed in the far north to supply mining and other settlements. After potatoes, cabbage occupies the largest area, followed by cucumber, tomatoes, onions and garlic, turnips, and carrots. Western Siberia is renowned for its mushrooms.

Fruit-growing is found on collective farms and private plots throughout the country. In Northern European Russia, apples, pears, plums, and soft berries are most common; while south of Moscow, cherries and apricots appear. Besides a wide variety of fruits, melons are important along the Volga below Kazan, and in the lower Volga candied and dried fruits are produced.

On the Black Sea coast and in Western Georgia citrus fruits, like oranges, tangerines, and lemons, have been grown since the twelfth century. Georgia harvested 710 million pieces of fruit in 1949, though a series of cold winters in the early 1950's reduced output. Azerbaydzhan has developed citrus fruits, notably in the irrigated lowlands around Lenkoran and Baku, and has olive-growing. Dagestan grows fruit and nuts in sheltered, irrigated valleys, while Armenia has persimmon, peaches, and apricots. The Megrinsk district is renowned for almonds. Much fruit is canned or otherwise preserved. Central Asia is the home of many varieties, like figs, grapes, peaches, pears, melons, quince, pomegranates, and apples. Alma Ata is known for its apples, while Samarkand cherries and sweet, seedless grapes are famous. Tashkent, Yangi Yul and Ordzhonikidze have important fruit gardens, also found amid cotton fields in the Fergana valley. Japanese and American varieties have been introduced.

Russian wines, not well known in the West, have some good qualities, but the bulk appear to be mediocre. In 1958 state farms produced 93 million gallons of wine and 33 million bottles of sweet champagne. The westernmost producer, Moldavia, has vines grown on well-drained slopes and river terraces sheltered from the wind by tree belts. Extension of vine-growing uses non-grafted European stocks, mainly nineteenth-century French strains, and local hybrids are being replaced. Emphasis is on sparkling wines in the central districts and cognacs in the north. Main centres are Kishinev, Kotovsk, Tiraspol, Beltsy, and Grigoriopol.

The lower slopes of the Crimean mountains produce notable white wines; but some of the best wines come from the southern shores, particularly the former imperial estate at Massandra near Yalta, and are full-bodied and rich, resembling madeira and tokay. Vines are also grown in the lowest reaches of the Dnepr and Dnestr and around Odessa and Kherson. The lower Don has extensive vineyards, with the best qualities from Tsimlyansk, noted for its champagne.

Caucasia is known for its red wine. Sparkling wines come from Northern Caucasia, from Anapa, Abrau-Dyurso, Gelendzhik, and Paskoveysk. Georgia has about 150,000 acres of vineyards, one of the oldest branches of the economy, and is the largest producer. The vines grow on well-drained terraces sheltered from cold winds, like those along the Iori and Alazan and in the upper and middle Kura and Rioni, with full, dark, and often dry wines from iron-rich lateritic soils. Sweeter wines and champagne-types are from around Kutaisi and Tbilisi, whose 'champagne factory' produces about a quarter of Soviet output. Dry wines come from Azerbaydzhan, though many of the vineyards grow table fruit and raisins, with Kirovabad and Shemakha as main centres. Nakhichevan is principally concerned with distillation, also common in Armenia, where Yerevan is known for cognac, liqueurs, sherry, madeira, and muscat. The main growers are the Araks, Razdan, and Khasakh valleys. Central Asian wines are particularly sweet, but nine-tenths of the grapes are used otherwise.

Tea traditionally came across the caravan routes from China, but was first grown in Transcaucasia about 1848, though predated in Central Asia by tea suited to local taste. Output is about 140,000 tons a year and another tenth is imported. It demands warm, humid conditions without frost or drought and grows on well-drained, acid soils allowing deep root penetration, with plenty of nitrogen. The red earths of humid Western Georgia are ideal — rich in hydrates of ferric oxide and alumina oxide, formed from weathered, andesitic tuffs. Introduced in 1932 into the Lenkoran Lowlands, tea suffers from the cold.

Tobacco is grown throughout warm, moist Southern Russia, but is generally poor in quality, though the best Crimean and Caucasian types are reputedly as good as Balkan tobacco. About a quarter of the sown area is in the Ukraine, principally in the Crimea. Georgia, Moldavia, and the Krasnodar region are also important, where air-curing is practised. The central black earth districts, Western Siberia, and the Volga, also grow *makhorka*, a strong, dark, pipe tobacco, which is tending to give way to yellow tobacco. Comparatively cheap but loosely filled, about 200,000 million cigarettes are annually produced in widely-scattered factories.

Fodder crops now occupy a quarter of the sown area compared to less than 3 per cent in 1913, resulting from improved rotations and elimination of the old three-field system. Annual grasses form the largest area. Lucerne, from seed raised in Central Asia, is sown on limey areas, while timothy gives a good yield on heavy cold soils. Esparto grass has also been grown. The richness of Kaliningrad *oblast* is stressed as the reward of long-term ley-farming practised by the former German owners.

Livestock farming has had an unhappy history since large numbers of animals were slaughtered or lost in the early stages of collectivization. A high, though falling, proportion of the animals are kept by collective farmers or workers on their private plots, but since 1958 measures have been taken to eliminate this as an obstacle to more efficient breeding of high-quality animals.[8]

Key factors in animal farming are availability of fodder and

TABLE 13

HARVEST AND YIELD OF SELECTED CROPS

Crop	Harvest, mill. tons *			Yield, Qu/ha[1]			Best Year 1953–1959	Harvest mill. tons	Yield, Qu/ha[1]
	1913†	1940	1959	1913†	1940	1959			
Wheat, Winter	8·3	14·5	26·4	10·0	10·1	15·2	H. 1958 Y. 1958	29·5	16·2
Wheat, Spring	18·0	17·2	42·7	7·3	6·6	9·4	H. 1956 Y. 1956	52·4	10·7
Rye	22·7	20·9	16·9	8·0	9·1	9·9	H. 1959 Y. 1959	16·9	9·9
Maize (dry grain)	2·05	5·1	5·7	9·4	13·9	16·0	H. 1955 Y. 1958	11·6	20·6
Rice	0·3	0·3	0·2	11·9	17·3	22·1	H. 1956 Y. 1959	0·25	22·1
Cotton	0·7	2·2	4·7	10·8	10·8	21·7‡	H. 1959 Y. 1959	4·7	21·7‡
Linen flax	0·4	0·3	0·3	3·2	1·7	2·3	H. 1956 Y. 1956	0·52	2·7
Sugar beet	11·3	18·0	43·9	168·0	146·0	159·0	H. 1958 Y. 1958	54·4	218·0
Sunflower	0·7	2·6	3·0	7·6	7·4	7·7	H. 1958 Y. 1958	4·6	11·8
Potatoes	31·9	75·9	86·6	76·0	99·0	91·0	H. 1956 Y. 1956	96·0	104·0
Vegetables	5·5	13·7	14·8	84·0	91·0	100·0	H. 1957 Y. 1957	14·8	101·0
Silage crops	..	5·2	170·4	..	..	..	H. 1959 Y. ..	170·4	..
Hay	..	67·8	80·4	..	..	..	H. 1958 Y. ..	85·2	..

H. = Harvest.
Y. = Yield.

[1] Quintal/ha = 0·795 cwts/acre.
* metric tons, 1 metric ton = 1·02 long tons.
† In present boundaries.
‡ From irrigated lands.

Source: *Selskoye Khozyaystvo SSSR*, Moscow, 1960, pp. 202–203, 208–209.

transport to consumers. Near to big towns, produce is largely fresh milk and meat, while in more distant areas there is conversion of milk into cheese, butter, and sour cream, while more meat is canned or reduced to extracts, and hides and fats become important. The transport factor is illustrated by the use of only 10 per cent of milk produced near large towns for butter-making, while over half the milk produced in Northern European Russia and Siberia is made into butter. Cheese is important in Caucasia, where Armenia produces Swiss-type, and in Western Siberia. Solution of winter fodder problems in Northern European Russia has allowed expansion of livestock-keeping. About three-quarters of the cattle are local breeds, like the Yaroslavl, Dom-

TABLE 14

NUMBERS OF LIVESTOCK IN SELECTED YEARS

Year	Horned Cattle	Of which Cows	Pigs	Sheep	Goats	Horses
	Million head					
1916	58·4	28·8	23·0	89·7	6·6	38·2
1928	66·8	33·2	27·7	104·2	10·4	36·1
1934	33·5	19·0	11·5	32·9	3·6	15·4
1941	54·5 (44·4)	27·8 (58·2)	27·5 (41·1)	79·9 (35·1)	11·7 (72·3)	21·0
1946	47·6	22·9	10·6	58·5	11·5	10·7
1950	58·1 (42·7)	24·6 (64·9)	22·2 (33·7)	77·6 (15·2)	16·0 (52·0)	12·7
1953	56·6 (41·6)	24·3 (59·0)	28·5 (45·2)	94·3 (14·5)	15·6 (72·2)	15·3
1960	74·2 (33·5)	33·9 (50·3)	53·4 (25·9)	136·1 (21·2)	7·9 (80·7)	11·0

Source: *Selskoye Khozyaystvo*, Moscow, 1960.

Figures in brackets represent holding by collective farmers and workers as a percentage of total stock.

shinsk, and Kholmogory, with their high-fat-content milk ideal for butter-making, also true of the Suksun and Tagil cattle of the Ural. The Western Siberian dairy industry developed by settlers from European Russia in the late nineteenth century had its centre, before 1914, in Kurgan, and used the rich fat milk of Siberian cattle for butter in an area where there was little demand for fresh milk. Useful new dairy stock has been raised in Transbaykalia by crossing Mongolian cows with European Simmenthaler. In 1959, milk output was 62 million tons and butter over 650,000 tons. Drier regions away from consuming centres concentrate on meat, tallow, suet, and hide production, though they also supply livestock for fattening and raising in other districts, notably for stall-feeding in the black earth lands. Meat-processing plants are found at Krimskaya (one of the largest factories) in Northern Caucasia, while others are Novosibirsk, Kurgan, Biysk, Tyumen, and Omsk, with Chita, Ulan Ude, and Borzya in Transbaykalia. Kazan has traditionally processed leather, soap, and tallow.

Sheep are raised mainly for wool and tallow, though mutton is eaten in Caucasia and Central Asia. Producing 335,000 tons (1959) of wool, concentration has been on improving quality. Before the Revolution, fine-woolled sheep were found in Kazakhstan west of the Irtysh : the bulk of the animals were coarse-woolled, fat-tailed varieties which have since been interbred with fine-woolled rams to produce a successful 'Kazakh long wool'. High mountain sheep have also been cross-bred to give a breed readily adaptable to long treks and poor pasture, now found mainly in the Alma-Ata and Taldy Kurgan districts. Two-thirds of Karakul products come from the poor steppe of Uzbekistan, notably around Bukhara, Samarkand, and in the Kashka Darya district. Gissar sheep are raised in Tadzhikia.

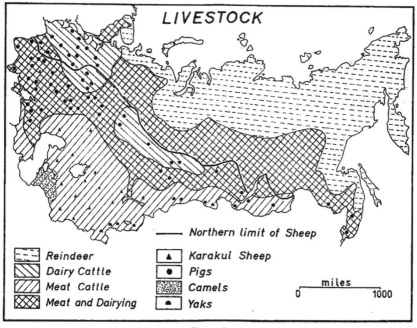

FIG. 36

Livestock. *Atlas Selskogo Khozyaystva SSSR*, Moscow, 1960, pl. 176-197.

With mechanization, the number of horses is declining. They have been bred traditionally in the steppe among Cossack groups, while Central Asian and Southern Siberian nomads have shown great skill. Even the Yakuts are known as horse-breeders. Breeds include European and Arabian strains, but throughout Central Asia and Siberia the small Mongolian horse is seen. Mules and donkeys are common in Central Asia and Caucasia, while camels are used in the sandier desert areas. Yaks are found in the Pamir and Altay.

Reindeer, only partly domesticated animals, are widespread in the tundra and tayga, where there are possibly about two million. Between the Pechora and the Ob they are raised by the Nentsy, by Lapps in the Kola Peninsula, and bred by the Koryaks and Chukchi in North-eastern Siberia. They provide food and raw material in the tundra and to a lesser extent in the tayga: their skin is valued for the property of remaining elastic even in great cold. In the tundra, reindeer draw sledges in harness, but in the tayga they are ridden or used as pack animals. In the Altay and Sayan, *maral* (Manchurian red deer) are kept on farms, and wild animals hunted for hide and horn, which has medicinal properties for the Chinese. Dogs are used for transport and herding in Northern Siberia, while in the Central Siberian tayga they are principally used for hunting. Siberian dogs are smaller, lighter, and faster than Eskimo dogs, though less strong.[9]

An early attraction to penetration into the tayga was its richness in furry animals. Long hunting has reduced the natural fauna in numbers and species, but it is now protected by closed seasons and nature reserves. Numbers of fur-breeding state farms have been established. In the tundra zone, notably from the northern islands, large numbers of birds are trapped.[10]

NOTES ON CHAPTER 6

1. Bergson, A. (Ed.), *Soviet Economic Growth.* New York, 1954.
   Gerasimov, I., 'Geographical Study of Agricultural Land Use', *Geographical Journal*, vol. 124, 1958.
   Atlas Selskogo Khozyaystva SSSR. Moscow, 1960.
2. Lyashchenko, P., *A History of the National Economy of Russia to the 1917 Revolution.* New York, 1949.
   Mavor, J., *An Economic History of Russia.* New York, 1925.
   Maynard, J., *The Russian Peasant.* London, 1943.
   Mirsky, D., *Russia — A Social History.* London, 1933.

Robinson, G., *Rural Russia under the Old Regime*. London, 1932.

Atlas Istorii SSSR. Vols. I-III. Moscow, 1949–1960.

3. Hubbard, L., *Economics of Soviet Agriculture*. London, 1939.

Jasny, N., *The Socialized Agriculture of the U.S.S.R.* Oxford, 1949.

4. Most important are *Selskoye Khozyaystvo SSSR — Statisticheskiy Sbornik*, Moscow, 1960, and *Atlas Selskogo Khozyaystva SSSR*.

Examples of farms were collected in Russia during 1960.

Report of the Central Committee of the C.P.S.U., 22nd Congress of C.P.S.U., Moscow, 1961, and the Draft Programme of the Communist Party of the Soviet Union discuss the future development of the country-side, notably the introduction of the agricultural town (*agrogorod*). The first of these has been built 35 miles south of Moscow on a 53,000-acre collective, though earlier attempts were made in the Northern Caucasus in the early post-war years but later abandoned.

Economic Survey of Europe, 1960. United Nations. Geneva, 1961.

Parsons, K. H., and Penn, R., *Land Tenure*. Madison, 1956.

It is worth noting that the labour force on collective farms varies from about 18 million in winter to 34 million in summer, particularly July and August, though maximum employment is in September and October in Central Asia.

5. Cherdantsev, G., *Ekonomicheskaya Geografiya* (three volumes). Moscow, 1957–1958.

Durgin, F., 'The Virgin Lands Programme 1954–1960', *Soviet Studies*, 1962.

Field, N., 'The Amu Darya — A Study in Resource Geography', *Geographical Review*, vol. 44, 1954.

Jackson, W., 'The Virgin and Idle Lands of Western Siberia and Kazakhstan', *Geographical Review*, vol. 46, 1956.

Malygin, E., *Zasukhi v SSSR*. Moscow, 1958.

Sotnikov, V., 'Farming Problems in the Zones of the U.S.S.R.', *Soviet Geography*, vol. I, 1960.

Suslov, S. P., *Physical Geography of Asiatic Russia*. San Francisco, 1961.

Thiel, E., 'Die Grundwasserverhältnisse Turkmeniens und ihre Beziehung zur Morphologie und Geologie des Landes', *Petermanns Mitteilungen*, vol. 95, 1951.

Grigoryev, A., 'Soviet Plans for Irrigation and Power — A Geographical Assessment', *Geographical Journal*, vol. 118, 1952.

'Reclamation in the Forest Belt in the U.S.S.R. in Europe', *Geographical Journal*, vol. 119, 1953.

Atlas Selskogo Khozyaystva SSSR.

Armand, D., 'Das Studium der Erosion in den Waldsteppen- u. Steppengebieten der UdSSR.', *Petermanns Mitteilungen*, vol. 98, 1954.

6. Cherdantsev, G., *op. cit.*

Nikishov, M., 'Experience in Distinguishing Agricultural Zones and Regions on the Agricultural Map of the U.S.S.R.', *Soviet Geography*, vol. 1, 1960.

Atlas Selskogo Khozyaystva SSSR.

Geograficheskiy Atlas dlya Uchiteley Sredney Shkoly. Moscow, 1960.

Atlas SSSR dlya Sredney Shkoly — Kurs ekonomicheskoy Geografii. Moscow, 1960.

7. *Selskoye Khozyaystvo SSSR.* Moscow, 1960.
   Atlas Selskogo Khozyaystva SSSR.
   Posevnyye Ploshchadi SSSR. Moscow, 1957.
   Jackson, W., 'Durum Wheat and Expansion of Dry Farming in the Soviet
   Union', *Annals Assoc. Amer. Geog.*, vol. 46, 1956.
      'The Russian Non-Chernozem Wheat Base', *Annals Assoc. Amer.
   Geog.*, vol. 49, 1959.
8. *Selskoye Khozyaystvo SSSR.*
   Atlas Selskogo Khozyaystva SSSR.
   Chislennost Skota SSSR. Moscow, 1957.
   Cherdantsev, G., *op. cit.*
   Kutafyev, S. A. (Ed.), *Rossiyskaya Federatsiya.* Moscow, 1959.
   Thiel, E., *Die Mongolei* (Munich, 1957) discusses the nature of Central
   Asian nomadism under modern conditions.
9. Tokarev, S., *Etnografiya Narodov SSSR.* Moscow, 1958.
10. Fisher, R., *The Russian Fur Trade, 1500–1700.* Berkeley, 1943.
   Kirikov, S., and Isakov, Yu., 'The Reserves of Wild Game and the
   Dynamics of Hunting and Its Prospects', *Soviet Geography*, vol. 1, 1960.

CHAPTER 7

Fuel and Mineral Resources

A MODERN industrial economy demands such a wide range of minerals that few countries have even a fair degree of self-sufficiency, but Soviet attempts to develop autarchy have been surprisingly successful because the country is richly endowed, although there are nevertheless serious deficiencies made up by imports, many of which come from other members of the Communist bloc. A particular problem is the unfavourable geographical location of several of the scarcer minerals in the more inaccessible parts of Siberia and Central Asia, where mining and transport are difficult in such harsh environments. The unequal distribution of mineral wealth has been an important hindrance in the establishment of the 'complex' and 'balanced' economies which the Soviets have sought to build in the country's major economic regions (Chapter 5).

Russia is fortunate in having large reserves of fuel minerals to which should also be added water power. Since the early post-war years, there has been a steady improvement in the petroleum reserves, though by Western standards *capita* consumption appears low. Until the middle 1950's, the U.S.S.R. depended largely on the supply of nuclear fuels from the Eastern European satellites, but here again home reserves are believed to have shown marked improvement.[1] Of the metallic minerals, the country is well supplied with large reserves of iron ore in deposits as yet unworked as well as in those currently exploited, where metal content is generally higher than in the Western world. There are abundant manganese and chrome, of which the Soviet Union is a major world producer. Adequate deposits exist of the rarer vanadium, titanium, and nickel; but the important alloy metals cobalt, molybdenum, and tungsten seem scarce. These metals unfortunately act as substitutes for one another in high-speed steels. Molybdenum is not found in Eastern Europe, but Russia can import supplies from Manchuria and North Korea. Cobalt reserves are low, though a little is mined in Eastern Germany;

DOI: 10.4324/9781003172048-7

and China, a major world producer, may augment supplies of tungsten.

Relatively little information exists about non-ferrous metals, but it seems certain that care and economy are needed in the use of available supplies and a special department of *Comecon* is occupied with co-ordination and economical employment of Soviet bloc resources. Copper, required increasingly for electrical use, is probably available in adequate quantities, though doubt has been expressed whether these will keep pace with rising demand. Lead is also at present adequate, but a changing pattern of deposit utilization is likely to raise costs as remoter and smaller deposits are worked. Zinc, usually found in association with lead, also comes from silver and copper ores in the Ural. Output appears satisfactory and some reserves lie in Eastern Europe. With fair reserves of bauxite, Russia can also call on valuable Hungarian deposits, while Eastern Europe as a whole contains a quarter of world reserves: aluminium alloys are frequently substituted for scarcer metals. Rising chemicals production increases demand for antimony, but there seems to be no supply problem. Barium, important in chemicals manufacture and oil-well construction, is mostly imported from the large German Mühlhausen deposit. Small deposits lie in inaccessible Central Asia. Beryllium, used in electrical engineering, appears to be readily available. Though mined in the remoter parts of Siberia, platinum metals seem to cover Soviet needs, as does the output of mercury and magnesium metals. Particular attention has been given to expanding output of columbium, zirconium, and germanium, all important for making high-speed and high-temperature steels and electronic goods. Although apparently adequate in size, deposits lie in the climatically trying parts of the Soviet Arctic and North-eastern Siberia. Despite the far-reaching economy and substitution of tin, output deficiency appears to have been offset by exploitation of remarkably inaccessible deposits in the mountains of North-eastern Siberia. Satellite production of these metals is small.

With rich gold deposits, it is often suggested that the Soviet Union now produces more than South Africa. Large reserves are available to modern recovery methods, but little indication has been given of silver output. Since the Second World War, new, rich diamond deposits, both placer and pipe, have been worked in Central Siberia. In several places, such as the Ural, there is the working of precious and semi-precious stones.

Sulphur, the basic mineral in a significant heavy chemicals industry, can be obtained in reasonable quantities, though there are technical and organizational bottlenecks. Before 1939, 45 per cent was produced from pyrites and 40 per cent from copper and coal by-products, with another 10 per cent from native sulphur. Development in Central Asia has possibly increased the importance of the latter. Phosphorus minerals, important for fertilizers, can be obtained from phosphate rock and apatite used directly in ground form, but mostly derived from complex superphosphates produced by treatment with sulphuric acid. Again, adverse location rather than shortage of reserves is the main hindrance. There is no lack of salt, though not all deposits are suitable for human consumption. Sodium salts, also readily available, are mostly associated with arid desert areas. Potash comes from the vast Solikamsk deposit, and there is no shortage of potassium minerals. Fluorspar is produced but it is also imported from East Germany and Korea. Asbestos and industrial refractories are abundant. Arsenic is not short !

The Russians are reluctant to publish figures of output, whether in terms of raw ore or metal content and do not always agree on which deposits are the most important. Furthermore, the situation is complicated by many rather doubtful estimates. Table 15 attempts to construct some idea of the possible output but it should be regarded with great caution. Russia is still in a pioneering state of knowledge of mineral resources. It is only since 1945 that the initial geological survey of the country has been completed. While new deposits of 'industrial' significance are constantly being claimed, mostly from the remoter parts, some of the deposits of European Russia are still incompletely known.

FEATURES OF MINERAL DISTRIBUTION

The richest mineral-bearing areas are the exposed roots of the older folded mountain systems, including the remains of the great shields, where deep-seated igneous and metamorphic activity is now revealed at depths accessible to modern mining. Many important deposits of lignite, coal, petroleum, and salt also occur in the almost undisturbed sediments overlying the great shields or deposited in geosynclinal structures. Gold and diamonds, for

TABLE 15
OUTPUT OF SELECTED MINERALS, FUELS, AND CHEMICALS

Mineral	Measurement			Output	Probable Share of World Output
					%
Antimony	s.t.	(1)	E.	6,600	10
Arsenic	s.t.	(2)	E.	5,000	n.a.
Asbestos	s.t.		E.	660,000	27
Barytes	s.t.		E.	140,000	4
Bauxite	l.t.		E.	3,445,000	13
(Aluminium: 745,000 s.t. = 15 per cent world output)					
Beryllium	s.t.		E.	110	Not more than 10
Cadmium	lb.		E.	1,035,000	Not more than 5
Copper	s.t.	(3)	E.	510,000	12
Fluorspar	s.t.		E.	210,000	10
Gold	T/Oz.		E.	11,000,000	24
Graphite	s.t.		E.	50,000	11
Gypsum	s.t.		E.	3,860,000	9
Lead	s.t.	(4)	E.	340,000	13-14
	s.t.	(3)	E.	350,000	
Mercury	fl.		E.	2,500	10
Magnesium	s.t.	(5)	E.	27,500	26
Phosphate Rock	l.t.	(6)	E.	6,495,000	16
Platinum Group	T/Oz.		E.	275,000	22
Potash	s.t.	(7)	E.	1,212,500	12
Silver	T/Oz.		E.	25,000,000	10
Sulphur	l.t.	(8)	E.	370,000	4
Tin	l.t.	(3)	E.	16,500	8
Zinc	s.t.	(3)	E.	38,000	1
Coal	m.t.		O.	374,933,000	19
Lignite	m.t.		O.	138,261,000	22
Coke	m.t.	(9)	O.	56,200,000	n.a.
Petroleum	m.t.	(10)	O.*	166,000,000	16
Electricity	million kwh.		O.*	327,000,000	n.a.
Oil Shale	m.t.		O.	14,000,000	n.a.
Iron Ore	m.t.	(11) (a)	O.*	118,000,000	21
		(b)	O.	61,770,000	27
Chromite	s.t.		E.	1,010,000	20
Manganese	m.t.	(12) (a)	E.	6,393,400	43
		(b)	O.	2,600,000	49
Molybdenum	lb.	(13)	E.	11,000,000	12
Nickel	m.t.	(14) (a)	E.	53,000	19
	s.t.	(b)	E.	64,000	18
Tungsten	s.t.	(15) (a)	E.	10,500	15
	m.t.	(b)	O.	5,700	34
Sulphuric Acid	m.t.	(16)	O.*	5,700,000	n.a.
Mineral Fertilizer	m.t.	(17)	O.*	15,300,000	n.a.
Salt	s.t.		E.	8,300,000	9
Soda Ash	s.t.		E.	1,873,900	11

instance, are found in placers of detrital materials eroded from ancient rocks. Petroleum is found in a wide range of structures, from the recently folded Caucasian mountains to Carboniferous and Devonian sediments lying in the older basins and warpings of the Russian Platform.

One of the most important mineral-producing regions is the 1,500-mile-long ancient Hercynian Ural system, folded and broken, whose exposed metamorphosed roots form low, rounded mountains. The more intensely dislocated eastern slope is particularly rich, so that it constitutes one of the greatest and most varied concentrations of mineral wealth in the world with almost 1,000 different minerals in 12,000 recorded deposits. Although some of the iron, copper, and gold deposits have been worked since the eighteenth century, the riches have been principally developed in the last forty years. The region is deficient, however, in coal, though it is found in small quantities and poor quality on both western and eastern slopes, while the physical nature restricts possibilities of hydro-electric power development. By using imported fuels, however, the association of iron ore and its alloy metals as well as the varied non-ferrous metals can be developed for industry. To the west, in the younger deposits of

---

Measurement

fl. = Flasks of 76 lb. each.
l.t. = Long tons = 2,240 lb.
m.t. = Metric tons = 2,205 lb.
s.t. = Short tons = 2,000 lb.
T/Oz. = Troy ounces.
O. = Official figures from Soviet sources (for 1961 where available *) or U.N. Statistical Yearbook, 1961.
E. = Estimate (mostly from Minerals Yearbook, 1960, U.S. Department of Interior, Washington, D.C.).

Notes

(1) Antimony content of ore and concentrate.
(2) Figure for 1947: 1960 data too meagre for reliable estimate.
(3) Smelter production.
(4) Content of ore at mine.
(5) Primary magnesium.
(6) All forms of phosphate rock.
(7) Potash in K_2O equivalent.
(8) Elemental sulphur.
(9) High-temperature coke.
(10) Crude petroleum.
(11) Iron ore (a) dressed ore, 65 per cent iron; (b) iron content.
(12) Manganese ore (a) grade unstated; (b) 30 per cent Mn content.
(13) Molybdenum ore and concentrate.
(14) Nickel (a) Metallgesellschaft A.G.; (b) Minerals Yearbook.
(15) Tungsten (a) 60 per cent WO_3; (b) WO_3 content.
(16) Sulphuric acid: 100 per cent H_2SO_4.
(17) Gross weight of mineral fertilizer.

the Russian Platform, large resources of salts and petroleum form the basis of a chemicals industry closely associated with Ural industry.

In the Kazakh Uplands, formed by roots of ancient mountains, good coking coal is found at Karaganda and Ekibastuz, and there are large reserves of non-ferrous metals, notably copper, as well as recently developed iron ore. The uplifted blocks of the Altay system have also been worked, particularly along the western slopes, where non-ferrous metals are mined and refined in conjunction with cheap hydro-electricity. More limited working, mainly for non-metallic minerals, is carried on in the lesser known Sayan. The northern edge of the mountains contains the rich coalfield in the Tom valley, the Kuzbass, lying in an old embayment of a Carboniferous sea. To the south, the Gornaya Shoriya country supplies iron ore. These rich producers are to be associated in future with the potentially great mineral wealth and large resources of hydro-electric power found in the southern part of the Central Siberian Uplands. Non-ferrous metals are already produced in the Aldan-Vitim country, while coking coal is to be mined in the Chulman field of southern Yakutia. Transbaykalia and the Far East supply scarce minerals such as molybdenum, beryllium, zirconium, and, until recently, most Soviet tin output. The coal and petroleum of the Far East have an important strategic value.

The complicated mountain structures of Central Asia contain non-ferrous metals, coal, and petroleum. From them the Russians are believed to obtain their nuclear fuels. From Transcaucasia come manganese and molybdenum. Petroleum is found in the periphery of the Tertiary orogenic belt in Jurassic and Cretaceous sediments. Until recently they supplied the bulk of Soviet petroleum output, but some of the most prolific fields are beginning to show signs of exhaustion.

The comparatively young and undisturbed deposits overlying the Russian Platform contain large resources of coals, lignite, petroleum, and oil shales, which contribute greatly to the 'fuel balance' of the European regions. Particularly important are the petroleum reserves in deposits ranging from Devonian to Jurassic age in the Ural-Volga oilfields, the major Soviet producer. Many of the metallic mineral deposits, such as iron ore, are associated structurally with the older horsts standing above the level of the concealed platform at depths which can be reached by modern mining. The ancient Azov-Podolsk shield contains much of the Ukrainian mineral wealth (iron ore, manganese, etc.). Along its northern edge, an arm of a Carboniferous sea gave rise to the rich Donbass coalfield. Another great downfold fed by ancient rivers from the north is marked by the high-quality lignites of the Moscow Basin. In Karelia, the ancient Baltic shield supplies iron ore, nickel, and several non-metallic minerals, while more recently there has been development of the Vorkuta coalfield in the Pechora syncline where petroleum has also been worked.

PROBLEMS OF MINERAL WORKING

Even by the eighteenth century Russia was an important producer of salt and pig iron and there was growing interest in the precious-metal deposits of Siberia. Iron was mined in the Ural and smelted with charcoal from the extensive forests as was the bog iron ore of European Russia. Even in prehistoric times, the petroleum seepages of Baku had been used for oil and bitumen. Investment of foreign capital in Russia in the nineteenth century began the exploitation of these deposits on a greater scale, notably the rich coal and iron ore of the Ukraine and the great petroleum deposits of Baku; but before the First World War foreign money was already being spent on some of the more promising Siberian deposits. The Soviet period has brought development not only of the rich deposits already worked but also the drive to open up less attractive sites even in the most inhospitable parts of the country, in the search for rarer minerals needed in building an extensively self-sufficient economy.

Besides the technical problems of the geological nature of the deposit and petrological occurrence of the mineral, development has been directly influenced by the physical environment of the site. The Soviet authorities have shown a ruthless determination to work deposits of essential minerals irrespective of such site problems, though this has often meant high cost and fluctuating production.

Climate has generally been more important than terrain problems in developing most sites. In Central and North-eastern Siberia, intense winter cold reduces the efficiency of machinery, while ores with a high water content may freeze hard, making handling difficult. Drilling of shafts, erection of surface machinery, and the disposal of water can be extremely difficult in *permafrost*, which may make open-cast working, particularly working by dredger, hazardous. In arid Central Asia, water for settlements and processing, hard to find in adequate quantity and in the correct quality, has often to be brought by pipeline and aqueduct over considerable distances. Intense heat, like intense cold, can adversely affect the operation of machinery, while sand and dust storms are an added hazard.

In both Arctic and desert mining, provision of labour is a difficulty overcome by resort to direction and forced labour. For highly-qualified volunteer personnel there are also inducements of better pay and other 'fringe' benefits. Special organizations have been charged with development of the most inhospitable areas, such as the Kolyma Basin, included in the vast area of North-eastern Siberia under the control of *Dalstroy*. In the war-time development of the Taymyr Polymetal Combine at Norilsk and the Vorkuta coalfield, large numbers of German and Polish prisoners of war were used. In order to reduce the burden of transport, as much food as possible is produced at the mine sites. The Russians have been pioneers in developing 'Arctic agriculture' (Chapter 6) and irrigation schemes in desert areas. The use of fresh food, despite its very high cost, is an important element in maintaining health.

Transport (Chapter 9), constitutes one of the most significant problems in mining in inaccessible regions. In Siberia, river boats are used during the short summer navigation period, but in the winter motor lorries run along the frozen rivers. The Northern Sea Route is also used in summer to evacuate products and to bring in supplies and workers. Mines have been linked to rivers by roads or even by light railways, which in some instances also join them to the main railway system or to the coast. Long branch lines have been constructed to large producers from the main railway system; for example, the railways around Karaganda and the lines to Leninogorsk and Zyryanovsk in the Altay, as well as the long and costly branch line built to the Vorkuta coal-mines. There have also been scattered reports of the use of aeroplanes and helicopters. Transport costs may be offset by installing heavy and costly facilities at the mines to treat raw ore so that the payment of carriage on waste material is avoided. Nowadays there seems to be a move towards treatment at the mines, with modern methods of ore benefaction used in Siberia and Central Asia, where mechanization has been extensively applied in the last decade, but many of the remoter mines still appear to be only primitively equipped.

COAL RESOURCES [2]

Coal supplies about 46 per cent of Soviet energy requirements, although it is being steadily replaced by other sources. The

TABLE 16

OUTPUT OF COAL (INCLUDING LIGNITE) BY COALFIELDS OR GROUPS OF COALFIELDS
(Millions of metric tons)

Field	1913 *	% †	1940	% †	1958	% †
Donbass	25·3	87.0	94·6	56·8	181·7	36·5
Moscow	0·3	1·0	10·0	6·0	47·2	9·5
Kuzbass	0·8	2·6	23·7	14·3	75·3	15·2
Pechora	—	—	0·4	0·2	16·8	3·3
Ural	1·2	4·2	12·0	7·3	61·0	12·4
Karaganda	0·09	0·3	7·0	4·2	24·3	4·9
Ekibastuz ⎫ Central Asia ⎭	0·13	0·4	1·7	1·0	⎰ 6·1 ⎱ 7·7	1·3 1·6
E. Siberia ⎫ Far East ⎭	1·21	4·2	15·3	9·2	⎰ 36·1 ⎱ 20·0	7·3 4·0
Georgia	0·07	0·2	0·6	0·4	3·0	0·6
U.S.S.R.	29·1		165·7		495·8	
Of which hard coal	28·5	98·0 ‡	145·0	88·0 ‡	353·0	71·0 ‡

* Excluding the Polish fields in tsarist empire.
† Percentage of total Russian output.
‡ Percentage of total output of all types of coal.

Sources: *Narodnoye Khozyaystvo SSSR v 1958 godu.* Moscow, 1959.
  Oxford Regional Atlas of U.S.S.R. and Eastern Europe. Oxford, 1960 (reprint), p. 52.

Soviet Union is considered to possess about one-fifth of the world's coal reserves, sufficient for many decades at present or substantially increased consumption. About a sixth of the reserves are lignite. There is evidence, however, that the Russians have been working inferior coals either because the very best seams are already nearing exhaustion in some areas or because the development of new fields to supply local customers and reduce rail haulage has meant working poorer coals. This has been reflected in the shifting geographical distribution of mining (Table 16). In 1917, the Donbass was the main producer, but it has fallen relatively compared to the newer fields of the Kuzbass and Karaganda, although quantitatively its output is now much greater as new pits have been opened in the eastern part of the field. Since the late thirties, new but mostly small fields have been opened in Siberia (Fig. 37). The loss temporarily during the Second World War of the coking coal of the Donbass hastened development of the rich but inaccessible Pechora field in North-east European Russia. The proportion of output from the Moscow lignite field has also greatly increased. The Soviets developed mining concessions on Spitsbergen, first made about 1912, to supply shipping and industry in the Barents and Kara Seas. Coal has also been imported from Polish Silesia for use in the Baltic region and around Leningrad.

Coal is mainly used in industry as a boiler fuel or as metal-lurgical coke, while a large but falling share has been used by the railways. Lignite and poorer coals are widely used for electricity generation, while both lignite and coal provide chemical by-products in the process of coking and gas-making. Domestic fuel is still mostly firewood. Nowadays there is careful conservation of coking coals, whereas even in the late twenties they were being used indiscriminately as railway fuel. The use of higher ash-content coking coals from the Donbass and the Kuzbass and the use of very high ash coals from the Ural (Kizel) and Karaganda has been reflected in increased consumption of manganese to remove sulphur from blast furnace charges. Mixing of metal-lurgical coking coals to give a medium-quality coke is now widely practised. Recent availability of good coking coals from Pechora has eased the situation, though these are mainly used in North-west Russia, replacing both long-haul Donbass and Polish coals.

The first large-scale exploitation was in the Donbass, still the largest producer, lying in a broad depression in the Russian Platform, once the arm of a Carboniferous sea, though its southern edge has been uplifted against the edge of the Podolsk-Azov shield. Since the Second World War, the possibility of mining to both west and east of the conventional limits on to the concealed field has been discussed. Mining tends to gather on the anticlinal crests, where the shallowest shafts may be sunk. The steepness of the folds is greater towards the north, but the southern part is complicated by thrusts and overfolds. Some of

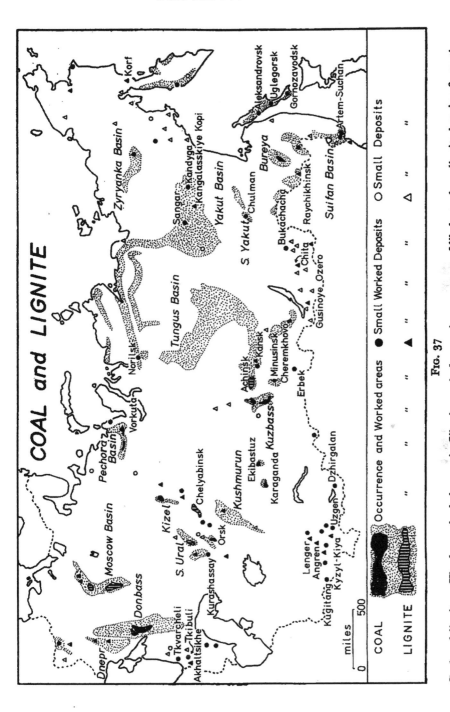

FIG. 37

Coal and lignite. The large shaded areas in Siberia mark the maximum extent of likely coal- or lignite-bearing formations, though these minerals are not expected to be found throughout these areas. Based primarily on Onika, D., *Ugolnaya Promyshlennost SSSR*, Moscow, 1958, fig. 23, and augmented from other Soviet sources for the smaller deposits.

the most intensive folding occurs in several brachy-anticlinal uplifts north of the main Rovenki-Nikitovka railway. Apart from gas, dust, and rock-fall hazards, the field is, however, comparatively simple to work, with generally shallow mines. There are about 200 coal partings, mostly in the middle and upper Carboniferous. Seams are usually about 1 foot 3 inches to 2 feet 5 inches thick, though a few are thicker than 6 feet. There are between thirty to forty workable seams (*i.e.* exceeding 1 foot 6 inches), but in a few places sixty seams may be worked. The ratio of the total thickness of coal seams to the total thickness of the coal-bearing formation is low (0·66 per cent).

The quality changes from north-west to south-east, with long-flame coal grading into anthracite, particularly around Shakhty. Only 5 per cent of the reserves are coking coal, much of which requires washing to remove excess ash and sulphur. The share of coking coal in the total output has tended to rise, but it has been achieved only by mining inferior grades. From the field comes 60 per cent of Soviet coking coal, consumed principally in the Ukraine, but large shipments of all coals are sent elsewhere in the U.S.S.R. as well as to East European countries. Mining began in the late eighteenth century : local coals were used at the Lugansk ironworks opened in 1797. It was not until railway construction in the 1870's that the field began to grow, though even at this time English coal was still cheaper in nearby Azov ports. At the end of the nineteenth century about half the output was used on the railways. Early capital investment was by France (about half), Britain, and Belgium.

The second coalfield, whose share of national output has steadily risen, is the Kuzbass, the principal supplier to West Siberian and Ural industry. Mining began in 1851, but development came only in the late 1890's with the building of the Trans-Siberian railway. In the first five-year plans, most Kuzbass coal was sent westwards to the Ural in exchange for iron ore. Much of the present output is still sent westwards to the Ural and even to the Volga region, while quantities are shipped to Transbaykalia and the Far East. The basin lies in a downfold between the Kuznetsk Ala Tau and the Salair Range, forming a broad but undulating lowland across which meanders the Tom. The coal measures were laid down in an arm of an Uralo-Siberian sea which gradually changed into a lake from lower Carboniferous times. The eighty-three workable seams have a total thickness of 547 feet, mostly in Permian series. The seams are very thick, some attaining 40 to 50 feet near Prokopyevsk. The most important seams are in the Kolchugino series, varying in thickness from 2 feet 5 inches to 4 feet 6 inches in the lower parts to 13 to 20 feet in the upper, easier to work by conventional methods than the very thick seams, where much unmined coal has often to be left, since to extract completely can substantially raise the unit cost. The seams are moderately folded and little faulted or broken, though around Anzhero-Sudzhensk reverse folds and imbricate structures are found. Kemerovo works coals of coking quality on the right bank of the Tom, which has given rise to a large coal by-products industry. Prokopyevsk mines lean and fat coals, with a considerable proportion cokable. Unfortunately the field here is strongly dislocated. Belovo, Ossinovka, and Leninsk-Kuznetskiy mine coals in the least disturbed part of the field. Coking coals are sent to cokeries at Novokuznetsk. A coal especially suited to distillation found at Barzas in the north-east of the field has been developed only since 1945, while bituminous schists found here produce synthetic petroleum. Adit mining in the south in the Tom-Usinskoye district has recently produced the cheapest coal.

The third field lies on extensive Carboniferous deposits in the heart of the Kazakh Uplands, with the main producing area around the town of Karaganda, though open-cast mining has started at Ekibastuz. The coals will coke but have a high ash content and are frequently mixed with richer Kuzbass coal for use in the Ural or Central Asia, while growing quantities are consumed locally. Poor location in an arid steppe is offset by cheapness of open-cast mining, which supplies a third of the output. The coal, chiefly lower Carboniferous or Mesozoic, comes from about thirty moderately inclined seams, many exceeding 25 feet. First worked before 1914 by an English company, the field was developed in the middle 1930's.

The newest field lies in North-east European Russia, in the Pechora Basin around Vorkuta and Inta. The mines, developed during the Second World War, send coal chiefly to North-west European Russia, while some is exported through Labytnangi up the Ob and Irtysh to Siberia. Despite the difficulty of sinking shafts and working the field under tundra conditions, individual mine output is greater than in the Donbass. The field, associated with the Ural and Timan ranges and deeply concealed beneath younger deposits, contains a wide range of coals, some of coking quality, though with a high ash content. In the Vorkuta series there are a hundred coal partings with a total thickness of 165 feet. Thirty seams exceed the minimum workable thickness of 2 feet. The degree of metamorphism decreases from west to east, where lignites are found.

A problem of using the immensely rich mineral wealth of the Ural, particularly iron ore, has been scarcity of good coal. Despite efforts to increase local output, the region still depends on imports from the Kuzbass or Karaganda. Fields occur on both the west and the east slope, but a large part of the reserves are in the Kizel field, known since the eighteenth century, which provides some coking coals, though these require to be mixed with better-grade imported coals to reduce the high sulphur content. Low-quality coals and lignites are used for electricity generation. The most important lignite field lies on the eastern slope around Korkino-Kopeysk, in a long *graben* extending a hundred miles south of Chelyabinsk. The seams are between 66 and 165 feet thick and exceed 500 feet in one place. The high-quality upper Triassic-upper Jurassic lignites are easy to mine open-cast, though shafts are used at Kopeysk. The field was started in 1910 near Chelyabinsk, but the principal centre at Korkino opened in 1935, and a wide range of coals is produced near Sverdlovsk. The Orsk-Khalilovo industrial district is connected with the working of the small deposit of good-quality (possibly coking) coal around Dombarovskiy.

The Moscow Basin lignite appears to be contemporaneous with Carboniferous bituminous coals of the Donbass. The deposits lie around the southern and western parts of the Moscow region, but the bulk of output comes from the south in Tula *oblast*. The lignite, with a higher than average calorific value, is used for electricity and gas generation, for the local chemicals industry, and has even been used on the railways. Coal brasses form a source of sulphur and liquid fuel is made from bog-head-type coals containing 12-20 per cent resin. Glass and refractory sands are found among the seams, which are comparatively undisturbed, about 6 feet thick, and lie at an average depth of between 150 and 200 feet, though they are deeper in the west and north. Unfortunately, there are many barren patches in the beds, and depth of mining is influenced by the water table. It is uneconomic to transport the lignite more than 60-100 miles from the mine. The principal mining centres are Novomoskovsk, Kaluga, Tovarkovo, Shchekino, Skopin, while underground gasification has begun near Tula. In the west, Safonovo and Selizharovo are shaft producers. Output from the northern mines at Borovichi moves to Leningrad. The field was first mined in 1855, but it was unimportant until after the First World War.

Several small fields are mined in European Russia, notably in the Ukraine where expansion of open-cast mining in the Aleksandriya field is planned. The Tkvarcheli and Tkibuli bituminous deposits in Transcaucasia are poor in quality, while lignite is found at Akhaltsikhe. Small deposits on the northern slope of the Great Caucasus are hard to work because of their structure.

In Central Asia, the western mines work Jurassic-Triassic lignites, while mines in the mountains around the Fergana valley produce varied types of coal. In the Uzgen field there is some coking coal, found also at Shargun and Kugitang, and at Dzhirgalan-Sovetskoye east of Issyk-Kul where transport is difficult. Much coal is used in electricity generation, as at Angren where underground gasification has been used. In Northern Kazakhstan, lignite mines are being developed at Kushmurun. Kurashassy coal is used at Aktyubinsk and Orsk, while Berchogur coal is used by the Aralsulfat chemical works.

Apart from the Kuzbass, there are many scattered fields in Siberia and the Far East. Soviet maps show large areas in Central and North-eastern Siberia where coal may be expected to occur. The Angara-Tunguska basin is worked near Norilsk for local smelters and electricity generation. Mining at Nordvik is used in salt extraction, and the small mines along the Lena supply river boats, settlements and Yakutsk. Bituminous coals come from Sangar and lignites from Kangalasskiye Kopi. The Zyryanka mines supply low ash and moisture lignite for the Kolyma nonferrous metal mines. The Northern Sea Route ships have used steam coal from Ugolnaya Bukhta and the poorer coal of Korf.

In Southern Siberia, the main field is Cheremkhovo-Irkutsk, where simple structure makes extraction among the cheapest in Russia. The field was opened in 1898 for railway use: it now supplies industry (including chemicals from coal by-products) and power generation. Coals of poorer heating value and higher sulphur content, mostly Permian, are mined open-cast at Chernogorsk in the Minusinsk Basin. Lignite comes from open-cast mines in the Yenisey and Kansk fields. Drift mines exist at Erbek in Tuva. Of eighteen coal and lignite occurrences east of Lake Baykal, eight are worked, generally near to railway lines, but climatic conditions hinder open-cast mining, while shaft mining is hampered by water infiltrating from *permafrost*. Chemicals are produced from the newly opened lignite field at Gusinoye Ozero. The planned metallurgical industry of Southern Siberia will depend on the Chulman coals, reported to be of excellent coking quality.

Strategical reasons governed mining development in the Far East: the first mines at Suchan served the Russian Far East fleet. The largest deposits are in a structurally complex and difficult field in the upper Bureya, to which a railway has been built. The coal will coke but has a high ash content. Along the Amur, anthracite is mined near Birobidzhan and lignite at Raychikhinsk, but as this disintegrates rapidly it is only used locally. Steam coals are produced at Suchan and Lipovskiy in the Suifan basin (suitable for distillation). Kraskino produces lignite. Three major deposits are found on Sakhalin, of which the western field is the main producer, with Tertiary coals metamorphosed into a type suited to coking. Good coals are also reported from the Nai-Nai river, but the island is one of the most expensive producers.

PETROLEUM, NATURAL GAS, AND OIL SHALE [3]

Although, in the early years of the century, Russia was the largest world producer, output was later overtaken by other countries. It seemed as though the Soviet Union would be unable to cover home requirements if petroleum were to be introduced on a large scale in industry and transport, but the rapid growth of production from the Ural-Volga fields has recast the position since the 1950's. There is now a growing use of oil and natural gas, as well as feverish development of petrochemicals. Petroleum supplies well over one third of Soviet energy and natural gas about 12 per cent. The geographical distribution of production is more restricted than coal, though large areas of the country contain geological structures associated with petroleum bearing (Fig. 38).

Modern development came in 1833 when petroleum was derived from deposits near the surface at Groznyy. With the inflow of foreign capital, commercial production really began in the 1890's, and by 1903 Russia was the world's largest producer, supplied overwhelmingly from the Baku field. After the Revolution, output in 1920 was only a quarter of 1901 and no

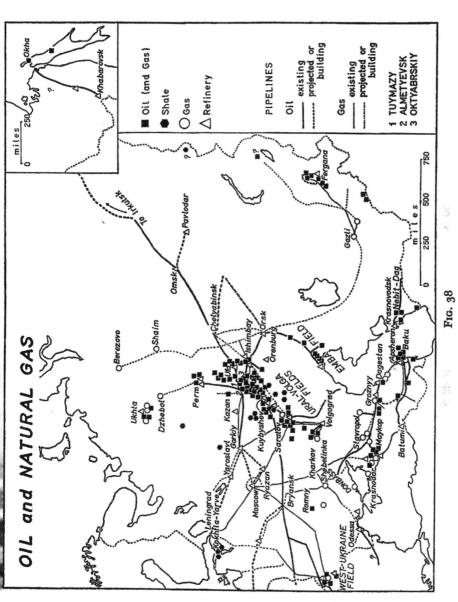

Fig. 38

Oil and natural gas. Much of Northern and North-eastern Siberia have been omitted to show the main fields at the largest possible scale, though these areas have small known deposits, some possibly worked. Based on sources listed in the bibliography for Chapter 7.

new fields were opened until 1929, with the discovery of oil at Chusovoy on the western flank of the Ural. By 1937, the Ural-Volga fields were beginning to develop, though they did not surpass Baku until 1953, and now supply almost three-quarters of Soviet output. After 1945, output increased by annexation of the Western Ukraine and the Japanese concessions in Sakhalin. Today, the Soviet Union is the second major world producer, with one-third of the United States' production.

TABLE 17

PETROLEUM OUTPUT IN MILLION METRIC TONS

Year	World Total	U.S.S.R.
1938	280·3	30·1
1951	550·0	42·0
1954	708·7	58·0
1958	938·3	113·5
1960	1,089·6	148·0
1965	1,500·0 est.	243·0 est.

Figures issued by Petroleum Information Bureau, London.

Technically, the Russian petroleum industry lagged behind the western world for a long period. Wells tended to be poorly drilled with outmoded equipment, frequently too shallow to reach the better horizons and there were too few exploratory and producing wells to get the best from a field. Output of necessary equipment and of barytes for preparing 'muds' was inadequate. Until the end of the Second World War, reliance was on natural pressure to operate wells, accounting for low recovery rates (15-25 per cent); but recently modern methods have been introduced, such as turbo and electric drills, repressuring, contour flooding, which improve the design of wells and the recovery rate.

The rapid change in the early 1950's in the geographical distribution of crude petroleum production was not accompanied by an equally rapid redistribution of refining capacity, producing an increased burden on transport of crude products to distant refineries. Attention has been given to the development of a new and well-equipped refining and petrochemicals industry in the Volga region and in Western Siberia, which is being linked to producers and consumers by pipeline, also well placed to cater for future development of recently discovered deposits in Northern Kazakhstan (notably in the Mangyshlak Peninsula) and in the

West Siberian lowlands. The Soviet Union includes about a quarter of the world's potentially petrol-bearing area and 45 per cent of its own area is composed of sedimentary formations, the greater proportion of which show petrological and structural features associated with petroleum bearing. The known and developed resources cover only a small part of this vast area: in 1946, 42 per cent of the reserves were in Azerbaydzhan, mostly in upper Tertiary strata, and some 30 per cent in the Ural-Volga fields, where they lay in Palaeozoic Permian-Devonian rocks. In 1956, exploration showed that over 80 per cent of the known reserves were in the Ural-Volga fields and only 10 per cent in Azerbaydzhan. The Palaeozoic rocks may also be the source of petroleum found in Jurassic salt domes in the Emba field, in the Northern Ukraine, and in Northern Siberia.

The distribution of oil and gas fields shows a striking preponderance in the western part of the country, so that Asiatic Russian producers are small and of local significance. The earliest field, and for a long time the most important, was the Apsheron Peninsula, where natural seepages ('eternal fires') were worshipped from early times and crude petroleum collected in pits was exported to Persia, India, and Mesopotamia. Development began in 1860, when distillation for illuminating oil started after an unsuccessful attempt to obtain photogen from local *kir* deposits. By 1863, Baku kerosene began to oust American from Russian markets and, despite wasteful exploitation and uncertain government policy, output mounted to a peak in the early nineteen-hundreds. From 1928 onwards the original producers were augmented by new wells, while the refining capacity of Baku and Batumi was extended and improved. Since 1945, possibly a quarter of output has come from offshore wells and complete settlements have been built on stilts several miles out in the Caspian.

The Baku fields are structurally part of the eastern termination of the Great Caucasus and the downfold of the Kura Depression. Occurrences in the Apsheron Peninsula are associated with some twenty complex anticlinal structures. Volcanic activity related to the structures is a continual threat to operations, notably at Lok Batan. The productive measures lie in Tertiary intercalated sands and clays between 4,000 and 5,000 feet thick. Light and heavy oils are found, the former mostly from diatomaceous beds. Water hazards operation in places, though by-products (bromine, iodine) are obtained from it. The presence of water is claimed to be the result of rapacious drilling in tsarist times. Petroleum is exported via the Caspian to the Volga, or sent by pipeline to Batumi for shipment to Ukrainian Black Sea ports or overseas.

Soviet initiative opened other producers in the Kura Depression. Neftechala (1926) and Pirsagat (1936) are important, while inland producers are Mirzaani and Naftalan (oil for medicinal purposes). Drilling is hampered by soft, crumbly rocks, complicated structures, and the aridity of the country.

Similar structures to Baku are found in the oilfields along the northern flanks of the Great Caucasus, where the principal fields are the Kuban and Groznyy-Dagestan. The oil is removed by pipeline to Black Sea and Caspian ports and to the Donbass. There are refineries at Tuapse, Krasnodar, Groznyy, and Makhachkala, which also refined Ural-Volga oil until capacity in the latter region was enlarged. More recently, natural gas from Stavropol has been piped to Central Russia. Development in the Groznyy fields began about 1893 and expanded rapidly. The fields are in sandstone reservoirs in Tertiary anticlinoria separated by broad depressions. After 1930, a similar field was developed at Malgobek and later in the Chechen autonomous district. The fields of the Dagestan littoral lie in two large Tertiary anticlines with dome-shaped structures, where the eastern fold has been most productive. At Izberbash and Achi-Su, overfaulted and fractured anticlinal domes are principal producers. The wells are shallow (900-2,200 feet) but slow producers. Seepages were worked after 1870 but the main production has come from Miocene sandstone since 1930. At Bereki, oil has been replaced by gushing, hot iodine-bromine water. Natural gas, found in the lowest horizons, is mostly exploited around Dagestanskiye Ogni.

The Kuban field, developed in the 1860's, draws oil from Oligocene and Miocene beds, while Pliocene beds yield oil in the Taman Peninsula and traces of oil and bitumen have been found in lower Cretaceous and Jurassic rocks. The reservoirs are usually in gentle monoclinal structures near Maykop, Apsheronskiy, and Khadiyy; but to the west oil lies in anticlinal folds. Light and heavy oil is worked, but in some fields 'edge' water is now a serious problem.

On the northern Caspian shore, the small Emba oilfield sends most of its high-quality crude by pipeline to the Orsk refinery for use in the Ural and Kazakhstan. The field developed about 1911 and expanded after the Revolution. During the Second World

War, a refinery and petrochemicals plant was built at Guryev. Local natural gas is used to fire electricity generators. The small scattered deposits are frequently associated with salt domes. Recently, further petroleum deposits have been discovered in North-west Kazakhstan.

The south-eastern shore of the Caspian contains the Nebit Dag and Cheleken deposits, worked before the Revolution. In 1959, interest shifted back to the coast with offshore drilling near Krasnovodsk. Cheleken lies on a faulted dome, but other workings are in anticlinal structures. Output is used locally or shipped away across the Caspian. The Krasnovodsk refinery and petrochemicals plant, using local and imported Baku crude, is linked to Nebit Dag by pipeline, which also carries refined products east to Ashkhabad. Like the Emba field, it is hampered by the hostile, arid environment.

The Central Asian oilfields are growing in importance, chiefly from twenty-one locations worked around the Fergana valley producing both oil and gas, used locally by railways and industry. Fergana is the main refinery. A small, expanding producer is near Termez, while a small field lies near Frunze. The fields are linked by short pipelines to consumers. At Gazli north of Bukhara, a large new gas field will be linked to Sverdlovsk by pipeline. Other indications of oil and gas have been found around Bukhara and large potential deposits are claimed from the Issyk-Kul region.

Siberia receives petroleum from western producers by rail or pipe, particularly the large Omsk refinery. New deposits have been discovered in the West Siberian lowlands: Berezovo and Shaim natural gas will be piped to the Ural, while oil is worked on a small scale between the Mulymya and Konda rivers. Oil showings are reported from Northern and Central Siberia, notably Nordvik, Maly Kheta, and Ust Voyampolka, but it is uncertain whether any are worked. Natural gas from Taas-Tumus at the Vilyuy mouth will be piped to Yakutsk. Sakhalin supplies the Far East and East Siberia from wells in the north-east of the island. Khabarovsk is the main refinery.

Russia annexed petroleum-bearing Polish territory on the northern slope of the Carpathians between Przemysl and Chernovtsy in 1945. Producing since 1860, these West Ukrainian fields had begun to show a slow exhaustion by 1918, for careful development was handicapped by their chequered political ownership. Borislav and Stryy are oil producers, but Dashava produces natural gas, now piped to Kiev, Moscow, Minsk, and Vilnyus, while it is also used locally in Lvov. Elsewhere, an extensive gas deposit at Shebelinka is used nearby in Kharkov.

Oil showings, known since the eighteenth century in Northeast European Russia, were developed only during the Second World War. The oil has a high pour-point and freezes at high temperatures. The first wells were on the Yarega river, but after 1942 development centred around Ukhta, while Voy-Vozh developed a carbon-black industry. Natural gas is also obtained near Voy-Vozh and Dzhebol. Ukhta refines oil, but most goes by rail to upper Volga refineries. The main producing horizons are in Devonian strata, and asphalt and bitumen are also found.

The Ural-Volga fields now form the most important oil-producing districts extending from the upper Kama around Perm to the lower Volga near Volgograd, lying between the west bank of the Volga and Ural foothills and covering some 190,000 square miles in which are found discontinuous oil-bearing structures. The petroleum is generally poorer than the Caucasian type, with a lower petrol fraction and high sulphur content, creating refining difficulties, but offset by a high gas fraction, until recently largely flared off. The bulk of production comes from Palaeozoic sediments overlying the crystalline platform and formed into broad anticlines. It has been generally necessary to drill deeper and through tougher materials than elsewhere. Over half the output comes from Devonian and Carboniferous strata. Oil was known in the region in the eighteenth century, but before the First World War lack of success in probing the overlying layers tended to divert attention to other fields. Oil was found in Permian rocks near Verkhne Chusovskiye Gorodki in 1929: in 1933, deeper drilling carried exploration into Carboniferous rocks, while after 1941 prolific Devonian strata were explored.

Increased demands for petroleum in the war and the difficulties of moving Baku and Turkmen oil to the front, coupled with the loss of North Caucasia to the Germans, added impetus to development of this field, so conveniently located in the heart of the country. Even in peacetime, the field lies fortunately at the centre of a triangle formed by the Moscow and the Donbass-Ukrainian industrial regions on the west and the Ural industries on the east. It is already well supplied with rail and water communication and can use the petroleum-handling facilities developed along the Volga to deal with Caucasian imports.

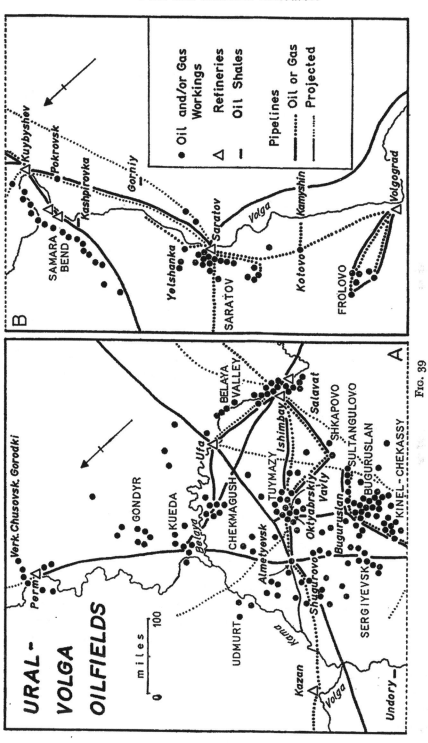

Fig. 39

Ural-Volga oilfields. Trofimuk, A., *Uralo-Povolzhye: novaya neftyanaya Baza SSSR*, Moscow, 1957, figs. 1-10, and augmented with later additions.

Containing 80 per cent of measured reserves, it has risen from producing 6 per cent in 1938 to 29 per cent in 1950 and to over 70 per cent in 1961, and about half the Soviet refining capacity is now in this region.

The most prolific areas around Almetyevsk in the Tatar A.S.S.R. and Oktyabrskiy in the Bashkir A.S.S.R. are the main centres: over half the Soviet output probably comes from within a radius of 150 miles of these centres, while the marginal fields around Perm in the north and Volgograd in the south are relatively limited producers. Along the Tatar-Bashkir border, the wells are deeply driven into Devonian and Carboniferous limestone and sandstone. In the Tatar A.S.S.R., the best wells draw from the rich reservoir of the Romashkinskiy anticline, first explored in 1934 but not opened until 1945 around Almetyevsk. During 1943–1944, the Shugorovo wells were developed, followed by centres such as Nurlat, Aktash (1950), and Aksubayevo (1944). Tatar oil costs a quarter of the Caucasian price. In Bashkiria, an oil gusher was struck near Tuymazy in 1937, coming from lower Carboniferous oölitic limestone; but since 1944, the Devonian strata have also produced. Oktyabrskiy, now a Soviet boom town, developed in 1948, and in 1953 the rich Devonian reservoir of Shkapovo was worked. Oil in the Belaya valley above Ufa was discovered in 1931 but had to await railway construction in 1935 before development. It is linked to refineries at Ufa and Orsk by pipeline, while Ishimbay and Salavat have refineries. The wells are shallow, drawing from porous, sandy Permian limestone and clays in broad, dome-shaped anticlines.

Anticlinal structures associated with the great Samara bend of the Volga were explored in the middle thirties and began to produce by 1937. In 1936, an oil of peculiarly low specific gravity was worked at Syzran and extended eastwards during 1938. Drilling westwards into Penza and Ulyanovsk *oblasts* began in 1951–1952, but the western boundary of the oil-bearing region is not yet known. Further south, west-bank producers have been mainly for gas. In 1942, the Yelshanka gas field was linked to Saratov electric power station, while in 1944 a pipeline was connected to Moscow. Kamyshin receives gas from Jurassic strata at Kotovo. Known since 1892 but opened in 1946, Frolovo sends gas to Volgograd. The occurrences are associated in the south with structural dislocations which form the Volga Heights.

Ural-Volga crude is refined at Volga refineries, like Gorkiy,

Kazan, Kuybyshev, Saratov, and Yaroslavl, which also handle Caucasian crudes, as well as new refineries near the fields, such as Ufa, Ishimbay and Salavat and Perm. Movement is by rail and by river, but also by a growing length of pipelines. The main emphasis has been on natural gas lines, but oil-carrying pipelines have been built to Omsk in Siberia, to Gorkiy, and will also extend to Moscow, and from the Ural-Volga fields to Eastern Europe and to a new oil-loading terminal at Ventspils. Oil shale reserves have been exploited only to a limited extent, though output has risen from half a million tons in 1937 to almost 12 million tons in 1957, supplying under one per cent of the nation's energy. The main worked deposits are around Lake Peipus at Kokhtla Yarve, Gdov, and Slantsy. Other deposits, not all worked, occur in the Ural-Volga oilfields (Gorniy, Undory, Kashpirovka, Ozinki) and in Orenburg *oblast*. They are also found on the Izhma river near Pechora, in Siberia and in Eastern Kazakhstan as well as recently discovered reserves in Kirgizia. The Peipus oil shales, intensively worked by Estonia after 1918, extend over 1,500 square miles with an average depth of 7 feet but may reach 33 feet. Oil content varies from 20 to 29 per cent. Distillation plants are at Kokhtla Yarve and Akhtme, with gas now piped to Leningrad and Tallin. Experimentally, the shales have been burned by railway locomotives. The Russians claim that working one million tons of Volga oil shale would yield 150 million cubic metres of high calorific gas, 40,000 tons of petroleum products, and sufficient semi-coke to generate 25,000 kwh. of electricity.

ELECTRIC POWER [4]

Tsarist Russia produced proportionally less electricity than other great powers. In 1913, generating capacity greater than 10,000 kw. was found only in Moscow, Leningrad, Kiev, Kharkov, Odessa, and Baku, all thermal plants, with 60 per cent generated by oil and the rest by coal. After the Revolution, a new start was made with the GOELRO, State Electrification Plan, envisaging large regional generating systems. Nowadays, about two-thirds of electric current is used in industry and mining, about a fifth by municipal and domestic consumers, and a twentieth by transport, though railway electrification will alter this pattern.

Russia possesses large resources of low-grade fuel suitable for thermal generation; for example, anthracite dust is used at

Slavyansk, Donbass, and waste from 150 sawmills is burned at Arkhangelsk and waste oil in Baku. Peat is consumed by specially built stations east of Moscow, while lignite is a common fuel and there is a growing use of oil and natural gas. Nuclear power has also been used. Thermal generators produce about 80 per cent of the current; but high-grade and long-haul fuel contributes only 10 per cent, and the Russians claim they can build a grate to burn fuel with up to 55 per cent moisture and 60 per cent ash and other undesirables.

While more costly and slower to build, hydro-electric stations produce more cheaply and do not involve long hauls of bulky fuels. At present, they supply about a fifth of Soviet electricity and use little more than a tenth of the claimed potential of 300 million kw. (annually giving 2,700 milliard kwh.) of the 1,500 major rivers. The basins of Asiatic Russia account for 81 per cent of this capacity (Lena and Yenisey 41·8 per cent, Amur 10·1 per cent, Ob 7·3 per cent, Central Asia and Kazakhstan 21·8 per cent), but the bulk of the potential may never be harnessed because of remoteness and immense physical difficulties. The wide, shallow valleys of the Russian plains with their soft substrata mean low head stations operating on long, low barrages impounding vast but shallow reservoirs that often involve resettlement from the inundated area. Resultant slight changes in the water table may have wide repercussions on agriculture and local fauna and flora. Few schemes have been conceived solely for power but are more commonly linked to plans for irrigation, water conservation, or navigation. Sites at narrow rocky gorges, mostly in Transcaucasia, Central Siberia, or mountainous Central Asia, pose fewer problems but generally offer less generating potential. Although electricity stations are widely scattered, several major groupings appear near important industrial areas, each containing various types of generators with differing technical characteristics and linked to local grids. Because of the difficulties of transmitting current over great distances, stations lie adjacent to consumers, and no national grid has yet been developed, though some inter-regional lines, such as between the Volga hydro-electric stations and Moscow, do exist. Such connections are important in adjusting fluctuating seasonal output from hydro-electric stations, while a national grid might make adjustments for inter-regional variations in daily load patterns.

Power for the varied manufacturing industries of the Central

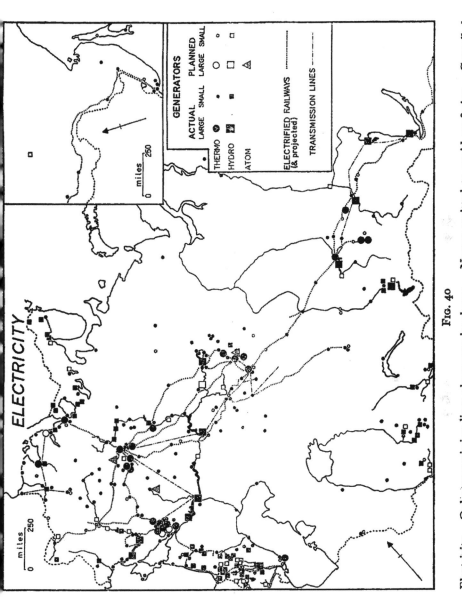

FIG. 40

Electricity. Only transmission lines above 300 kv shown. Names omitted to avoid confusion. Compiled from sources listed in bibliography to Chapter 7. Electrification of railways taken from Nikolskiy, I. V., *Geografiya Transporta SSSR*, Moscow, 1960, and Slezak, J. O., *Breite Spur und weite Strecken*, Berlin, 1963.

Industrial region comes from both thermal and hydro-electric generators. Moscow city has two main stations which have turned from Donbass coal to piped oil and gas. South of the capital, large lignite-burning stations lie at Kashira, Novo-moskovsk, Shchekino, Cherepets, and Safonovo, while a gas-turbine generator using underground gasification of lignite works near Tula. Some peat is used south and west of Moscow, but the main stations lie to the north and east, notably Shatura, Ivanovo-Komsomolsk, Yaroslavl, and near Gorkiy. The hydro-electric stations are smaller units of the Great Volga project: Ivankovo, Uglich, and the 33-Mw. Rybinsk station (including the large reservoir). Gorkiy, besides thermal generators, has been served by the 400-Mw. Gorodets plant since 1956.

To reduce long hauls of fuel, Leningrad uses peat and shale-oil gas, notably at the Dubrovka and Red October stations. The 56-Mw. Volkhov hydro-electric station, completed in 1926, was one of the earliest in Russia and serves the southern shore of Lake Ladoga, while former Finnish stations at Svetogorsk and Raukhiala lie on the north shore. A hydro-electric barrage is to be built at the Falls of Narva.

The Ukraine is served by stations along the Dnepr, on the Donbass coalfield, and in most big towns. The Donbass stations are large and designed mostly to burn waste fuel such as dust and low-grade coal. Most important are Donetsk and nearby Kurakhov and Zugres (Zuyevsk); Shtergres (Shterovsk); the North Donets plant at Lisichansk, and at Mironovo and Slavyansk, completed since 1945. Power is also taken from Rostov-na-Donu. The electro-metallurgical industry of the Dnepr bend is supplied by thermal stations at Dnepropetrovsk and Dneprodzerzhinsk and by the large hydro-electric plant of Dneproges (near Zaporozhye) at the southern end of the rapids where the Dnepr crosses the hard crystalline shield, begun in 1932. One of the earliest major Soviet projects, owing much to American engineers, it was blown up in 1941 before the advancing Germans and reconstructed after the war to an enlarged capacity of over 600 Mw. A 312-Mw. barrage was completed in 1955 lower down the river at Kakhovka; lying in the flood plain in the arid Black Sea littoral, it also supplies water for irrigation as well as aiding navigation. Other schemes include Kremenchug (625 Mw.) — already complete — and a 250-Mw. station at Dneprodzerdzhinsk, as well as Kanev.

Fuel and Mineral Resources

The Ural depends largely on thermal stations using local or imported fuel, as few streams are suitable for hydro-electric stations. Current is also imported from adjacent regions. The industries of the eastern slope are served by a group of thermal stations near the Chelyabinsk lignite field, and coals around Sverdlovsk power other stations. All big towns, like Nizhniy Tagil, Serov, and Magnitogorsk, have their own stations. The Kizel coalfield on the west has large generators, while oil-burning stations lie in the Belaya valley. On the west, hydro-electric stations lie along the Kama river, but on the east, Alapayevsk is the only hydro-electric plant. A regional grid exists.

The small stations in the headstreams of the Great Volga hydro-electric cascade were completed before the war, while Gorkiy (Gorodets) was finished after 1945. On the Kama, Perm was also completed in 1954 and Votkinsk is advanced. A station on the Ufa near Krasnyy Klyuch is projected. On the Volga, the 2,100-Mw. Kuybyshev station, standing on comparatively hard substrata in the Samara bend, sends 60 per cent of its current to the Moscow region. The 2,500-Mw. Volgograd barrage, where the Volga flows into its braided lower course, will supply 40 per cent of its current to Central Russia and 30 per cent locally. The associated project of the Volga-Don Canal and irrigation works added the 164-Mw. Tsimlyansk station in 1954. Two more projected stations, each over 1,000-Mw. capacity, are planned at Cheboksary and near Astrakhan, while work has also begun near Saratov. The large reservoirs formed behind the barrages (mostly over 1,000 square miles in area) will help to regulate the flow of the river, provide water for irrigation and aid navigation. A scheme is also planned to divert water from the northern rivers, Vychegda and Pechora, into the Volga. The new landscape of the Volga valley has required considerable resettlement and adjustment to communications, while it is already reported to have affected (not always beneficially) fauna and flora.

Until the Second World War, electricity in Siberia was supplied mainly by thermal generators in large southern towns and in a few isolated mining centres, but railway electrification and industrial development demand increasing supplies. The main concentration of thermal stations is in the Kuzbass, at Novokuznetsk, Kemerovo, Belovo, and the 400-Mw. Yuzhkuzbassgres at Kaltan. To these will be added the 1,300-Mw. Tom-Usinskoye station and the large Nazarovo plant, augmenting supplies from

Achinsk, Kansk, and Itat. The coal-fired Irkutsk and Cheremkhovo plants have been supplemented by the 600-Mw. barrage
near Irkutsk on the Angara, one of the fastest-flowing Siberian
rivers which shows little seasonal fluctuation in volume since it is
fed by Lake Baykal acting as a compensating reservoir. The
Angara flows through a series of rocky gorges providing excellent
sites for barrages, though the rapids at these points hinder navigation. A 'cascade' of six stations is planned, of which the 4,500-
Mw. Bratsk station at the Padun rapids is already being built
and plans exist for the Ust-Ilim station. It will ultimately be
possible to sail from Lake Baykal to the Yenisey, on which a large
barrage is being built at the Shumikha gorge, 25 miles south
of Krasnoyarsk, where the river is only 850 yards wide. On the
Ob, a 400-Mw. station has been built at a barrage 20 miles
above Novosibirsk, and a second station at Kamen (630 Mw.)
will prevent spring flooding and provide summer irrigation water
in the Kulunda Steppe. The Irtysh stations, serving the metal-
refining industries of the Gorno-Altay district, include the 322-Mw.
Ust-Kamenogorsk station and the partly completed Ust-Bukhtarma
barrage which will greatly extend the area of Lake Zaysan.

Construction of these stations presents numerous technical
problems. Most sites are remote and need branch railways to
be built to transport cement and constructional materials;
settlements for workers have to be built; electricity has to be
generated locally by diesel generators or brought by overland
lines; at some places, for example at Bratsk, *permafrost* has
created constructional difficulties, while there are problems of
intense winter cold and freezing of the river, besides the danger
in spring of sudden thaw. As most reservoirs flood forested land,
it is necessary to clear away the trees before the dam fills, otherwise floating timber might jam the turbines.

The rivers flowing swiftly from the Central Asian mountains
provide many suitable sites for hydro-electric stations, while the
dams created store irrigation water, though they need to be
larger than in Siberia in relation to the turbine size in order to
hold sufficient for the low-water season. The largest (660 Mw.)
hydro-electric plant at the Farkhad dam on the Syr Darya
supplies current to the Begovat steelworks. Near Tashkent, two
large stations on the Chirchik power the local electro-chemicals
industry, and in the Fergana valley the Shaarikhan station, one
of a series planned, is already operating. In Kirgizia, a 112-Mw.

station is being built on the Naryn, besides the Orto-Tokoy station at the Boom gorge on the Chu, 8,775 feet above sea level, while there are smaller stations along the river. In Tadzhikistan, a number of low-capacity diesel generators were augmented by hydro-stations on the Varzob in 1937 and at Khorog in 1949. Most large towns and industrial centres also have thermal generators, using oil, lignite, or coal, with underground gasification at Angren. In Turkmenia, diesel generators are the main suppliers (*e.g.* Ashkhabad, Mary, Krasnovodsk), though a small hydro-electric plant was built on the Murgab at Gindukush in 1917. In Central Kazakhstan, the Karaganda coalfield is served by the large Kargres plant.

Petroleum and water power are the main generators in Transcaucasia. The fast-flowing rivers provide admirable sites for cascades of stations, like the Rioni. A part of the Khrami cascade is also complete, while the Samgori irrigation scheme on the Iori and Kura generates power. Oil is used in Baku, but now power is also drawn from the Mingechaur barrage. In Armenia, six of eight stations planned for the Razdan are already complete, where the steady fall of 440 yards in 90 miles is used for the head of the stations, from which water is distributed to irrigation channels.

Russia has actively developed nuclear fuels, but until the early 1950's appeared to depend on raw materials from satellites: from the Czech-German borderland at Jachymov and Aue, from Hungary, and possibly also Romania.[5] The first Russian workings were at the extensive but low-grade Central Asian deposits at Tyuya-Muyun, Maylisay, and the Kara Tau. Small but good-quality betafite deposits exist in Transbaykalia, possibly at Slyudyanka, and in the Aldan shield, but thorium deposits are limited and there are no significant monazite sands, though they are known in Krasnoyarsk *kray* at Tarak and Kazachinskoye. There are well-endowed deposits of beryllium and graphite for reactor coolants. As world production shifts into poorer deposits, Russia's initial disadvantage may be reduced, so that it may be worth mining phosphate rocks with 0·01 per cent U_3O_8 content.

The main research centre is at Dubna near Moscow, while another centre is in Siberia, possibly near Novosibirsk. It is thought the desert between the Irtysh and Lake Balkhash as well as the Soviet Arctic may have been the site of test explosions. Russia claims the first atomic power station (5 Mw.), opened in 1954 at Obninsk near Kaluga. Atom-powered electricity generators are reported from the Volga at Ulyanovsk and from a 420-Mw. station at Voronezh, while a 900-Mw. station has been built at Beloyarsk near Sverdlovsk: rumours of several generators in Siberia are unconfirmed. The atom-powered ice-breaker *Lenin* has already proved itself on the Northern Sea Route. With such large resources of water power and cheap fuel minerals, the development of atom power for electricity genera-tion is possibly less pressing in Russia than in the United Kingdom.

RESOURCES AND DISTRIBUTION OF NON-FERROUS METALS [6]

Widely scattered non-ferrous metal deposits are associated par-ticularly with the older crystalline basement rocks or the remains of the older mountain systems. As a major producer, the Hercynian Ural structures stand out, while other important regions are seen in the Kazakh Uplands and the older rocks of the complex Altay-Sayan. Central Siberia, the Transbaykal region, and mountains of North-eastern Siberia are rich in non-ferrous minerals, which are also found in the varied but com-plicated mountains of Central Asia, notable suppliers of rarer minerals, and some come from the ancient basal rocks of the Russian Platform and the Baltic and Podolsk-Azov shields (Fig. 41). Many deposits contain ore matrices with two or more metals. There are indications that poorer and remoter deposits in difficult environmental conditions are having to be increasingly worked in order to maintain a high level of self-sufficiency.

Copper is no longer usually imported on a large scale, as development of deposits in the eastern regions has eased the internal supply position. The earliest producers were the Ural mines, notably Karabash and Baymak; but during the inter-war period, Kazakh output rose, particularly at Kounradskiy and Dzhezkazgan, claimed to be the second largest deposit in the world. Ural output has also increased from the large Degtyarsk mine and at Gay near Orenburg. During the Second World War, mining in the Taymyr Peninsula was stepped up, while Almalyk in Central Asia was worked. There are also several small and medium-sized producers in the Altay country. Most deposits are poor in metal, while flagging technical standards led to poor recovery rates. Smelting is usually done near the mines.

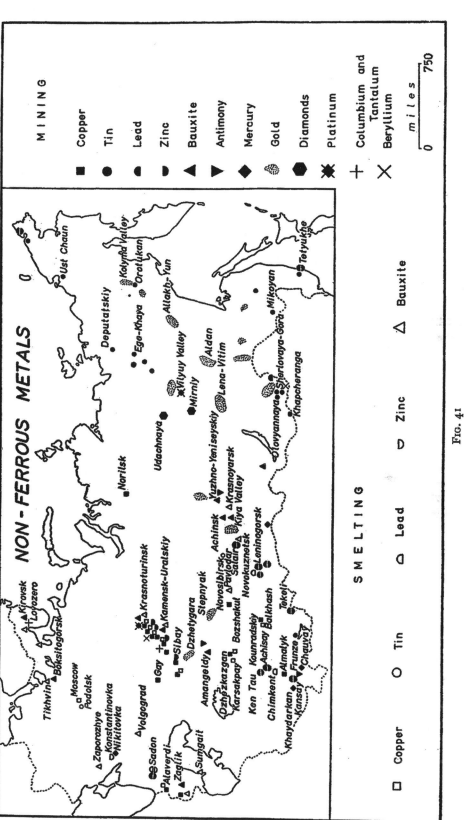

FIG. 41

Non-ferrous metals. It is unlikely that all deposits shown are worked. Only main smelters shown. Sources listed in bibliography to Chapter 7.

Reserves are primarily in the copper content of mines producing other metals. Old tailings may also contain recoverable quantities, but perhaps a third of present consumption comes from scrap. Russian attempts to economize in the use of copper are interpreted as a sign of the unsatisfactory future supply position.

Lead and zinc frequently occur together. Before 1941, home needs were covered by production, but heavy demand for lead during the war may have exhausted the principal mines at Achisay and Kolyvan. It is not certain whether new workings have offset this loss. The most important producers are the Ural and the medium-sized mines of the Altay, especially near Leninogorsk. A newer area has been opened in the Ken-Tau in Central Asia, while significant producers lie at Sadon in Caucasia and Tekeli in Kazakhstan. The importance of the Tetyukhe mines in the Sikhote Alin has risen and the ore is sent by light railway to a smelter at Tetyukhe-Pristan before shipment by sea. Except the Ural deposits, all contain lead and zinc, while several also contain cadmium. Smelting is done locally, though the largest works appear to be Chimkent and Leninogorsk, where zinc output was increased by erection of a smelter from German reparations.

Tin reserves have been increased by prospecting in North-eastern Siberia. Production began in 1933 at Olovyannaya, Khapcheranga, and Sherlovaya Gora in Transbaykalia. Later, between 1938 and 1941, mining and smelting developed in North-eastern Siberia at Ege-Khaya, Vankarem, Endybalsk, Deputatskiy, and Imtondzha. Mines were also opened in Eastern Kazakhstan as well as in the Sikhote Alin in the Far East. Two smelters away from deposits are Podolsk near Moscow (using imported materials) and Novosibirsk. Some North-eastern Siberian deposits and the workings at Mikoyan in the Khingan mountains are placers from weathered igneous rocks and several are claimed to be of high quality.

Aluminium is produced by complicated electrolytic processes from clayey bauxite. Production, begun in the early five-year plans, suffered heavily during the war because of its concentration in western European Russia. From 1932, the main supplier was a smelter using Volkhov electricity and ore from Tikhvin-Boksitogorsk, but later smaller and richer deposits were worked in the northern Ural, near Krasnaya Shapochka and Kamensk, while more southerly mines also opened. Since 1945, Jurassic lacustrine bauxite from Northern Kazakhstan at Amangeldy has been refined at Pavlodar. Deposits of nepheline near Achinsk will be used at Krasnoyarsk and near Irkutsk. Bauxite found at Yuzhno-Yeniseyskiy and near Angarsk was supplied to a wartime factory at Novokuznetsk. Available hydro-electric power and large nepheline and similar deposits in the Kola Peninsula are a potential source. In Caucasia, Sumgait uses Zaglik bauxite, and cryolite for all plants is synthesized at Polevskoy in the Ural.

Platinum ores, principally in the Ural, once made Russia the world's largest producer, but these dunite masses may be nearing exhaustion. Norilsk has important but lean ores, and an unworked body is reported from the Vilyuy river in Central Siberia, while Ural and Yakut gold placers also contain platinum and until recently were probably a chief source. Derivation may now be from by-products of other non-ferrous metals. Antimony comes mainly from Central Asia, where there is local smelting, from Turgay in Kazakhstan and Razdolinsk in Siberia. Small deposits are known in the Amur valley. It is also produced as a secondary metal at the Ukrainian Nikitovka mercury mines, now difficult to work in the lower horizons. During the German occupation, Khaydarkan and Chauvay in Central Asia became important for mercury. Russia is a major world producer of antimony and a moderate producer of mercury. Columbium and tantalum resources appear to be good. Lovozero in the Kola Peninsula is claimed to be one of the world's largest deposits. Zirconium is found in small deposits in the Azov coast and the Kola Peninsula, but a large though little-known deposit is reported from the Far East Zeya valley. Izumrud emerald mines are the main source of beryllium, but it also occurs in Eastern Siberia at Sherlovaya Gora and in scattered deposits elsewhere.

Commercial gold-mining began in the Ural in 1814 and was later followed by mining in Siberia. Most important are the placers of the Lena-Vitim and Aldan fields worked since the late nineteenth century and now highly mechanized. Lodes and placers are also worked in Transbaykalia. Fields occur in the Ural and around Dzhetygara and Stepnyak in Kazakhstan. Deposits are worked on the east bank of the Yenisey. In Eastern Siberia, the Allakh Yun field was developed during the war years. Silver, associated mostly with polymetallic ores as a secondary metal, comes from main producers in the southern Ural and the Altay.

Diamonds were discovered in 1829 in the Ural. In 1954,

large deposits in kimberlite pipes were discovered in Central Siberia, where several score are now known, but the most important are Mirniy, Udachnaya, and Aykhal on the Sokhsolookh river. There are also placer workings. The stones, generally small but with gems up to 32 carats, are graded and sorted at Nyurba. Amber is worked in open-cast pits on the East Prussian coast, chiefly at Yantarnyy. Precious stones come from the Ural and Altay-Sayan.

RESOURCES AND DISTRIBUTION OF IRON ORE AND ALLOY METALS [7]

Iron ore is widely scattered over the country, but the major producers are in the Ural and the Ukraine (Fig. 42). Their output is proportionally larger than their reserves, so that a shift in the near future is expected, particularly into Kazakhstan, already a small producer with large reserves and well located to serve Ural, West Siberian, and Karaganda industry. The Kuzbass is also favoured by the substantial deposits in the south of the Central Siberian Uplands. During the Second World War the Asiatic deposits, including the Ural, were particularly important. Most current production comes from magnetites (39 per cent) and haematites (38 per cent), though half the reserves are in marginally useful bog ores (15 per cent of current output) and ferruginous quartzites.

TABLE 18

OUTPUT OF IRON ORE

	1940	1945	1950	1955
European Russia	70·4	25·9	55·5	59·8
Ural	27·9	68·9	39·0	34·8
Siberia and Kazakhstan	1·7	5·2	5·5	5·4
Total	100·0	100·0	100·0	100·0

Source: *Zhelezorudnaya Baza Chernoy Metallurgii SSSR*, Moscow, 1957.

The Ukrainian Krivoy Rog deposits, the product of Pre-Cambrian metamorphism, are the largest single producer. They contain over three-quarters of Ukrainian reserves and lie in a field about 80 miles long and 2-3 miles wide in the upper reaches

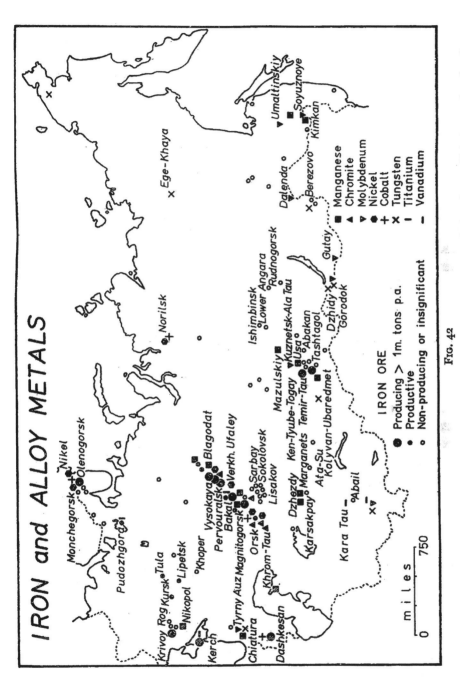

FIG. 42

Iron and alloy metals. Several iron deposits in Southern Siberia are not yet worked and are still being explored. Some remoter deposits of alloy metals may not be actively worked. Sources as listed in bibliography to Chapter 7.

of the Ingulets river. The ore is taken from both open-cast pits and shaft mines, but deep working is increasing in order to reach beds lying deep in the sharply compressed synclinal structure. The best ores contain over 60 per cent iron, but between 40 and 50 per cent is now most common. In 1955, about 37 per cent of the 40 million tons produced came from the northern suburbs of Krivoy Rog. Most ore is sent by rail to Donbass smelters. As poorer qualities are worked, dressing is more commonly used. A small producer of poorer quality but inexpensive to work is the Kerch Peninsula, where the ultimate iron content of the ores probably now exceeds Krivoy Rog. Lying in synclinal structures on the northern and eastern side of the peninsula, the watery and powdery phosphoric ores are among the few in Russia suited to the Thomas process. The main centre is on the north coast where ores containing 39 per cent iron are mined at Kamysh-Burun.

Central European Russia contains several small deposits and the large potential reserves of the sideritic ores and iron quartzites of the Kursk Magnetic Anomaly, explored in the 1930's but only worked since 1950, mostly by open-cast methods. The ores average 33 per cent iron but it is intended to concentrate them before shipment. The mines near Tula and Lipetsk, where bog ores are underlain in places by lignite, provide raw materials for local industry. The Khoper Valley ores near Balashov are not worked because of the very high phosphorus content.

Growth of heavy industry near Leningrad and shipment of ore to Poland and Central Germany has increased interest in Karelian and Kola ore, where some deposits were worked by Peter the Great. In the Kola Peninsula, magnetites and titano-magnetites are known at Yena, Kovdor, Olenegorsk, and at Afrikanda near the main railway to the south, where open-cast mining began in 1951. Most useful are the magnetites with a titanium-vanadium content and about 28 per cent iron, at Pudozhgora on the eastern shore of Lake Onega and in central Karelia at Mezhozero and Kostamuksh.

Famous since the eighteenth century, the rich Ural iron deposits, some containing chrome or nickel in natural alloy or in close proximity to alloy metal deposits, have been widely developed to feed the region's heavy industry. In 1955, three-quarters of the 34·8 million tons of ore mined were magnetites or martites. Almost half the output was from Magnitnaya Gora, a 'mountain of iron', although the best ores in the upper layers are largely exhausted, worked in a series of terraces on the western face. Output is expected to decline in the next decade and the Magnitogorsk steelworks will rely more on Kazakh Sokolovo-Sarbay ores. About an eighth came from Bakal, consisting of five major deposits from which ore is concentrated before shipment, and output will increase as railway facilities improve. A little less came from Vysokaya Gora, the main producer in the Kachkan group, linked to the main lines by narrow-gauge railways, which sends its ore to Nizhniy Tagil and Serov. Blagodat mines (also in the Kachkan group) are open-cast. Just north-east of Nizhniy Tagil, Lebyash uses shafts but the extra cost is offset by its nearness to the consumer. It has been worked for 200 years and the ore produces a good vanadium pig. One of the oldest mining districts is Pervouralsk-Alapayevsk, whose best ores contain up to 58 per cent iron. The most recent deposit to be worked is Orsk-Khalilovo, where ore contains chrome, nickel, and cobalt, which supplies the new Novo-Troitsk works.

In the early five-year plans, availability of ore was regarded as a serious bottleneck in the development of Western Siberian heavy industry, but intensive prospecting has revealed deposits sufficient for local needs and future expansion. Until recently the main source has been the Altay and Kuznetsk Ala-Tau, notably Gornaya Shoriya. Most of the latter ores contain lead and zinc and up to 40 per cent iron. The biggest mines are at Tashtagol, opened in 1941, but expanded rapidly since 1950. With completion of the Novokuznetsk-Abakan railway, easy movement of Abakan ores to the Kuzbass is possible. Detailed knowledge of the ores of the Angara-Lower Tunguska is not yet complete, but they will be used to expand Siberian heavy industry. Ores in the south-east of Chita *oblast* and in southern Yakutia may be used to serve a projected steelworks east of Lake Baykal.

Several deposits surveyed since 1945 in Kazakhstan will serve the Ural region. Some, like the Kustanay group, are said to be potentially among the greatest in Russia, though the ores are not particularly rich. The incompletely explored Sokolovo-Sarbay, Kachar, and Kurzhunkul ores have been worked already and Ayat has rich reserves of brown haematite. The Lisakov ores need complicated enrichment but are suitable for the Thomas process, so that a slag rich in phosphorus will be produced as a by-product. They are well served by railways. The Ata-Su ores, largely haematites containing 53-60 per cent iron and large amounts of manganese, supply the new Temir-Tau steelworks near Karaganda. Some ores in the Ken-Tyube-Togay deposit contain cobalt.

Limited reserves exist in the Far East, mostly in the northern Little Khingan and the Zeya-Selendzha region. The Khingan ores (30-36 per cent iron) go to Komsomolsk steelworks.

The widest range of alloy metal deposits is found in the Ural, while Caucasia has molybdenum and manganese. Central Asia has little iron but vanadium and tungsten and small deposits of molybdenum. The tungsten and manganese deposits of the Altay are possibly too small for Western Siberian needs, while European Russia is deficient in alloy metals apart from manganese. Karelia has some titanium and the Kola Peninsula contains nickel and cobalt, which are also worked under arduous conditions in Northern Siberia. Small molybdenum and tungsten mines are worked in Eastern Siberia despite inaccessibility.

Russia is one of the world's largest manganese producers, mostly in Nikopol in the Ukraine and Chiatura in Transcaucasia, though the latter is harder to work because of deep over-burden. Until the Second World War, Ural deposits, principally at Polunochnoye in the north and Uchaly in the south, and Kazakh deposits at Marganets and Dzhezdy were not intensively worked. In Western Siberia, the Usa river ores are in wild country recently opened by railway construction, but the Mazulskiy mine lies near the Trans-Siberian railway. It is uncertain whether the Turochak deposit is worked.

Chrome comes principally from the Ural and from the Khrom-Tau-Khalilovo-Kempirsay group, on the borders of the Ural region and Kazakhstan, reputedly one of the largest in the world. Many chrome deposits are only suited to refractory use.

Attempts have been made to extract cobalt as a by-product of other ores. Until 1940, output appears to have been limited to the Caucasian Dashkesan mine. Principal producers are now Khalilovo and Verkhne Ufaley (Ural), the Kola Peninsula (associated with nickel and augmented by post-1945 territorial annexations), and Norilsk. Tuva also produces some. Prospecting appears to be mostly in North-eastern Siberia.

About a third of nickel reserves lie in the Monchegorsk-Pechenga area of the Kola Peninsula. Additional sources may be revealed by study of ultrabasic intrusions at Norilsk, in the Ural,

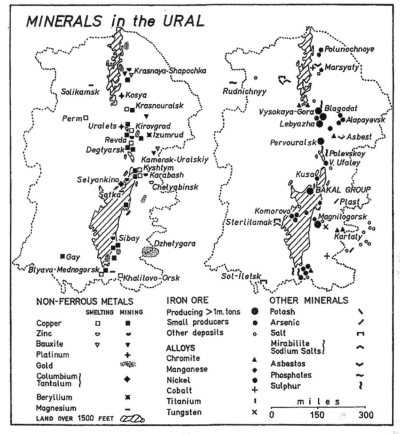

FIG. 43

Minerals in the Ural. Compiled from sources used in Figs. 41, 42, and 44. This map augments these figures where scale demanded excessive generalization of the Ural area.

and in Tuva. Orsk-Khalilovo is probably the main source at present.

Tungsten is imported chiefly from China. Two wolfram mines operated near Chita before 1917, though Gorodok may be exhausted. In the Buryat republic, it is found on the Dzhidy river. Prospecting may augment the Transbaykalian supplies. Small mines also lie in Central Asia and in the Altay. During the war, important discoveries were made in North-eastern Siberia at Ege-Khaya and in the Chukchi Peninsula, but both are difficult to work under rigorous physical conditions and remoteness. The main supplier is Tyrny Auz in the Great Caucasus.

Molybdenum comes from Tyrny Auz (three-quarters of the reserves) and from Umaltinskiy in the Far East. Large potential sources are the copper ores of Lake Balkhash as well as similar deposits in Transcaucasia, though it is possible that these are not suited to the preparation of metallic molybdenum. Molybdenum is also reported from Gorodok and the eastern Kuznetsk Ala-Tau. Smallness of the Central Asian deposits makes them difficult to work. Titanium is mainly from naturally alloyed pig from Kusa, Ural, for long possibly the only worked deposit. Pudozhgora in Karelia is also now used. Vanadium came chiefly from blast-furnace slag from the Kerch Peninsula, but it is derived from the Kara Tau deposit in Kazakhstan, where it occurs with mica.

MINERALS USED IN THE CHEMICALS INDUSTRY

These minerals form an exceptionally diverse group and include by-products of other branches of mining. Russia is fortunate in having a good resource base for heavy chemicals in sulphur, derived mainly from pyrites and other sulphur-bearing minerals. Native sulphur from Kuybyshev on the Volga and from Gaurdak and Shorsu in Central Asia is only a secondary source. An important source is from by-products of copper tailings at Blyava, Degtyarsk, and Polevskoy in the Ural or from coal brasses in several areas. Pyrites are known in Siberia, but the small, lean deposits are not worked.

Common salt (sodium chloride) is available in large quantities; but Siberia and Northern European Russia are the only regions with a deficiency — no significant deposits are known east of Lake Baykal. In the arid lower Volga and Central Asia, large numbers of seasonal salt pans, replenished annually, are a potential source, but few are pure enough for human consumption. The main producers of this type are Lakes Elton and Baskunchak: the latter is fed by streams, each moist spring, and usable salt is found in the top 20-30 feet. Lake Tavolzhan near Pavlodar is also used, but it is heavily contaminated with magnesium salts. In Central Asia, Lake Dzhaksykluch near Aralsulfat is a big producer, while Beyuk Shor in Transcaucasia provides iodine-contaminated salt. Solar evaporation is used on sea water along the Black Sea and Azov coasts. Large rock-salt reserves occur in almost undisturbed Permian beds near Solikamsk, though

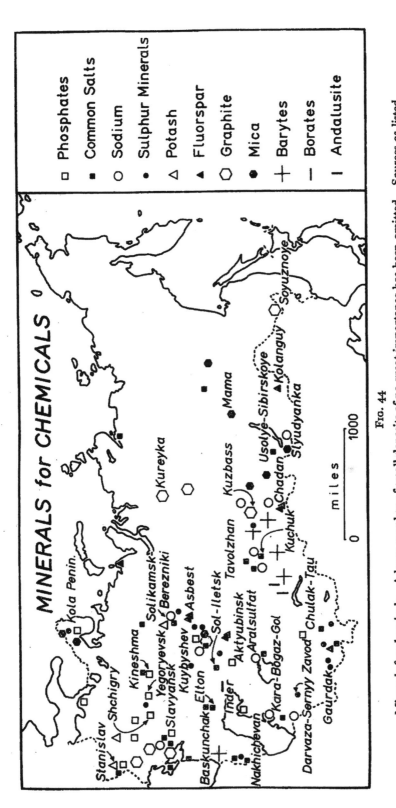

MINERALS for CHEMICALS

Phosphates □
Common Salts ■
Sodium ○
Sulphur Minerals ●
Potash △
Fluorspar ▲
Graphite ⬡
Mica ⬢
Barytes +
Borates —
Andalusite ∣

Kola Penin.
Stanislav
Shchigry
Kineshma
Yegoryevsk
Slavyansk
Solikamsk
Berezniki
Kuybyshev
Asbest
Sol-Iletsk
Baskunchak
Elton
Tinder
Nakhichevan
Aktyubinsk
Aralsulfat
Kara-Bogaz-Gol
Darvaza-Sernyy Zavod
Gaurdak
Chulak-Tau
Tavolzhan
Kuzbass
Kuchuk
Chadan
Usolye-Sibirskoye
Mama
Kolanguy
Soyuznoye
Slyudyanka
Kureyka

miles
0   1000

Fig. 44

Minerals for chemicals. A large number of small deposits of no great importance has been omitted. Sources as listed in bibliography to Chapter 7.

heavily contaminated; there are salt domes around the northern Caspian littoral, but only Sol-Iletsk is worked (for human use). Rock salt is also worked at Artemovsk in the Ukraine (human use), at Fergana, and at Nakhichevan. Salt springs supply Slavyansk in the Ukraine; while a usefully located but high-cost producer is the brine wells at Usolye Sibirskoye (suitable for human use) in Siberia. Sodium salts are available in the large mirabilite deposit of the natural evaporating pan of the Caspian Kara-Bogaz-Gol and Lake Kuchuk in Western Siberia and in lakes near Achinsk, Minusinsk, and Selenga in Eastern Siberia; though actual production appears to come mainly from Solikamsk-Berezniki, Slavyansk, and the Donbass or most recently from Sterlitamak in the Ural.

The most suitably located deposits of phosphate rock and apatite are in Central European Russia (Yegoryevsk, Fosforitnyy, Kineshma, Bryansk, Shchigry) and in the Ukraine (Khmelnitskiy, Izyum and in Chernigov *oblast*). Significant for the 'virgin lands' are the Kazakh workings at Aktyubinsk. For Central Asia, the discovery in 1936 of promising Kara Tau deposits, now processed at Chulak Tau, was important. The principal source for European Russia, despite the long rail haul south, is the Kola Peninsula's unlimited quantities of apatite which prove more suitable as concentrate than the Central Russian lower-grade phosphate rock. Basic slag is also used. Siberia is, in general, deficient in these rocks.

Potash comes chiefly from Solikamsk where a thick layer of potassium-bearing carnallite overlies a thinner bed of related sylvinite. Significant deposits of kainite-langbeinite were gained in 1945 in annexed Polish territory near Stanislav. It is unlikely that small Central Asian deposits are worked, except perhaps Gaurdak.

Arsenic is available readily in the Ural (Plast, Dzhetygara) and in smaller scattered deposits in Siberia and Central Asia. Barytes are limited to small deposits in the Kuzbass and Kazakhstan. Two major deposits of fluorspar exist: near Dushanbe and in Transbaykalia at Kolanguy, both remote from consumers. Some is reported from Amderma in European Russia. Borates come mainly from Lake Inder on the Ural river. Low-content muds are also worked in the Crimea and Taman Peninsula. There are adequate refractory materials. Magnesite in short supply from Satka in the Ural is frequently replaced by dolomite.

Metallic magnesium has developed since 1945 with dismantled German works now sited in the Ural. There are immense deposits of andalusite in the Kazakh Uplands (Semiz Bugun, Aktash, Kounrad), and at Lovozero in the Kola Peninsula, both in a harsh, inaccessible location. Asbestos is present in large, high-grade deposits. The Asbest open-cast mine in the Ural is the largest working, but there are several smaller workings. During the war, an important discovery of asbestos was made in Tuva *oblast*, near Chadan. Mica (several forms) is mainly in Eastern Siberia. Slyudyanka, which may be nearing exhaustion, and Mama are the main workings, but the Aldan country also has deposits, including amber mica. Vermiculite is obtained from the Ural. Graphite is available in large quantities and has been worked from the 1930's, although the quality is not high. Main workings have been at Dnepropetrovsk and Zavalyeyskoye (contaminated with calcite). The Staryy Krim deposit is small but important for large-flake graphite. The Ural is a growing producer, while ultimately lower Yenisey deposits at Naginskoye and Kureyka will be important and the Far Eastern Soyuznoye deposit lies conveniently near the Trans-Siberian railway. Preparation is chiefly in the Ukraine. Gypsum is plentiful from European Russia at Artemovsk, Ust-Kama, Gorkiy, and in the Ural near Kungur. Deposits are widespread in other regions but deficient in Western Siberia and the Far East. Kaolin and fire-clay are chiefly from European Russia, though prospecting in Siberia and the Ural was undertaken in the war years. There is no shortage of raw materials for cement.

NOTES ON CHAPTER 7

1. Soviet statistics on mineral production are very few and mostly extremely incomplete. Material has to be pieced together from scattered estimates and semi-official figures. These are summarized in Table 15. Some information is contained in the *Oxford Regional Economic Atlas of the U.S.S.R. and Eastern Europe* (Oxford, 1960, revised edition); Shimkin, D., *Minerals — a Key to Soviet Power* (Harvard, 1953); Pounds, N., and Spulber, N., *Planning and Resources in Eastern Europe* (Bloomington, 1957); Kruger, K., *Das Reich der Gesteine* (Berlin, 1957); *Promyshlennost SSSR — Statisticheskiy Sbornik* (Moscow, 1957). Occasional reports also appear in *Wirtschaft des Ostblocks* (Bonn), *The Financial Times*, and the various publications of the United Nations.

2. Fritz, F., 'Die Kohlenerzeugung in Russland', *Glückauf*, Heft 5, 1957.

Hoel, A., 'Coal Mining in Spitzbergen', *Polar Record*, vol. II, 1935–1938.

Buyanovskiy, M., 'The Prospects for Development of the Pechora Coal Basin', *Soviet Geography*, vol. I, 1960.

Markova, F., and Nalivkhin, D., *Geologiya Sovetskoy Arktiki*. Moscow, 1958.

Onika, D., *Ugolnaya Promyshlennost SSSR*. Moscow, 1958.

Prigorovskiy, M., *The Coal Resources of the U.S.S.R.* Moscow, 1937.

Zasyadko, A., *Toplivno-Energeticheskaya Promyshlennost SSSR, 1959–1965*. Moscow, 1959.

3. Fraser, D., 'Volga Oil', articles in *Petroleum Times*, 1957–1958.

Hassmann, H., *Oil in the Soviet Union*. Princeton, 1953.

Hooson, D., 'The New Pattern of Soviet Oil', *Geographical Review*, vol. 49, 1959.
    'The Middle Volga — An Emerging Focal Region in the Soviet Union', *Geographical Journal*, vol. 126, 1960.

Keller, A., *Neftyanaya i Gazovaya Promyshlennost v poslevoyenniye Gody*. Moscow, 1958.

Shabad, T., and Lydolph, P., 'The Oil and Gas Industries of the U.S.S.R.', *Annals Assoc. Amer. Geog.*, vol. 50, 1960.

Trofimuk, A., *Uralo-Povolzhye: novaya neftyanaya Baza SSSR*. Moscow, 1957.

4. Details of the construction of new electrical power projects have been published piecemeal in the Soviet press. A useful if incomplete map was published in *Pravda* on 2nd November 1960.

Krzhizhanovskiy, G., and Veyts, V., *Yedinaya Energeticheskaya Sistema SSSR*. Moscow, 1956.

Tsunts, M., *Velikiye Stroyki na Rekakh Sibiri*. Moscow, 1956.

Some statistics of electricity production appeared in *Narodnoye Khozyaystvo SSSR v 1958 godu — Statisticheskiy Yezhegodnik*. Moscow, 1959.

5. 'A Guide to Soviet Power Stations', *Nuclear Power*, October 1958.

'Nuclear Power', *Financial Times Survey*. London, April 1956.

Kruger, K., *Unser Wissen über die UdSSR*. Berlin, 1956, contains additional but unconfirmed information.

Pounds, N., and Spulber, D., *op. cit.* chapter iii.

The Siberian centre for academic nuclear research may be the new 'Science Town' outside Novosibirsk.

6. *Atlas dlya Sredney Shkoly — Kurs Ekonomicheskoy Geografii*. Moscow, 1959.

Shimkin, D., *op. cit.*

An article on Soviet non-ferrous metal production by J. Kowalewski appeared in *Optima*, December 1959, and in the *Mining Journal*, November 1960.

Shabad, T., *The Soviet Aluminium Industry*. New York, 1958.

7. United Nations survey of world iron-ore supplies, 1956.

CHAPTER 8

Industrial Geography

ECONOMIC policy pursued since the Revolution has made Russia one of the great industrial powers, in which industrial self-sufficiency has often been at the expense of other sectors of the economy. Contemporary geographical distribution of industry bears traces of the pre-revolutionary pattern adapted to fit Marxist-Leninist doctrines, whose application to industrial location and regional planning has produced a much wider territorial spread. Everywhere there has been an attempt to develop a 'complex economy' to further local as well as national self-sufficiency, itself associated with the desire to reduce the burden on an inadequately developed transport system. This has brought industrial development in unfavourable locations under adverse environmental conditions in the eastern regions, where there have sometimes been pressing strategic needs and where the Russians have shown a ruthless ability to assemble workers and equipment against great physical difficulties far beyond the fringe of pioneer settlement. The cardinal rôle of industry has attracted people to industrial settlements, which have been made the foci of the territorial-administrative system, and has led to the peopling of pioneer regions with a dominantly 'urban' population. The new eastern centres have been engaged typically in basic heavy industry and mining far from their consumers, the diverse manufacturing towns of European Russia, while many of them have developed the close spatial association of related industries, the *kombinat*, favoured by Soviet planners. Nevertheless, many projects in the eastern regions remain incomplete or unrealized. The bulk of industry is still found in the broad, settled wedge of country across European Russia where some of the greatest but least publicized achievements lie. The European Centre, the Ukraine, and the Ural contain over half the invested national capital, 60 per cent of the labour force, and most of the great industrial towns. Moreover, numbers of works evacuated during the Second World War to the Volga, Ural, and Siberia from western districts have returned to their old sites.[1]

    DOI: 10.4324/9781003172048-8

HISTORICAL ASPECTS OF INDUSTRIAL
DEVELOPMENT [2]

The traditional isolation of Russia has enforced an extensive dependence on its own resources: the village remained largely independent of the outside world until late in the nineteenth century, when over 80 per cent of the population was still agricultural. By the thirteenth century, crafts and manufactures not usually found in villages had begun to develop at the seats of princes, at the sites of great annual fairs, or even near the rich monasteries. Their products passed through the fairs to the surrounding countryside or into the fluctuating mediaeval trading channels. Rising Muscovite power attracted much of the manufacture and trade of the country to Moscow and nearby towns, like Mozhaysk and Tula, where in 1632 a Dutchman started iron making; but there were also important fairs and manufacturing towns along the main river routes. Leadership was also given by powerful families such as the Stroganovs, whose estates at Solvychegodsk employed 15,000 people to work bog iron and salt.

The reforms of Peter the Great in the early eighteenth century gave a new impetus and he drew in foreign artisans who introduced the factory system, developing mainly industries to supply government needs; but territorial spread and colonization increased the potential market for consumer goods, so far chiefly supplied by village artisans. S. Peterburg, Moscow, and the newly acquired Riga, with its enterprising German population, became the main industrial centres. Apart from many small forges, there were seventeen ironworks in Muscovy in 1695, none in the Ural. By 1725, fifty-two ironworks had been started, thirteen in the Ural, where they were larger and better equipped than elsewhere. The state Ural ironworks produced about a fifth of Russian output and another fifth came from the private works. With abundant iron-ore, charcoal, and water power, the Ural industry flourished until the new techniques of coal smelting were brought to the Donbass in the nineteenth century. During Catherine the Great's reign Russia was the largest iron producer in Europe. Towards the end of Peter's reign, silk, velvet, ribbon making, and china, glass, and brickworks had also made their appearance.

Industry employed both freemen and serfs. The latter were particularly important and the feudal system was readily adapted without great modification to industrial needs. In some instances, forcible movement of serfs was conducted to provide labour for new works in the Ural and in S. Peterburg. Industry grew most extensively in the northern forest belt, where serfdom had been reduced to payment rather than the personal service characteristic of the richer agricultural areas, and some serfs even managed to become *entrepreneurs*. Feudalism proved, however, unequal to the demands of rising technical standards and was strained by the rapid industrial expansion during the Napoleonic wars, when Russia was cut off by the 'continental system' from a wide range of imported semi-finished and manufactured goods. Even before the emancipation of 1861, the majority of industrial workers were already freemen.

Peasant industry sold chiefly to merchants who could not own serfs and who dealt in the great fairs. It showed considerable regional specialization and was most developed in the poorer regions where climate and soil conditions hampered agriculture and where the form of feudalism allowed the peasant more choice in his activity. In the forest regions, wooden goods, baskets, and bark shoes — common peasant wear — were made; in the Ural, peasants specialized in iron goods and there were whole villages of blacksmiths on the Volga, around Gorkiy, and where bog iron was worked, making wire, knives, padlocks, agricultural and other tools, and even guns and castings. Very small, delicate iron castings became a speciality of Zlatoust, while Volga villages monopolized the making of nails. There were even villages specializing in precious metals and jewels. In the flax-growing districts of the North-west, the weaving of narrow linen strips for leg wrappings was entirely a peasant industry.

One of the earliest modern factory industries to serve the almost unlimited market for cheap goods was making cottons from imported raw materials, concentrated particularly around Moscow, S. Peterburg, and in Narva and the German-dominated Baltic provinces. Growth was rapid after 1830, aided by falling international cotton prices and protected from severe foreign competition by tariffs. Big factories with economy of scale rapidly outstripped smaller units: in 1835, 484 factories employed 47,000 people, but fifteen years later, 536 factories had 110,000 operatives. By 1860, factory-made cottons surpassed homespun and greatly exceeded the output of wool and silk, while competing fiercely with linen. About this time spinning mills were opened and importation of yarn declined. British capital and management were important in the development of the textile industry.

The emancipation of 1861 released a new interest and vigour, reflected in the upsurge of industry and strengthened after 1870 by an inflow of foreign capital. As well as to Moscow, S. Peterburg, and the Baltic provinces, this investment began to flow into the Ukraine, while some even went to the Volga and into Siberia, where state interest remained predominant, and foreign money developed the Caucasian oilfields. In face of growing Swedish and British competition, using improved techniques, the Ural iron industry had not kept its earlier promise, though it remained the main Russian producer until almost the end of the nineteenth century, when it was outpaced by the Donbass. After 1861, the loss of cheap serf labour made it hard in the Ural to haul ore and charcoal over frozen rivers in winter, especially as inroads into the forests increased the distance to the furnace, and the most easily worked ores were used up. Use of coke in smelting, with greater productivity and cheapness, put the Ural, deficient in suitable coal, at a disadvantage compared to the rich coal and ore resources of the Donbass and Krivoy Rog and to the Dąbrowa area of Poland, both also nearer to large and assured markets. By 1889 there were seventeen large works in the Donbass at a time when Russia was the world's third largest pig producer.

By 1913, the area around Moscow had the most marked concentration of industries (Fig. 45), accounting for a third of the industrially employed and over a third of the value of industrial output. Moscow was also the country's main commercial centre, a function not lost even when the capital had moved to S. Peterburg. It enjoyed a nodal position in the best-developed part of the country and lay near to all the great fairs, which remained singularly important in Russian commerce. Its large reserve of skilled labour was later to prove important in the pattern of Soviet industrialization. Engineering, with British and German participation, was already established in Moscow and at the large Kolomna railway works, opened in 1862, as well as in the iron-smelting towns of Lipetsk and Tula, where samovars were made. The textile industry lay concentrated in and to the east

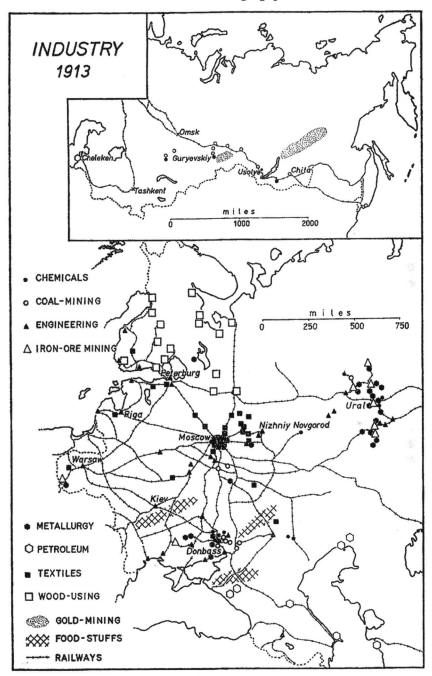

INDUSTRY
1913

Omsk

Cheleken   Guryevskiy

Tashkent   Usolye   Chita

m i l e s
0   1000   2000

• CHEMICALS

○ COAL-MINING

▲ ENGINEERING

△ IRON-ORE MINING

m i l e s
0   250   500   750

Peterburg

Urals

Riga

Nizhniy Novgorod

Moscow

Warsaw

Kiev

• METALLURGY

⬡ PETROLEUM

■ TEXTILES

□ WOOD-USING

GOLD-MINING

FOOD-STUFFS

RAILWAYS

Donbass

FIG. 45

Industry, 1913. *Atlas Istorii SSSR*, Moscow, 1954, Vol. III, pl. 9, and *Geograficheskiy Atlas*, Moscow, 1954, pl. 149, 150, 154, 155, 156, and augmented from sources listed in bibliography to Chapter 8.

of Moscow, where Ivanovo was the country's third largest industrial employer. The region claimed 65 per cent of all spindles and 71 per cent of all weaving frames in 1910 and was dominated by cotton cloths. Woollens, using imported and native raw wool, were made at Kolchugino; but the older eighteenth-century industry at Voronezh and Kazan still remained more important. Coal was brought by rail and water from the Donbass, though Tovarkovo and Podnoye near Tula were producing 30,000 tons of lignite by 1913. Nizhniy Novgorod (Gorkiy), where the Sormovo works opened in 1849, built Volga boats, railway equipment, and other machinery. To the south-west, at the junction of Bryansk-Bezhitsa, an eighteenth-century charcoal iron industry turned to rolling rails and building railway engines and wagons from Donbass raw materials.

TABLE 19

PIG IRON OUTPUT IN RUSSIA

Year	Total *	Ural *	%	South *	%	Bituminous Coal Output *
1867	17,028	11,084	65·1	56	0·3	26,700
1887	37,389	23,759	63·5	4,158	6·5	110,100
1897	114,782	41,180	35·8	46,349	40·4	683,900
1902	158,618	44,775	28·2	84,273	53·1	1,005,200
1913	283,000	55,800	19·7	189,700	67·0	2,213,800

* Million puds (1 pud = 16·38 kg.).
Source: Feigin, J. G., *op. cit.* p. 147.

S. Peterburg and the Baltic towns accounted for about 15 per cent of industrial employment and output. Industry in S. Peterburg, based on imported raw cotton, wool, jute, rubber, and British coal and using Ural and Donbass iron, manufactured high-quality machinery, some electrical goods, chemicals, and rubber, with an emphasis on articles for military or naval use, but there was also output of clothing, piece goods, shoes, and domestic articles. Over three-quarters of the workers were in firms employing more than 500 people. Most raw material imports were consumed locally, but imported manufactured articles were distributed throughout the country as the railway network grew. The German minority developed industry in the

Baltic lands where Latvia was the main centre, with shipbuilding, engineering, and chemicals in Riga. The largest single industrial enterprise in the empire, employing 12,000 workers, was the Estonian Krengolm textile mills at Narva.

Almost a fifth of the industrially employed were in the Ukraine, which supplied more than a fifth of industrial output. The rapid development of heavy industry in the last quarter of the nineteenth century owed much to the Welshman Hughes at Yuzovka (Donetsk) and to Pastukhov at Krasnyy Sulin. Originally, local brown ores were used, but completion of the Donbass-Krivoy Rog railway in 1884 provided large quantities of rich ore. By 1913, well over two-thirds of the raw iron and steel and over half the heavy engineering output was in this region, such as the big Lugansk railway works (opened 1896) and the shipyards at Nikolayev and Kiev. The Donbass supplied over 80 per cent of Russian bituminous coal, used locally or shipped by rail or water even to the Ural and Arkhangelsk.

Despite its eclipse by newer and more productive techniques in the Donbass, the Ural charcoal iron smelting expanded production of good quality pig. Apart from iron-ore, gold, copper, and precious stones, few of the Ural's varied minerals were worked, though some salt and oil seepages were exploited on the western slope. Manufactures included river boats on the Kama at Votkinsk and firearms at Perm and Izhevsk, while Zlatoust still produced fine castings as well as instruments.

The agricultural colonists brought to Siberia by the new railway provided a basis for the development of manufacturing and the processing of agricultural products. Coal began to be mined for railway use in the Kuzbass, and even at Karaganda and Ekibastuz, on a very small scale, to supply the small smelter at Spasskiy Zavod. Mining, particularly for gold, was dominated by the state but a little foreign money was also involved, while precious stones were sought in the Altay and some salt came from Kulunda. Small ironworks opened at Guryevsk and Petrovsk-Zabaykalskiy. In the Far East there was some industry in Vladivostok, with nearby coal-mining, to supply the navy.

By 1913, Caucasia was one of the most important mining regions of Russia, with the large oilfields as major world producers financed by the Nobel group, and ore was also worked at Chiatura. Apart from petroleum and some indifferent cotton, the Central Asian territories incorporated in the latter part of the nineteenth century contributed little. Yet one of the most important industrial producers lay in the Polish Kongresówka, where there was a large textile industry, notably at Łódź, and the small iron and steel industry on the Dąbrowa coalfield which used local coal and ore from the Jura Krakówska.

The Revolution weakened Russian industry with the loss of the relatively advanced Poland, Finland, and the Baltic countries, and war had spread destruction widely in the western provinces. Further, great damage was done in the Civil War, so that by 1920 the country was faced with the almost complete renewal of the industrial fabric. Expediency demanded that industry in already established areas should be rehabilitated before new areas were developed, for which equipment was bought abroad with resources sorely needed at home. With the grandiose plans for industrialization envisaged by the State Electrification Plan (1922) the first stage was to be the building of the Ural-Kuzbass Combine as a second metallurgical base and symbol of the new era.[3] Even before the Revolution, attempts had been made to carry Kuzbass coal along the Ob and Tobol to the Ural to modernize smelting there; now coking coal, deficient in the Ural, would be sent by rail from the Kuzbass in return for iron ore, since Kuzbass resources were thought poor and inadequate. In the Kuzbass, coke batteries were opened at Kemerovo in 1924, and in 1932 the first modern steelworks was completed at Novokuznetsk. In the Ural, 1931 marked the opening of the great Magnitogorsk works and there was modernization of older plants. Lying a thousand miles apart, the weakness of the Combine lay in the inadequate railways to handle greatly increased traffic. From the middle thirties, the emphasis changed and the Ural and the Kuzbass began to develop independently.

Problems to overcome in creating a diversified and more balanced industry in European Russia, where the foundations were already laid, added to preoccupation with making the Ural-Kuzbass Combine work, left little for surmounting the much greater problems faced in the Asiatic regions whose vast emptiness, harsh climate, widespread *permafrost*, and the sparseness of population and communications made it simplest to take their raw materials to established industrial centres for manufacture. The need for coking coal in the Ural made it prudent to develop the Karaganda coalfield in the arid steppe of Kazakhstan, while nearby non-ferrous metals also attracted mining. Karaganda town was built in 1928 and later linked to the main-line railway system. The deep impression left by defeat in the 1905 war with Japan and the weakness of communications with the Far East brought strategic industrial development; as a beginning, iron and steel making was started in the new town of Komsomolsk-na-Amure, though progress was slow; coal-mining expanded near Vladivostok and in the Bureya valley, while mining of non-ferrous and ferrous metals started, notably in the Sikhote Alin; and in North Sakhalin, petroleum production was pushed. Like all eastern regions, it depended heavily on manufactured goods brought from European Russia. Mining also developed in Transbaykalia, where Ulan Ude became an important railway engineering town and other industries developed around Irkutsk, Cheremkhovo, and Krasnoyarsk; but it was the West Siberian towns, Novokuznetsk, Omsk, and Novosibirsk, which became the industrial focus of Siberia. By the Second World War, using extensive forced labour, mining and industrial towns were appearing even in Arctic Siberia.

GEOGRAPHICAL DISTRIBUTION OF SELECTED MAJOR INDUSTRIES

Iron and Steel Industry

Iron and steel output has risen from 4·2 million tons of pig iron and 4·3 million tons of steel in 1913 to 66·2 million tons of pig and 91·0 million tons of steel in 1965 : Russia now makes over a fifth of world steel. There has also been a marked shift in production eastwards, shown in the table below.

Together, the Ural and Ukraine supply over 80 per cent of the pig and almost three-quarters of the steel. The third main producer, Western Siberia, makes less than a tenth of both metals and the Centre only a twentieth. As might be expected, steel production is more widely scattered than pig production.[4]

Iron and steel making, requiring large quantities of bulky, low-grade raw materials and process water, tends to be attracted to these sources. To produce one ton of pig iron, Russian works need between 1·75-2·17 tons of iron ore and flux and between 0·6-1·02 tons of fuel. The iron content of ores and their physical properties appear to be deteriorating, with a rising consumption of ore per ton of pig as coal consumption declines. Manganese, easily available, is freely used. The rising proportion of coal coked at iron and steel plants, with its advantages of by-product output, adds a locational pull towards coal sources, though in the drier parts of Southern Russia, Central Asia, and parts of Siberia, water supplies become a major locational factor: at Temir-Tau near Karaganda, large reservoirs hold a supply for the dry season. Steel making is both closely associated with ore smelting and with

TABLE 20

RUSSIAN PIG IRON AND STEEL OUTPUT, 1913–1955
(Million metric tons)

	Pig Iron		Steel	
	1913	1955	1913	1955
R.S.F.S.R.				
North-west	—	—	0·2	0·9
Centre	0·2	1·9	0·3	2·6
Volga	—	—	0·1	1·5
N. Caucasus	—	—	0·3	0·9
Ural	0·9	11·9	0·9	16·4
W. Siberia	—	2·4	—	3·9
E. Siberia	—	—	—	0·4
Far East	—	—	—	0·3
Ukraine	2·9	16·6	2·4	16·9
Uzbekistan	—	—	—	0·2
Kazakhstan	—	—	—	0·2
Georgia	—	0·4	—	0·6
Azerbaydzhan	—	—	—	0·4

Source: *Promyshlennost SSSR — Statisticheskiy Sbornik*, Moscow, 1957.

major engineering districts where there is much scrap (Fig. 46). In some districts, steel making is dependent on imported pig or small, local, high-cost pig production to augment imports of scrap. Soviet sources frequently complain that planning has been unequal and large cross-flows of various forms of semis occur. Many new plants established to satisfy Soviet regional planning concepts have proved uneconomic.

The Ukraine, the largest producer of both iron and steel, exports a large part of its output of pig for conversion and a wide range of steel semis to manufacturing industry in other regions.

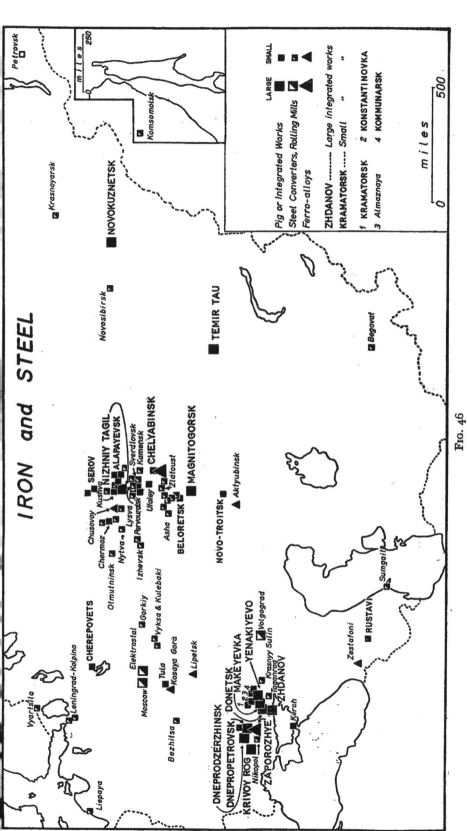

FIG. 46

Iron and steel. Full cycle integrated plants undertake ore smelting, steel conversion and processes such as rolling or drawing. Sources as listed in bibliography to Chapter 8.

The principal works lie in the western Donbass coalfield where they originally used local blackband ores, fortunately both nearest to the modern ore supply from Krivoy Rog and above the main coking coal deposits. Limestone is locally available for the furnaces, which use the heaviest charge of all regions, and the manganese ore used so freely in them comes from the important Nikopol deposit.

Makeyevka is a major centre, with the Kirov works — one of the biggest in the Donbass — and the Kuybyshev pipe plant. Donetsk has blast furnaces and heavy rolling mills, while Yenakiyevo also has large blast furnaces. Donetsk and Makeyevka together produce half the metal smelted in the Donbass. Smaller plants are Konstantinovka (rolled sheet) and Kramatorsk, which is noted for electric steels and which converts local scrap from engineering shops and does foundry work. There are several large cokeries, notably Gorlovka. In the eastern coalfield, coal-mining and engineering are more important, though Lugansk has pipe-rolling. Blast furnaces and cokeries work at Kommunarsk which also turns out complicated sections. Almaznaya makes special qualities of pig iron from Krivoy Rog ore. There are open-hearth furnaces and rolling mills at Krasnyy Sulin and Taganrog makes tubes and pipes. On the coast, away from the coalfield, Zhdanov has one of the most modern Russian plants, Azovstal, using Gorlovka coke, Yelenovka limestone, and a mixture of Krivoy Rog and Kerch ores. The rolling mills are among the largest in Europe, and nearby plants make pipes, armour plate, rails, and sections. The low-iron and high-phosphorus Kerch ore, when mixed with additional phosphates, is suitable for Thomas steel, useful in welded work.

Krivoy Rog sends ore to the Donbass and to other parts of the country, but it also imports Donbass coal and coke to smelt fragile ores locally. On the Dnepr banks, between the Krivoy Rog ore and Donbass coal, Dneprodzerzhinsk smelts but Dnepropetrovsk concentrates on open-hearth conversion. Zaporozhye, based on cheap Dneproges electricity, is principally an alloy steel producer, using local pig and alloys, besides specializing in cold rolling. Other centres, including Nikopol, turn out a wide range of sheets, pipes, and rolled goods.

The Ural possesses little coal suitable for coking but has large reserves of good-quality iron ore and many of the metals usually alloyed with iron. Eclipsed by the Donbass in the late nineteenth century, its output since the Revolution has climbed to equal that of the Donbass and other Ukrainian producers, a feat achieved by modernizing old plants and building some of the largest and most modern integrated works. The works lie mostly on the eastern slope near major ore deposits and along the main-line railways by which they receive their coal and coke. Of the metal exports, mainly rolled steel, over a third goes to the Centre, almost a quarter to Siberia and roughly a tenth is taken by the Volga and another tenth by Central Asia and Kazakhstan.

On the north, Serov uses Kemerovo coke and Rudnichnyy ore and it has still a charcoal-fired blast furnace, while there is a subsidiary plant at Severouralsk, specializing in high-grade steel and ferro-alloys. Further south a group of works lies around one of the largest integrated works in the country, Nizhniy Tagil. Ore comes from nearby mines at Vysokaya Gora, Lebyazhka, and Blagodat, as well as Tagil-Kushva and Alapayevsk; but the alumina content of some complicates smelting and restricts furnace size. Coal and coke are mainly from the Kuzbass. Products include heavy rolled sections, wire, plates, and thin sheets. Small smelters use local ore, Kuzbass coke, and some charcoal at Nizhnyaya Salda and Verkhnyaya Salda; and the open-hearth furnaces of the former supply the latter with steel for billets and angles. Alapayevsk makes plates, Nizhnyaya Tura roofing sheets, and Teplaya Gora foundry work.

A cluster of works around Sverdlovsk depends mainly on local engineering waste for conversion. The Verkhne Isetsk works makes steels for the electrical industry; Pervouralsk and Kamensk-Uralskiy have specialized pipe-making equipment, while nearby Bilimbay provides raw steel. The Polevskoy tin-plate works are the largest Soviet suppliers of metal for food containers. Nizhniy Sergi is a small full-cycle producer and Staroutkinsk a charcoal smelter. A mixture of old and new works found around Chelyabinsk (which first made steel in 1943 and has postwar blast furnaces) produces high-grade steels for machine tools from Bakal ore and coal and coke from the Kuzbass and Karaganda. Bakal ores are used for ferro-silicon and other alloys come from local chrome and tungsten. Nearby lignite-burning power stations supply power for electric furnaces, while process water is taken from the river Miass. Zlatoust, an important older works, makes special steels for instruments and general engineering from imported pig. Steel made from Komorovo-Zigazinskiy ores at

Beloretsk is rolled locally and also twenty miles away at Tirlyanskiy for roofing sheet. The Chebarkul works, specializing in high-quality precision castings, use machinery brought in 1941 from European Russia. Asha, Satka, Katav-Ivanovskiy, and Yuryuzan are small works.

Magnitogorsk, built 1929–1931, is one of the largest Russian works and stands in the wooded steppe at the foot of Magnitnaya Gora. Ore comes from the mountain, but the best, remarkably pure oxidized magnesite has been used and ore is now usually treated by washing, crushing, and wet magnetic separation, with about 86 per cent of the charge sintered. Growing ore supplies are coming from new mines in Northern Kazakhstan. Flux is from the nearby Agapovsk limestone, while coal and coke come from the Kuzbass and Karaganda, according to the smelting process. Large coke batteries treat more than 500,000 tons of coal monthly. Process water is taken from a barrage across the Ural river, while electric current comes from a local thermal generator. Besides the blast furnaces there are open-hearth furnaces, blooming mills, rolling mills, billet mills, and a plate mill, as well as continuous sheet production. In 1956, the works turned out 6·25 million tons of steel ingots and 5·5 million tons of basic pig iron; by 1960, it was to have made 8·5 million tons of the 14·5 million tons of steel produced in the integrated Ural works. The output, distributed widely over the country, is claimed to be twice as cheap as the best Ukrainian mills. Further south lies the new integrated Novo-Troitsk plant on the Ural river near Orsk, completed after 1950, and using Khalilovo chrome-nickel-iron ore and Dombarovskiy and imported coal; while the Aktyubinsk ferro-alloy plant uses Kurashasay coal and Khrom-Tau ore.

Several small works lie on the western slope of the Ural, mostly in the upper Kama basin. The most important, the integrated Chusovoy works (founded 1879) use local Kizel coking coal and Kuzbass coke, while Kusa titano-magnetite yields ferro-vanadium and vanadium slag. It produces angle iron, channels, sheets, and tin plate. Lysva (founded 1784) makes wire, thin sheet, galvanized iron, and tin plate, a combination typical of the older works. Perm also has rolling mills, while small producers include Dobryanka (sheet and plate), Maykor (pig), Nytva (sheet?), and sheet mills at Izhevsk and Votkinsk.

Western Siberia, third in order of regional production, developed during the first five-year plans with its principal works in the Kuzbass coalfield, where Novokuznetsk, the second largest in Russia, uses local coal and ore brought from the Tashtagol-Shalym mines to the south, which supply up to 85 per cent of the charge. Originally, ore from Magnitogorsk was used, but in the next few years the change to Altay-Sayan ore will be complete. Its well-equipped rolling mills produce a wide range of bars, billets, sheets, and rails. When complete, it will have the world's largest continuous sheet mill. There is also a ferro-alloy plant, while nearby at Guryevsk chrome-silicon steel for aircraft is made. A second large works is being built here to use ores brought from Abakan along the new Novokuznetsk-Abakan railway. Novosibirsk has hot and cold rolling mills, working imported steels from both the Ural and Kuzbass.

The Centre, a close rival to Western Siberia, serving the specialized industry grouped around Moscow, is a high-cost producer offset by low transport charges to nearby consumers. At Tula and Lipetsk, local ore is smelted with Donbass coke, mostly for foundry work but also for continuous cast steel. Lignite-burning electric generators supply current for furnaces making ferro-silicon at Lipetsk and ferro-manganese at Kosaya Gora. At several large engineering works in and around Moscow, scrap conversion is found. Moscow also makes foundry pipe and electrically welded goods. Bezhitsa is also a small scrap converter. Yelets supplies limestone to Lipetsk and Tula as well as to other regions.

Subsidiary producers are found in the upper Volga, where Kulebaki experimented with coke from peat in 1897 and now makes steel bars, beams, and channels, also made along with small diameter pipe at Vyksa. Gorkiy produces a wide range of plates, bars, heavy sheet, and continuous cast steel. Leningrad (including Kolpino), using imported iron and coal plus local scrap, covers the special demands of local engineering with varied production of bars, billets, plates, sheets, and pipes, as well as wire and cable. The Karelian Vyartsila electric steel plant was taken from Finland in 1945 and there is a nail and sheet works at Liepaya in Lithuania and some small production at Riga. Much of the requirements of the North-west and Baltic regions will be covered by the new Cherepovets works sited on the shore of the Rybinsk reservoir and using Vorkuta coal and Karelian ore.

Large ore reserves, chiefly at Ata-Su, and coking coal at

Karaganda and Ekibastuz have been the basis for development of the Temir-Tau steelworks on the banks of a reservoir on the Nura river, where water has been a major locational factor in the dry steppe. Rails and constructional materials for mining are the main products. A ferro-alloy plant has been planned at Pavlodar. The Kazakh Lisakov ores suited to the Thomas process are for use in a projected steelworks at Barnaul. To use available scrap, a steel conversion works was built in 1943 in Uzbekistan at Begovat near the Farkhad dam, which supplies electric current. It also converts imported West Siberian and Ural pig and turns out small rolled products. In Transcaucasia, using local coal and Dashkesan ore, the Rustavi steelworks, commissioned in 1955, concentrates on rolling and drawing, while its pig is used by the Sumgait pipeworks, opened in 1952. Cheap hydro-electric current, Chiatura manganese, and other ores provide the base for the Zestafoni ferro-alloy plant. In Eastern Siberia, the old Petrovsk-Zabaykalskiy works does foundry work, using local materials in small blast furnaces. Designed to use Bureya coke and local ores, it is uncertain whether Komsomolsk in the Far East does in fact smelt ore, but its tin-plate works serves the Pacific fishing industry. Using Angara-Ilim and other ore and southern Yakut coal, several projects have been suggested for Central Siberia. Outside the Ural, the Ukraine, and the Kuzbass, most plants are relatively small, though Soviet propaganda likes to suggest that they are all major bases of heavy industry.

Engineering Industry

Engineering produces a wide and diverse range of articles from a large selection of raw materials, but labour forms a major element in cost, and inherited skill, design, and development play an important rôle. Because its products have a high value in relation to their bulk and thus stand transport well, economies can be made by mass production and association regionally of different branches, though some sections are governed by special site considerations. A considerable part of the industry appears, however, to be in small and relatively inefficient plants. Even some 'showpieces' visited by the author leave much to be desired by Western standards. It is markedly concentrated: the most diverse patterns are found around Leningrad, Moscow, and Gorkiy, while in the European South there is considerable heavy

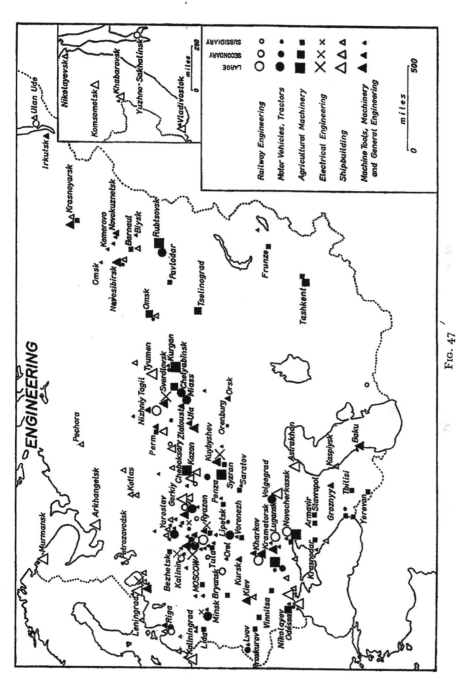

Fig. 47

Engineering. Many smaller centres are primarily repair and maintenance plants, while large numbers of the smallest plants have been omitted. Sources as listed in bibliography to Chapter 8.

engineering, with smaller concentrations in the Ural and Western Siberia. Most of the smaller, scattered centres are concerned more with maintenance than production. The Centre produces, for example, a quarter of the industrial boilers, two-thirds of the excavators and over three-quarters of the motor vehicles, while Leningrad turns out half the hydro-electric turbines. Engineering in eastern regions was given considerable stimulus by the war-time migration of plants, and though many returned after the war to their old sites, branch works were commonly left in the east. Products tend to be tied to local requirements, like mining dredges and gold-working machinery in Eastern Siberia. The regional spread of engineering may be accelerated by industrial decentralization introduced in 1957, while the patterns may also be affected by agreements on international specialization between Soviet bloc countries through *Comecon*.[5]

Manufacture of boilers, turbines, diesel engines, and electric motors is frequently associated with branches which incorporate them in their products. The principal producers include Moscow, which builds engines for road and rail vehicles and industrial purposes; Leningrad, where marine engines and hydro-electric turbines are made; Gorkiy, with motors for road vehicles and river vessels; and Kharkov, Riga, Kuybyshev, Sverdlovsk, and Taganrog, noted for its boilers. The ball-bearing industry, whose products are used in many sections of engineering, is principally concentrated in Moscow, Saratov, Kuybyshev, Sverdlovsk, and Minsk.

Railway engineering is widely scattered, and before the Revolution some large railways built their own locomotives and wagons, while smaller ones had extensive repair facilities. Russia has always bought equipment abroad, so that some works grew in association with foreign plants, like the Lugansk works affiliated to the Chemnitz works in Germany. After the Revolution, many smaller builders were restricted to repair works and building concentrated in larger units, while a further change has taken place with the replacement of steam by diesel and electric locomotive construction since 1956. Works not formerly linked to railway requirements have been organized to make parts for diesel and electric locomotives. The two largest works, now producing diesel locomotives, are Lugansk and Kharkov, while the former third plant at Kolomna may now do other work. Novocherkassk is the principal producer of electric locomotives, while Gorkiy builds diesel locomotives. Leningrad ceased main-line locomotive building about 1930, though it and Votkinsk, Kambarka, Kaluga, and Murom build narrow-gauge and small industrial locomotives. In 1938, Ulan Ude works opened in Eastern Siberia to supply locomotives, while the evacuated Lugansk works built locomotives between 1943–1945 at Krasnoyarsk, which has since done repair work. Riga builds diesel and electric rail-cars, while Mytishchi deals with the Moscow suburban and underground railways. The largest wagon builders are Nizhniy Tagil, Kanash, Kalinin, and Kolomna (coaches); but Bryansk concentrates on the growing fleet of refrigerator cars and Zhdanov on tank wagons. Narrow-gauge wagons are made at Petrozavodsk. Special facilities for the Sakhalin railways, using mainly Japanese equipment, are found at Yuzhno-Sakhalinsk. Major maintenance centres are Tbilisi, Astrakhan, Chelyabinsk, Tashkent, Kizil Arvat, Novosibirsk, Voronezh, Michurinsk, and Leningrad, as well as Penza and Kalinin. Among many towns supplying railway materials, Tula builds mechanical track-laying and maintenance machines.

In 1965, Russia produced 616,000 cars and trucks and 355,000 tractors. Over three-quarters of the cars and lorries come from the Central Industrial region, with Moscow and Gorkiy as the largest producers, and subsidiary production from the Volga, Ural, Byelorussia, the Ukraine, and Georgia. Tractors and earth-moving machinery manufacturers are more widespread, though dependent on motors and electrical equipment from the Centre. Since 1924, Moscow has built buses, lorries, and passenger cars and Gorkiy turns out lorries, landrover-type vehicles, and large saloons, with buses made nearby at Pavlovo. The Yaroslavl works, brought from Riga in 1917, is a smaller producer; Kirzhach makes lighting and ancillary equipment, while instruments are produced in Vladimir. Tractors are made in Vladimir, Tambov, and Lipetsk. Minsk builds Euclid-type dumpers and lorries. Kharkov makes tractors and Lvov, buses, though assembly is also done at Odessa, Dnepropetrovsk, and Rostov-na-Donu. Zaporozhye has begun building baby cars. On the Volga, Kuybyshev makes electrical and other parts, while Ulyanovsk started assembling lorries during the Second World War, and one of the largest tractor plants, opened in 1930, is at Volgograd. In the north, Petrozavodsk supplies tractors for the lumber industry.

Wartime shifts started motor-vehicle building at Miass in the Ural, extended the tractor works at Chelyabinsk and developed tractor building at Orenburg and Krasnoufimsk, all wartime tank builders. Motor-car assembly plants operate at Omsk and Novosibirsk, while Kurgan assembles buses. The Altay tractor works at Rubtsovsk was a wartime shift of the Kharkov plant. In Transcaucasia, Kutaisi assembles and builds lorries. Though motor-vehicle manufacture is underdeveloped in the eastern regions, aircraft are probably mostly made there out of strategic considerations in a largely 'footloose' industry, but very little reliable information is available.

Shipbuilding is mostly concerned with river boats, and some of its busiest yards lie along the Volga. Originally the largest builders were in the upper reaches where wood was available, but a change to iron vessels drew yards nearer to the metal supplies from the Donbass. The main ship and marine-engine builder is Gorkiy and there are yards at Rybinsk, while in the lower reaches Krasnoarmeysk and Astrakhan, which also builds Caspian vessels, are important, besides several ship repair yards strung along the river. On the Kama, Perm and Votkinsk are shipbuilders; the Dnepr is supplied from Kiev and Nikopol; the northern European rivers are served by Arkhangelsk (N. Dvina), Kotlas (Sukhona), Syktyvkar (Vychegda). The great river systems of Siberia are usually based on one or two main centres; for example, ships built in the big yards of Tyumen can reach the whole Ob-Irtysh system, but repair work is done at Tobolsk, Omsk, Novosibirsk, and Barnaul. The Yenisey is supplied from Krasnoyarsk, but for the Lena there are yards at Kachuga, Peleduy, and Yakutsk. Ulan Ude is the centre for the Selenga and Lake Baykal, with Komsomolsk for the Amur-Ussuri basin. Some vessels appear to be prefabricated and supplied in sections for assembly on smaller rivers and lakes, or coasted to them in the summer sailing season. Barge building is generally more widespread than powered vessel construction.

Many sea-going vessels come from foreign yards, but others are built chiefly at Leningrad and Nikolayev. On the Black Sea, yards (mostly repairs) lie at Odessa (whalers), Kherson, Osipenko, and Tuapse. On the Baltic, Petrokrepost builds fishing vessels and there are facilities at Riga, Klaypeda, and Tallin; East Prussian yards taken in 1945 appear to concentrate on naval work; in northern waters, yards are well equipped at Arkhangelsk and Murmansk. Vladivostok is the main Far Eastern base, though fishing vessels are handled at Petropavlovsk-Kamchatskiy and Nikolayevsk. The Caspian receives ships built at Astrakhan (notably tankers) and Kaspiysk.

Very little farm machinery was built before 1917, but the present nation-wide distribution is a reflection of the mechanization of agriculture and of the varying local needs for specialized types, with works found in both the main metal-producing districts as well as the principal farming areas. In the south, Rostov-na-Donu builds a large share of the grain combines and turns out a fifth of all agricultural machinery, though large works are also found at Zaporozhye (maize harvesters) and Taganrog. In the Western Ukraine, Odessa builds sugar-working machinery, Kirovograd, sowers for the mixed farming belt, and there are works at Vinnitsa and Proskurov. The Northern Caucasian farming areas get machinery from Armavir, Stavropol, Krasnodar, and other towns. The Volga has big works at Syzran and Kazan, with others at Saratov and Chapayevsk. Numerous important plants lie in Central European Russia: Voronezh and Kamenka (sowing machines) serve the black earth lands, while machinery for working potatoes, flax, rye, and vegetables is made in Tula, Ryazan, Smolensk, and Lyubertsy (one of the earliest producers), and Bezhetsk is primarily concerned with flax harvesters. Gomel and Lida are the main Byelorussian works. In the Ural, Chelyabinsk makes tractors and ploughs; Perm produces silage cutters and threshers; and there are several smaller centres. In Siberia, the large Rubtsovsk works builds a range of ploughs and cultivators; Kurgan and Omsk turn out a similar range and the latter along with Barnaul has built combine harvesters since 1956. Kurgan also makes dairy machinery. Krasnoyarsk serves Eastern Siberia, but all the works have contributed equipment to the new Kazakh grainlands, which are supplied by a new works at Tselinograd, to be augmented by another factory at Pavlodar. Tashkent, Chirchik, and Frunze serve the special needs of Central Asian oasis cultivation and the tea, citrus fruit, and vine growing of Transcaucasia are supplied from plants at Batumi, Tbilisi, Poti, and Yervan.

Machine tools and electrical equipment manufacture, dependent on great skill and complex design and development, is well suited to the older industrial areas where skilled labour is present and raw materials are brought in from other regions; but despite the concentration in these areas in European Russia, a movement of new plants into Siberia and the Ural can be seen and a further eastwards shift will doubtless take place. Many small towns possess engineering shops primarily concerned with repair and maintenance. About a quarter of the output of machine tools probably comes from the Centre. Mining machinery is made in the Donbass at Gorlovka and Novokramatorsk and nearby at Kharkov and Lugansk; in the Ural at Votkinsk, Kopeysk, and subsidiary centres; in the Kuzbass at Prokopyevsk and at Kutaisi in Transcaucasia. Rolling mills, smelting equipment, and heavy engineering machinery are produced in the Donbass, notably at Novokramatorsk and Lugansk; at Moscow and Elektrostal in the Centre; in the Ural at Sverdlovsk (mostly with ancillary electrical equipment) and Orsk; Novosibirsk, Krasnoyarsk, and Irkutsk are producers in Siberia, including machinery for gold and diamond working. Petroleum-working machinery is made in Baku, Armavir, and Groznyy, at Ufa-Chernikovsk (Ural) and at Podolsk near Moscow. Peat-working equipment is supplied from Kalinin and Ivanovo. Works in the Ukraine, the Ural, and around Moscow supply chemical plant, though this will ultimately also come from Siberia. Moscow and Kiev make automated process equipment and other important manufacturers of various machinery are Leningrad, Gorkiy, Kuybyshev, Sverdlovsk, Novosibirsk, Ryazan, Minsk, Kharkov, and Kramatorsk. Specialized production includes printing machinery from Rybinsk; sewing-machines at Tula and Podolsk; woodworking and gas equipment from Gorkiy, which along with Kineshma also turns out paper-making machinery; biscuit-making, flour-milling, and baking equipment comes from Voronezh, Kursk, Orel, and Shebekino, while other food-processing plant is made in Minsk, Odessa, Kiev, and Kharkov and Caucasian towns. Textile machinery is built in the textile manufacturing areas east of Moscow, notably Ivanovo, Shuya, and Klimovsk, but also in Moscow and Leningrad. Electrical goods are made principally in Moscow and Leningrad, but very heavy equipment is supplied from Sverdlovsk and Novosibirsk, and domestic electrical equipment and radios are made in Kalinin and Ryazan, Gorkiy and Kuybyshev and other Volga towns. Precision instruments come from Moscow, Zlatoust, Kurgan, Tomsk, and Kharkov (notably cameras), but Cheboksary makes cinematograph installations.

Chemicals Industry

A large and diverse chemicals industry is essential for a modern industrial economy : until recently the Russian industry has been relatively small and specialized, concentrating on strategic needs, but it is to expand more rapidly in size and in range of product than other industries in the 1959–1965 plan. Particular emphasis on artificial fertilizers is linked to the need to increase substantially agricultural output, while there will also be big increases in the output of synthetic substances, dyes, and detergents, as well as sulphuric acid. There are contrasts in the locational pattern between heavy processes, usually sited near to bulky raw materials, and light chemicals and pharmaceuticals, sited near to consumers and centres of skill and research. Heavy processes are found particularly in the Donbass, the Ural, and Western Siberia, with smaller producers in Central Asia, Caucasia, and Eastern Siberia. Heavy processes in the Centre depend on waste, by-products, and raw materials found locally or brought from other regions. The Centre, notably Moscow, and Leningrad have the bulk of the light and pharmaceutical branches, though these are found in other European Russian towns and on a limited scale in Siberia.[6]

The Central European region contains several older, specialized branches, such as perfumes, cosmetics, soapmaking, as well as heavier branches. Cosmetics, perfumes, and pharmaceuticals, typical of Moscow, are also found in Kursk and Voronezh, while Pereslavl-Zalesskiy supplies photographic film. Zagorsk and Kolomna turn imported latex into rubber, but potato alcohol is supplied to the big synthetic rubber plants at Yaroslavl, Yefremov, Tambov, and Voronezh. Analine and other dyes are made around Moscow and at Tambov. The Moscow lignite field, augmented by Ural pyrites, provides raw materials for sulphuric acid at Novomoskovsk and Voskresensk, and calcium carbide is made at Lipetsk. Kola apatites and phosphates from Yegoryevsk, Shchigry, and Dmitrovsk-Lgovskiy are made into artificial fertilizer ; there is a large superphosphate works at Voskresensk and Novomoskovsk, which uses lignite for nitrogen fertilizer production. Plastics are made at Lyubertsy, Mytishchi, Klin, Pereslavl-Zalesskiy, and Vladimir, while synthetic fibres come from Klin, Orekhovo-Zuyevo, and Shchelkovo, which has a sulphuric acid plant. The Centre is the main producer of synthetic fibres, though oil and natural gas imported by pipeline will provide a

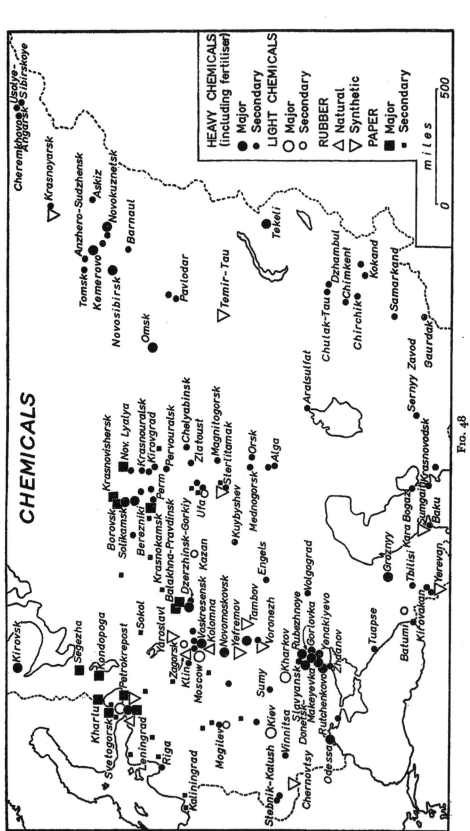

CHEMICALS

HEAVY CHEMICALS
(including fertiliser)
● Major
● Secondary

LIGHT CHEMICALS
○ Major
○ Secondary

RUBBER
△ Natural
▽ Synthetic

PAPER
■ Major
▪ Secondary

0   500
miles

FIG. 48

Chemicals. Though widespread, the industry comprises mostly small units, notably in the light chemicals and pharmaceuticals branches. This map should be compared with Fig. 44. Compiled from sources as listed in bibliography to Chapter 8.

further source of synthetics production. The Dzerzhinsk group of works near Gorkiy make fertilizers, toxic substances, synthetic fibres, caustic soda, tars, and organic solvents.

Even before the Revolution, Leningrad imported raw materials for chemicals manufacture. It uses both imported latex for natural rubber and potato alcohol for synthetic production. Mineral fertilizers are made from Kola apatites and sulphuric acid from Ural pyrites, while pharmaceuticals, soap, cosmetics, plastics, celluloid, viscose fibres, and paints are produced.

In the South, the Donbass and the Dnepr bend are important producers of heavy chemicals. The earliest Donbass works were makers of soda and superphosphates, but there is now a large and varied coal by-products industry, with benzol, ammonia, tars, nitrogenous substances as well as dyestuffs. The main centres are Gorlovka, Rubezhnoye, Lisichansk, Donetsk-Makeyevka, Rutchenkovo, and Yenakiyevo ; while Slavyansk and Artemovsk are important for sodium salts and there is by-production from the Konstantinovka zinc smelting, which also uses coal brasses, local and Ural pyrites to be one of the largest sulphuric acid producers. Sulphuric acid is made in Odessa, Vinnitsa, and Sumy (which makes superphosphates from Kola apatites and Central Russian phosphates) ; Stanislav potash is worked at Stebnik and Kalush. Light chemicals are made at Kharkov, Kiev (along with pharmaceuticals), and at Odessa, where there are also plastics ; synthetic fibres come from Chernigov and rubber from Chernovtsy.

Diverse mineral resources, metallurgy, and mining provide a wide range of raw materials in the Ural which now produces only a little less mineral fertilizer than the Ukraine, based on the immense potash and associated deposits of Solikamsk-Berezniki, one of the largest centres of heavy chemicals in Russia. Berezniki is a major sulphuric acid producer, with subsidiary centres using pyrites, coal brasses, flue gases, and even copper tailings brought from Balkhash, at Kirovgrad, Krasnouralsk, and Mednogorsk. Coal by-products come from Gubakha on the Kizel coalfield and from imported and local coals in coke ovens at Nizhniy Tagil, Magnitogorsk, and Chelyabinsk ; Pervouralsk makes chemicals from chromite and methyl alcohol, acetone and cellulose from wood are produced in Asha, Zlatoust, and Novaya Lyalya, while the Ural is also the main centre for asbestos.

Growth of the Ural-Volga oilfields has brought petroleum refining to the Volga-Kama basins and the planned expansion of petrochemicals to replace 'agricultural' sources for raw materials. Kuybyshev is an ethylene producer; Ufa and Sterlitamak are growing centres, with synthetic rubber at the latter based on a petrochemicals base. Other centres are Perm, Saratov, Syzran, and Volgograd. Petrochemicals are produced from Emba oil at Orsk and development will take place away from oil and gas fields as pipelines develop.

Western Siberian chemicals are derived primarily from coal by-products in the Kuzbass at Kemerovo and Novokuznetsk: plastics, drugs, and dyes come from Kemerovo, Novosibirsk, and Anzhero-Sudzhensk, with artificial fibres from Barnaul and Leninsk-Kuznetskiy. Omsk has a growing petrochemicals industry using piped Volga oil, and Novo-Blagoveshchensk consumes salts from the lakes of the Kulunda Steppe. In Eastern Siberia, Krasnoyarsk already has a synthetic rubber works, using ethyl alcohol from cellulose hydrolysis, and makes artificial fibres. At Kritovo, piped supplies of Volga oil will be used for petrochemicals. The Cheremkhovo coals are well suited to chemicals production and there are also local supplies of limestone and gypsum: the main centre is the salt-mining town of Usolye-Sibirskoye, but Angarsk has petrochemicals associated with oil refining. There have been reports of synthetic petrol production from dismantled German plants now working in the Kuzbass and Cheremkhovo fields. Both Western and Eastern Siberia have a considerable potential for development of chemicals extraction from their immense forest resources.

In Kazakhstan, in the first five-year plan, the Alga-Aktyubinsk works were opened to make sulphuric acid from Ural pyrites in order to treat local phosphates, while boracic acid was made from the Inderborskiy deposit. Aralsulfat, working local deposits, is a major sodium sulphate producer. In 1937, mineral fertilizer manufacture started from the Kara Tau deposits processed at Chulak Tau by the side of a reservoir on the Tamdy river; these phosphates also go to the Kokand superphosphate works and to the works at Dzhambul which uses sulphuric acid from Tekeli and Achisay. The large Chirchik works in Uzbekistan make nitrogenous fertilizer by electrolysis of water and atmospheric nitrogen. At Temir-Tau in Kazakhstan, a wartime calcium-carbide plant has been associated with development of synthetic rubber making.

Transcaucasian producers are associated with local petroleum and non-metallic minerals. Synthetic rubber is made from petrochemicals from Baku refineries, while calcium carbide is used for it at Yerevan. Tbilisi produces plastics; paints and lacquers come from Kutaisi. In Northern Caucasia, Tuapse and Groznyy are associated with petrochemicals.

Wood-using Industry

Almost a third of the land area of the Soviet Union is forested, mostly by coniferous stands, and contains between a fifth and a third of world timber reserves; but because of inaccessibility and its virgin nature, particularly in Northern Siberia and the Far East, almost two-thirds of the timber is mature or over-ripe, with large areas in Northern and Western Siberia of poor forest interspersed by bog. Around three-quarters of the production and consumption of timber is in European Russia; the annual cut of

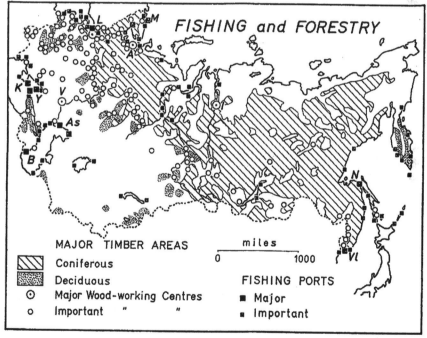

FIG. 49

Fishing and forestry. Large areas of the forests may not be suitable for exploitation, while many small areas of high-yield forest in European Russia are too small to show at this scale. Only principal fishing ports are shown. Forestry based on *Atlas SSSR*, Moscow, 1962, pl. 90–91, and *Geograficheskiy Atlas*, Moscow, 1954, pl. 111. For sources of fishing information, see bibliography to Chapter 8. A: Arkhangelsk; As: Astrakhan; B: Baku; I: Igarka; K: Kerch; L: Leningrad; M: Murmansk and outports; N: Nikolayevsk; V: Volgograd; Vl: Vladivostok; Y: Yeysk.

timber exceeds annual growth in the Centre, Byelorussia, the Ukraine, the Volga and the Baltic region, in contrast to less than eight per cent of annual growth in Siberia.[7]

Central European Russia produces about a fifth of the total timber output, notably from Kirov, Kostroma, and Gorkiy *oblasts*, and has a large and diverse wood-using industry augmenting its consumption by imports from other regions. Almost a fifth comes from the Ural, mainly supplied from Perm and Sverdlovsk *oblasts*. In the north are conifers, but the southern Ural has outliers of deciduous forest, though in many parts the best stands were eaten into by the charcoal-smelting iron industry. There is a big dispatch of timber to other regions along the Kama-Volga waterway. The European North supplies over a tenth, chiefly from Vologda, Arkhangelsk *oblasts* and the Komi A.S.S.R., with good stands within easy reach of the Northern Dvina and its tributaries. Some is exported abroad from Arkhangelsk, where there are many sawmills, also found along the Vologda-Arkhangelsk railway, and around Kotlas and Syktyvkar. Inta on the Usa river supplies Vorkuta with pit props. Leningrad *oblast* and Karelia supply not quite a tenth, but there are sawmills, prefabricated house makers and other plants along the Leningrad-Murmansk railway and Lakhdenpokhla has a big plywood factory. Leningrad, with many wood-using factories, also exports wood, as does Murmansk. Only about 7 per cent of Soviet timber comes from Western Siberia but Eastern Siberia produces about 13 per cent. Some is exported via the Ob-Irtysh waterway, but the main exporting port is Igarka on the Yenisey. Timber comes from the southern fringe of the forest belt and from the Altay and Sayan foothills, but construction of large reservoirs at hydro-electric stations has necessitated clear-felling of their inundation areas. Wood-using industries are sited chiefly along the Trans-Siberian railway at Novosibirsk and Krasnoyarsk, but match combines operate at Barnaul and Biysk. Novosibirisk and Barnaul also produce turpentine, camphor, and resins; Tyumen has a big plywood factory and Kurgan, wood-chemicals, also produced at Kansk; prefabricated houses are made in Eastern Siberia. The Far East supplies probably a twentieth of the timber and provides species not found elsewhere, with its sawmills, plywood factories, and house-building plants situated along the Trans-Siberian railway and at Vladivostok and Nakhodka.

The Ukraine, Byelorussia, the Volga and the Baltic republics, the main deficit regions augmenting local production by substantial supplies from other regions, together supply less than a fifth of the total. Sawmilling is particularly developed on the Volga, along which so much timber is floated, and Volgograd handles timber moving to the Ukraine and Northern Caucasia. These regions are marked by their production of furniture, matches, plywood, and prefabricated houses. Central Asia and Kazakhstan, poor in forests, import most of their wood from Siberia and the Ural or from Caucasia, which also sends wood to European Russia.

About half the timber is used for building and wooden houses; another quarter goes for pit props; the remainder is used for paper, cellulose, matches, and small quantities for laminates. Poor timber and waste are sold for domestic fuel, though this use may form a big share of production in peripheral regions. Selective felling does not appear to be used in the main forest regions; in many agricultural districts, afforestation is undertaken for land reclamation, dune stabilization or shelter belts. Felling is primarily a winter occupation, so that timber can be floated downstream on the spring thaw.

Paper and card are made predominantly from wood. Over a quarter of production is from the North-west, with big mills around Lake Ladoga and at Kondopoga and Petrokrepost; over a fifth is from the Ural, chiefly Krasnokamsk, Borovsk, Novaya Lyalya, and Krasnovishersk (for special-quality papers); Ufa also makes some paper. Not quite a fifth is from the Centre, notably near Gorkiy at Balakhna-Pravdinsk, which receives wood from the Unzha, Northern Dvina, and Karelian forests. Almost a tenth comes from the Baltic republics and a little less from the Far East, where the small mills on Sakhalin are reputedly uneconomic.

Textile Industry

The textile industry is markedly concentrated in the Central Industrial region and has shown little shift in its major location patterns since before the Revolution, so that it truly lies neither near its raw material sources nor near some of its most important markets.[8] The Ukraine, Siberia, and the Volga-Ural regions obtain the greater part of their textiles from the Centre, which supplies almost 80 per cent of the cotton cloth, over two-thirds of the woollen cloth and even three-quarters of the 'silk' (mostly artificial) cloth and the same share of linen. The North-west, the

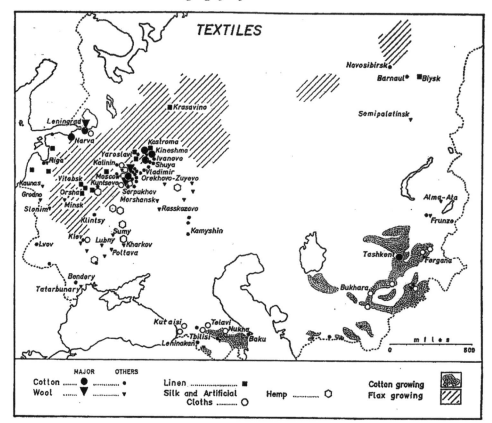

FIG. 50

Textiles. Sources as listed in bibliography to Chapter 8. Areas of cultivation based on maps in *Atlas Selskogo Khozyaystva SSSR*, Moscow, 1960. Owing to space limitations, several smaller producers east of Moscow have been omitted.

Baltic republics, and Byelorussia supply about 12 per cent of the woollens and a fifth of the linen cloth; but they contribute only about a twentieth of the cotton cloth, a share similar to the new factories in Central Asia, which is the centre for locally produced natural silk and contributes the bulk of Russian raw cotton, though smaller quantities of both come from Transcaucasia; North-west European Russia and Byelorussia grow flax. Russia nevertheless still imports raw cotton from abroad as well as wool.

The big concentration of cotton mills lies east of Moscow, particularly along the moist valleys of the Klyazma and its tributaries: these include Ivanovo (coarser counts), Vladimir, Kovrov, Shuya, and Kineshma, while nearer to Moscow, itself an important producer, lie Orekhovo-Zuyevo, Mytishchi, Noginsk, Pavlovskiy

Posad, and several other towns. Yaroslavl also makes cotton goods. In tsarist days, weaving greatly exceeded spinning, but production is now more balanced between the two branches. Leningrad is also a significant producer. Of the mills in the Baltic republics, Narva and Tallin are important; in the Ukraine, Poltava, Lvov, and Kirovograd are producers and Bendery in Moldavia. Modern mills at Krasnoarmeysk and Kamyshin serve the Volga; in Siberia, Barnaul is a growing textile centre. The principal mills in Central Asia are at Tashkent, but smaller units work in Frunze and Fergana, while there are small plants in Transcaucasia at Leninakan, Gori, Kirovabad, and Baku. These newer mills appear to be primarily concerned with yarn production, while weaving and finishing remain concentrated in the Centre.

Flax, the traditional Russian fibre, is widely grown in the northern parts of the Russian Platform, where there are numbers of small scattered retting and linen works. The largest producers with the biggest mills are the Yaroslavl and Kostroma districts east of Moscow followed by the Byelorussian works, notably Vitebsk, Vysochany, and Orsha. Vologda is a significant producer and there are mills at Pyarnu, Birzhai, and Yelgava in the Baltic. Small production of fibre and cloth comes from Biysk and Shadrinsk in Siberia.

The small woollen industry is relatively widely dispersed, but the major producers are concentrated in and around Moscow, which produces itself a third of the cloth output: the main mills are Kuntsevo, Pavlovskiy Posad, and Zagorsk, besides several smaller centres, and there are worsted mills at Ivanovo. Other centres are mostly small and include Bryansk, Borovsk, and Klintsy, with pre-revolutionary mills at Morshansk and Rasskazovo; a group of mills lies near Barysh in the Volga region and Leningrad is also a producer. Modern mills at Kiev, Sumy, Kharkov, Lubny, and Kremenchug supply a diverse assortment of products in the Ukraine; Byelorussian mills are Grodno, Minsk, and Slonim; and Tatarbunary serves Moldavia. Riga and Kaunas also have some production, while mills have been developed at towns in Caucasia near to raw materials. The Alma Ata mills in Central Asia, along with Frunze, Dushanbe, and Mary, use improved qualities of local wool. Central Asia produces about a fifth of Soviet washed wool; raw wool is washed in the Ukraine and in the Centre, which supplies a quarter in contrast to its two-thirds of cloth output. A washing plant at Khadkal on Ubsu-Gol in Mongolia supplies raw wool to the Soviet Union. Rug and carpet-making is a traditional industry in Central Asia and Transcaucasia, though execution and quality are usually considered inferior to India and Persia.

'Silk' cloth, much from artificial fibre, comes from the Centre, notably Moscow, Kuntsevo, Kalinin, and Naro-Fominsk, while Leningrad produces slightly more silk cloth (again much artificial) than the Central Asian republics. Kiev is the Ukrainian centre; while Kutaisi, Tbilisi, and Telavi are new works and Nukha an old centre in Transcaucasia, which produces raw silk, though the main producer is Uzbekistan in Central Asia. Local hemp is made into cord at Orel, Kharkov, Kirovograd, and Smolensk; imported jute is used for sacking in ports such as Odessa, Leningrad, Murmansk, and Arkhangelsk, while locally grown jute is used at Kirovabad and Frunze. Although the Centre dominates the artificial-fibre cloths, a widespread manufacture is envisaged for the future. Moscow and Leningrad remain the biggest single centres for knitted goods and hosiery, though over half the output comes from outside the R.S.F.S.R.

Fisheries

The substantially expanded Russian fishing industry accounts for less than a tenth of the world landings with a catch of about 2·54 million tons annually, of which 500 thousand tons come from fresh water. Distribution over long distances to interior consumers accounts for smoking, drying, or freezing of about 1·7 million tons, to which may be added 225,000 tons canned and 40,000 tons reduced to fish meal. A great deal now comes from the Russian-landed catch and the big import of foreign fish (notably herring) has fallen.[9]

Landings are dominantly at North-western ports, closely followed by Far Eastern ports, while the Volga ports land fish from the river, but particularly Caspian fish at Astrakhan; heavy landings from the warm, shallow Sea of Azov are made. Summer fishing accounts for most fish landed in the Northern ports. The Caspian, Aral Sea, and Lake Balkhash are the main sources of fish for Kazakhstan; Uzbekistan also takes fish from the Aral Sea. In Siberia, a rich and diverse fresh-water fauna is available, but a considerable proportion of the catch taken by natives probably goes unrecorded.

Far Eastern waters are fished by both natives, who show great skill in using primitive equipment, and by mechanized fleets. Species of river salmon form the biggest catch and are mostly canned or smoked; salted salmon eggs (red caviar) are a delicacy. Herring and sardine, also exported in processed form, are taken in shallow bays around Vladivostok and Sakhalin. Pacific cod comes mainly from Kamchatka, the Commander Islands, and Anadyr Bay. The west coast of Kamchatka, notably Khairyuzovo Cape, is renowned for crabs, mostly for canning, caught when they come into shallow water in spring. Claims of overfishing have led to Japanese fleets in these waters being restricted by the Russians. Whaling is conducted in the Bering strait by ships based on Vladivostok, though ships from European ports as well as here now take whales in Antarctic waters. Trepang and seaweeds are also exploited. The much-diminished hunting of seal, walrus, and other sea creatures is now mainly around the Chukchi Peninsula.

In the Caspian, the broad expanse of the northern fresh-water basin, warmed by the sun and fed with nutrients brought down by the Volga and Ural, has a rich fish fauna of types also found in the Volga. Particularly important are sturgeon, salmon, beluga, whitefish, and herring in several forms; but the Caspian roach (*vobla*) is taken in the greatest quantity. About three-quarters of the catch is made round the Volga mouth and handled at Astrakhan, well placed for dispatch to major consuming centres by good transport, and with adequate labour and nearness to tin plate and to the Baskunchak salt deposits. However, the annual catch is now only half that of 1913 as fishing has been adversely affected by the falling level of the Sea, which has reduced the shallow-water area in which anadromous and semi-anadromous species breed. Hydro-electric barrages have also affected migration of species in the Volga, so that new spawning grounds have to be provided; the decreased inflow of fresh water has brought rising salinity as well as reducing the amount of nutrients carried into the Sea. In the deeper, southern basins, the catch is largely sea pike, sea perch, and sprats, and the falling level has had less influence. Seal-fishing is now conducted from the Dagestan coast and from Bautino on the Mangyshlak Peninsula.

Atlantic and Arctic waters mix in the north to provide a rich plankton diet on which fish abound, and the North Atlantic Drift keeps the western areas ice-free in winter. The main banks in the Barents Sea are Central Rise, Goose Bank, the Rybachye Peninsula, Kildrin Island and waters about Novaya Zemlya. Finmark waters and Bear Island are also fished. The catch, largely cod, sea perch, and herring, is handled mostly at Murmansk and subsidiaries at Sayda Guba, Teriberka, and Port Vladimir, where a great part is processed or canned. Cod-liver oil extraction, fish meal and glue-making are also developed. Arkhangelsk trawlers work the White Sea in summer and shift to the Murman coast in winter. Coastal dwellers hunt seal, chiefly based on Mezen, and the White Sea herd is estimated at several million head. Some salmon-fishing is conducted in the Ob and Yenisey estuaries. At times, British vessels have been allowed to fish to within three miles of the coast instead of the twelve-mile limit claimed by the Soviets.

The warm, shallow northern waters of the Black Sea and the waters of the almost enclosed Sea of Azov, fed by the many rivers flowing to them, abound in fish. The Black Sea is, however, not quite as rich as the Sea of Azov, and no animal life is found below 600 feet. Its main fisheries are on the shallow shelf along the Ukrainian *liman* coast. The catch in the Danube delta is handled at Izmail. Bream, chub, sturgeon, and kefal are the main species, with carp from the marshes of the Kuban delta.

Russian and satellite boats land catches of Baltic, Atlantic, and North Sea herring, Atlantic cod and other species at Soviet Baltic ports. Flatfish are taken in local waters : Tallin is known for its pilchards, Riga Bay for sprats, and eels are extensively caught. Most ports have freezing and processing, notably Baltiysk, Klaypeda, Riga, Liepaya, and Ventspils, as well as a new base on Khuima Island. Russian fishing fleets with their own supply and mother ships are common now in the North Sea and Atlantic waters and have even fished the African coasts.

MAJOR INDUSTRIAL REGIONS

Examination of the major industries shows that despite their apparent wide scatter over the country, they are primarily associated with a few favoured regions which appear on the map as marked agglomerations of industry, and outside them industry has primarily a service and maintenance character or is formed by lighter branches such as food processing and consumer goods.

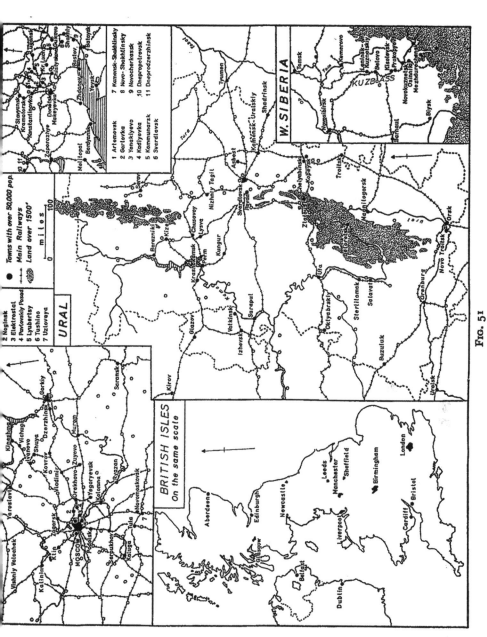

Fig. 51

Industrial regions. *Atlas Mira*, Moscow, 1954.

The geographical pattern of industrial distribution represents the product of the struggle between the centripetal tendencies of industry in seeking peculiarly favoured sites and the centrifugal forces created by Soviet planning attempting to scatter industry broadcast across the country. The bulk of Soviet industry lies in the Ural and European Russia, which together probably contain about three-quarters of the total, with Moscow-Gorkiy, the South (the Ukraine and Eastern Donbass), and the Ural as the three largest concentrations, but smaller groupings lie around Leningrad and in the Volga basin. Outside European Russia, Western Siberia is the most important group though comparatively small; other groups of industry occur in Northern Caucasia, Transcaucasia, and the Central Asian oases, with very small industrial districts in Eastern Siberia (the Cheremkhovo-Irkutsk region) and the Far East in the Ussuri valley-Vladivostok areas. Some thirty large and many small industrial towns lie outside these groups.

While these agglomerations stand out on the map of Russia, they are not like the closely associated industrial towns and landscapes such as South Lancashire, the Potteries, or even the more dispersed German Ruhr. Even the industrial towns immediately around Moscow are spread over an area larger than South Lancashire and the West Riding industrial area together and lie as far from those around Tula as London is from Birmingham; Leningrad and its industrial satellites form an area as large as industrial South Wales; the Donbass and its associated industrial towns cover an area larger than the whole of the English Midlands; but the Ural industrial region is greater in area than England and Wales together.

The European South contains three sub-regions : the Donbass coalfield, the Dnepr bend, and the iron-mining districts of Krivoy Rog. The region is closely linked to the big industrial town of Kharkov on the north and the metallurgical industries of Zhdanov and Kerch on the south, as well as to the lower Volga, Northern Caucasia, and the varied industries of Kiev. The raw materials base is the rich and varied coals of the Donbass (notably supplying good coke), the high-quality iron-ore of Krivoy Rog and the poorer qualities of Kerch, limestone for fluxing, plus local resources of manganese, salt, mercury, and natural gas (Shebelinka). Pipelines bring Caucasian petroleum and natural gas : electric power is taken from large thermal generators in the Donbass and other big towns and from hydro-electric stations on the Dnepr, at the Tsimlyansk Barrage and the big Volgograd station on the Volga. Water, both for human and industrial use, is a serious deficiency in the dry countryside of the Donbass: local supplies are taken from underground and from rivers, augmented by piped water from the Donets.

The association between raw material and fuel resources of the Donbass, the Dnepr bend, Krivoy Rog, and Kerch has led to the development of a major iron and steel industry, with emphasis on ferro-alloys and electric processes in the Dnepr bend. By-products of coking plus imported and local raw materials are the basis of the well-developed heavy chemicals industry. Availability of raw metal and a large local market has attracted engineering, notably to the towns on the edge of the coalfield or beyond, like Lugansk and Kharkov, some of which are producers of national importance.

The local railway network, particularly well developed in the coalfield, is linked by trunk lines to other regions, with three arterial lines to the Central Industrial region. To cope with the especially heavy freight traffic, electrification has been undertaken on the most important routes. Large quantities of coal and coke and iron-ore are shipped by rail and water to other regions; and the region is one of the major national suppliers of semi-finished metal goods, like pipes, sheet, and girders, heavy chemicals and heavy engineering goods such as railway locomotives, boilers, and agricultural machinery, as well as precision articles. It draws mainly foodstuffs and consumer goods from other regions and imports quantities of northern timber via the Volga-Don Canal, but some raw materials and semi-finished metal goods not produced locally are also imported, particularly from the Centre, the Ural, and Transcaucasia.

The Donbass is a landscape of mines and factories set amid towns and workers' settlements. The mining settlements tend to cluster in valleys near to water along the slopes of the Donets Ridge, on whose flanks many collieries lie. More open sites on the exposed interfluves above the straggling valleys and erosion gullies have been selected for modern mines and factories, despite greater difficulty in obtaining water. Widespread gullying in many parts has cut the open steppe into a rough topography, with a scarcity of flat sites and requiring more earthworks and steeper gradients along railways than usually common in Russia. Much has been done to improve the industrial towns by addition of modern buildings, parks, and greenery.

The Central Industrial region may be subdivided between Moscow and its immediate suburbs, the textile towns of the Klyazma basin, the lignite-mining district of Tula, and the industrial node around Gorkiy. Fuel is provided from lignite on the south and peat on the east, piped supplies of oil and natural gas from other regions, along with imported Donbass coal. Electric power is generated by local thermal stations as well as brought from the hydro-electric barrages on the Volga. Nearly all the other raw materials required by industry have to be brought from other regions. The soft water of the Klyazma is an important locational factor in the textile industry. With large reserves of highly skilled labour, nearness to markets and to research and development, using imported raw materials or locally produced steel from scrap or imported pig, the region has the largest national concentration of varied engineering industries; and it is the centre of Soviet consumer-goods industries. Chemicals, using local by-products or imported materials, are primarily associated with lighter branches, like pharmaceuticals, photographic film, or cosmetics.

Transport is provided principally by railway, as throughout Russia, and eleven main lines radiate from Moscow, though there is a growing network of roads and short-distance road haulage. Moscow is linked by river and canal to the Volga-Kama waterway. It imports nearly all its raw materials from every part of the country, to which it sends a diverse range of manufactured articles, while the region remains the technical and organizing centre of Soviet industry despite recent decentralization; and some forty million people live within about 300 miles' radius of Moscow. Industrial plants are scattered throughout the city, but engineering and chemicals are notably found in Khimki and Nogatino, adjacent to the Moscow river and Volga canal, or in Lyubertsy and suburbs along main railway lines from the Donbass. Many of the older industrial towns of the Centre are unplanned and squalid, like Podolsk and Serpukhov; and around Tula and Shchekino industry is spreading into farmland, even within sight of Tolstoy's idyllic home, *Yasnaya Polyana*.

There is not an even spread of industry in the immense Ural region, but several clusters occur. On the eastern slope, in the north around Serov, there are mining and metallurgical towns, with another larger group centred on Nizhniy Tagil, and a cluster of towns lies around the transport and engineering 'capital', Sverdlovsk; further south, clusters include Chelyabinsk, Magnitogorsk, and Orsk-Khalilovo; on the western slope, there is an industrial sub-region in the upper Kama, near Berezniki-Solikamsk, on the Kizel coalfield and at Perm; further south, industrial and mining towns lie along the Belaya valley south of Ufa-Chernikovsk. The eastern slope is richer in minerals and many towns here are situated on a belt of intense mineralization; on the eastern slope, there is a contrast between the northern 'forest' towns and the southern towns on the steppe fringe.

The Ural is particularly rich in iron-ore, ferro-alloy metals, non-ferrous metals, and non-metallic minerals. The magnetite iron deposits of the eastern slope are richer than the sedimentary deposits of the west, but along the western foot are important petroleum and salts deposits. Unfortunately, there is a deficiency of good coals, particularly those suited to metallurgical coke, though the small deposits of both coal and lignite are used to generate electricity. By its physical form, the Ural offers little possibility of hydro-electric generation, though stations on the Kama and Volga now send current to the region. Petroleum is received from the Volga fields and from Emba by pipeline, and natural gas will shortly be piped from Western Siberia (Berezovo-Shaim) and Gazli near Bukhara. Great quantities of coal for metallurgical use are received from the Kuzbass and Karaganda and rising amounts of iron-ore are sent from Kazakhstan.

In the first five-year plans, the development was closely linked to Western Siberia, but recently the two regions have taken more independent courses. Heavy iron and steel production, typical of the region, comes from vast, modern, integrated plants, like Nizhniy Tagil, Chelyabinsk, and Magnitogorsk, and from five small integrated works. Heavy chemicals have been associated with ferrous and non-ferrous metallurgy and use both local and imported raw materials besides coke oven by-products. Engineering developed in the thirties, but the greatest impetus came from wartime evacuation from the west and the need for armament manufacture well behind the front: its products are now mostly heavy mining and rolling mill machinery, the heavier forms of transport equipment and even industrial electrical installations. Paper is made in the north using wood from the immense forests.

There is a thin but basically sufficient network of railways, some electrified, and three east-west trunk routes join it to Western Siberia and the Volga basin. A sparse network of not very good roads exists between towns and the railways are the main means of inter-urban communication. The region exports large quantities of pig for steel conversion and semi-finished steel goods, non-ferrous metals, and chemicals and paper, besides a growing volume of heavy engineering equipment. It imports coal (mostly for coking), petroleum, natural gas, and such semi-finished metal products not made locally, besides engineering equipment, consumer goods, and particularly foodstuffs.

On the eastern slope, industrial towns lie either in relatively broad valleys amid low, rounded, forest-covered ranges, like many sites on the western slope, or on the more open country of the Trans-Uralian peneplain; south of Chelyabinsk, the countryside becomes drier and reservoirs have had to be built on some rivers before industrial development could take place. Many newer works and settlements rise amid virgin forest or steppe and sometimes stand near to great open-cast workings from which they get their raw materials.

The industrial towns of Western Siberia lie mostly in the valley of the Tom, on the Kuzbass coalfield, but there are big industrial outliers at Novosibirsk, Barnaul, and Rubtsovsk. The raw materials available locally are coal of coking quality and iron-ore in the Gornaya Shoriya, and there are large reserves of timber. The Altay contains non-ferrous metals and there are nearby resources of tungsten. Coal and lignite are used for electricity generation, but current will also come from hydro-electric barrages under construction in Southern Siberia. Petroleum from the Ural-Volga fields is imported by pipeline and diminishing quantities of iron-ore are taken from the Ural. The region is a producer of coal and semi-finished iron and steel goods widely exported throughout Siberia and Soviet Central Asia. The coke ovens have provided chemical by-products which are being extended by petrochemicals. With local semi-finished metals and supplies from the Ural, engineering has grown at Novosibirsk and the Kuzbass towns and a big tractor and farm machinery works operates at Rubtsovsk. The textile industry of Barnaul works raw cotton brought from Central Asia, and Siberian timber is used for constructional purposes and extraction of resins and other substances. Communications are provided by the focusing on the region of the trunk Trans-Siberian, South Siberian, and Turksib railways, while it is linked to Northern Siberia by the rivers Ob and Yenisey. Novosibirsk, one of Russia's largest towns, has grown since the building of the Trans-Siberian railway in 1894, and the large iron and steel town of Novokuznetsk was raised in virgin countryside in the late nineteen-twenties. The other towns have grown out of old Russian trading centres founded in the eighteenth and nineteenth centuries and often remain typical Siberian wooden towns.

The industrial structure of the Volga is formed of a widely scattered yet related industry, based on availability of petroleum natural gas and salts; it can easily obtain timber and potassium minerals from the Kama basin and products of the metallurgical industry of the Donbass and Ural as well as the manufacturing industries of the Centre. Coal, brought from the Donbass and even the Kuzbass, is being replaced as a fuel by local petroleum and by growing availability of locally generated hydro-electricity. The region is developing as a centre of electro- and petro-chemicals, though processing timber (floated along the river) and foodstuffs are important. Using local limestones, it is a major cement producer, and engineering is found in most of the large towns lining the high right bank.

Linked by canal to Moscow, the Black Sea, and the Baltic (at present being enlarged), and with access to the Caspian, the north-south route of the Volga-Kama waterways is one of the busiest in Russia, while several main east-west trunk railways cross the region and others terminate on the river's west bank. It thus handles traffic associated with its own economy and also transit movement between other regions.

Industry around Leningrad has its main emphasis on manu-facturing from imported raw materials from both within and without the U.S.S.R. It imports coal from the Donbass and Vorkuta and even from Poland, and uses local peat and oil shales to generate electricity, which also comes from hydro-electric stations. Petroleum is brought from the Ural-Volga field and Ukhta. Apart from converting its own steel from imported pig and local scrap, its raw metal requirements come from the Ural, the Donbass, the Centre, and increasingly from the new Chere-povets plant. Leningrad produces a wide range of machinery and transport equipment, chemicals, textiles, and consumer goods. A large port of international significance, it is well served by railways radiating to all parts of the country and linked by canals to the White Sea and to the Volga. The factories are scattered around the outskirts, along the main navigable channels of the Neva delta, in the harbour on the south-west, and in small industrial towns nearby such as Gatchina and Kolpino. The consumer-goods industries of this region are even more pronouncedly developed in the Baltic republics in towns such as Riga.

The industries of Northern Caucasia and Transcaucasia, both associated with the Volga region via the Caspian and the Ukraine across the Black Sea, bear similar features. The common major resource is petroleum, though coal, iron-ore, ferro-alloy, and non-ferrous metals and non-metallic minerals enrich Transcaucasia. Both regions have a large petroleum-refining capacity which until recently treated large quantities of crude petroleum from other fields. Taking limestones from the Great Caucasus, Novorossiysk is one of the largest cement producers in Russia, and there is an extensive industry processing the grain and animal products of Northern Caucasia and the citrus fruits, tea, and grapes of Transcaucasia. Engineering, mainly to serve agriculture, using raw metals from the Ukraine, has been developed in Northern Caucasia. Iron and steel making, using local coal and ore, and ferro-alloy production have been developed in Georgia to provide metal for the Transcaucasian pipe, tube, and petroleum engineering industry. Cotton, wool, and silk grown in Transcaucasia, and a long native tradition, have been harnessed to the development of a small, modern textile industry.

Central Asian industry has been the product of Soviet times, using ferrous and non-ferrous ores and non-metallic minerals developed in the nineteen-thirties, as well as local coal and petroleum and natural gas. Hydro-electric stations at irrigation barrages provide electric current. Heavy industry is best represented by chemicals, principally the production of mineral fertilizers for use on intensively-farmed irrigated lands. Engineering developed during the Second World War, supplied with metal from Siberia, though pig from the Kuzbass and local scrap are used for steel conversion at Begovat. Cultivation of cotton and silk have been the basis of a modern textile industry associated with an old native tradition, with large mills at Tashkent. The principal industrial centres are the republic capitals, all served by railways, but a grouping of industry is seen on the lower foothills of the Fergana valley and in the chemicals producers on the desert fringe. Krasnovodsk refinery has been developed as an important petrochemicals plant. By the Turksib railway, the region is closely linked to Western Siberia, from which raw metal, coal, and manufactured goods as well as Altay wheat are received in exchange for cotton, chemicals, and products of oasis agriculture.

In Eastern Siberia, the Cheremkhovo-Irkutsk region uses local coal and salt or minerals imported from nearby producers, plus locally generated hydro-electricity. It makes mostly heavy chemicals and engineering articles, owing some of its importance to its position on the main trans-continental railway. The Amur-Ussuri region is also a producer of semi-finished goods, notably with concentration on wood by-products. At Vladivostok there is marine engineering and local supplies of steel come from Komsomolsk; Khabarovsk has engineering, oil refining of Sakhalin oil, ferro-concrete and civil engineering, and the building of river boats. Food processing, particularly canning fish and crabs, is an important element in the regional economy. It is highly dependent on imported raw materials, capital equipment, and consumer goods, brought along the Trans-Siberian railway from the west, but has a considerable economic potential awaiting development.

NOTES ON CHAPTER 8

1. The problems and attitudes of oviet industrial locational theory are discussed in :
Dobb, M., *Soviet Economic Development since 1917*. London, 1953.
Feigin, J., 'Standortverteilung der Produktion', etc., *op. cit.*
German, F., 'Economic Geography in the Soviet Union', *Bull. Material on Geography of U.S.S.R.* Nottingham, 1960.
Hooson, D., 'Recent Developments in Soviet Geography'. *Annals Assoc. Amer. Geog.* vol. 49, 1959.
Vasyutin, V., 'Razmeshcheniye Proizvoditelnykh Sil SSSR v shestoy Pyatiletka', *Izvestiya Akademii Nauk*, 1956.
Classical theory is to be found in Marx's writings and in the memoirs of Lenin, V. See also references in note 12 of Chapter 5.
2. Balzak, S. (Ed.), *Economic Geography of the U.S.S.R.*, New York, 1949, surveys the scene at the outbreak of war in 1939.
Dobb, M., *op. cit.*
Feigin, J., *op. cit.*
Komar, I., *Ural.* Moscow, 1959.
Kulischer, J., *Russische Wirtschaftsgeschichte.* Jena, 1925.
Lyashchenko, P., *History of the National Economy of Russia to the 1917 Revolution.* New York, 1949.
Mavor, J., *An Economic History of Russia.* 2 vols. New York, 1925.
Miller, M., *The Economic Development of Russia, 1905–1914.* London, 1926.
Portal, R., *L'Oural au XVIIIe siècle.* Paris, 1950.
Voznesensky, N., *The Economy of the U.S.S.R. during the Second World War.* Washington, 1948.

3. Holzman, F., 'The Ural-Kuznetsk Combine', *Quarterly Journal of Economics*, 1957.
4. Buyanovskiy, M., 'On the Question of Location of Iron and Steel Plants in Kazakhstan', *Soviet Geography*, vol. 2, 1961.
   Clark, M., *The Economics of Soviet Steel*. Harvard, 1957.
   Komar, I., *op. cit.*
   Livshits, R., *Razmeshcheniye chernoy Metallurgii SSSR*. Moscow, 1958.
   'Steel in South-West Russia', *British Iron and Steel Federation*, 1955.
   'The Russian Steel Industry', *British Iron and Steel Federation*, 1957.
   'The Russian Iron and Steel Industry', *Iron and Steel Institute, Special Report 57*, 1954.
   Pravda, 11.9.58, 'Tretya Metallurgicheskaya Baza SSSR'.
5. Alexandrowicz, C., *Comecon: Soviet Retort to the Marshall Plan*. London, 1950.
   Omarovskiy, A., 'Changes in the Geography of Machine Building in U.S.S.R.', *Soviet Geography*, vol. I, 1960.
   Price, J., and Le Fleming, H., *Russian Steam Locomotives*. London, 1960.
   Rakov, A., *Lokomotivy Zheleznikh Dorog SSSR*. Moscow, 1955.
   Skobeyev, D., 'Promyshlennost Moskvy', *Voprosy Geografii*, no. 51, 1961.
   Mikoyan, A., 'Einige Probleme des Welthandels', *Handelsblatt*, Hamburg, 20.5.60.
   Useful supplementary information appeared in the *Financial Times Review of the Soviet Union*, June 1961, and the supplement to the *Manchester Guardian* on trade with Russia, 7.12.60. Production figures appear annually for certain selected engineering goods.
   Atlas Sredney Shkoly — Kurs Ekonomicheskoy Geografii.
6. The Russians claim to be fifth world producer of plastics and sixth world producer of artificial fibres, but second for artificial fertilizers. Material is based largely on a series of articles on the chemicals industry which appeared in *Pravda* during 1958 and 1959.
   Lydolph, P., 'U.S.S.R. Chemical Industries (Abstract)', *Annals Assoc. Amer. Geog.*, vol. 50, 1960.
7. Gorovoy, V., 'The Timber Industry of Northern European Russia', *Soviet Geography*, vol. 2, 1961.
   Cherdantsev, G., *op. cit.*
   Rodgers, A., 'Changing Locational Patterns in Soviet Pulp and Paper Industry', *Annals Assoc. Amer. Geog.*, vol. 48, 1958.
   Atlas Selskogo Khozyaystva SSSR. Moscow, 1960, map. 59.
8. Cherdantsev, G., *op. cit.*
   Kutafyev, S. (Ed.), *Rossiyskaya Sovetskaya Federativnaya Respublika*. Moscow, 1959.
9. *Yearbook of Fishery Statistics*. F.A.O. Rome, annually.
   Atlas Selskogo Khozyaystva SSSR, Moscow, 1960, contains maps of the Soviet fishing industry.
   The Russians have tried breeding specimens of fish in new waters. Pacific *gorbusha* have been transferred to Atlantic waters and this variety of salmon has since been caught in the North Sea and adjacent Atlantic waters. See *Scottish Fisheries Bulletin*, no. 14, March 1961.
   In 1960, landings of fish amounted to 3,538,000 tons.

CHAPTER 9

Geography of Transport

THE great dispersion of industry, mining, and, to a lesser extent, agriculture, has depended upon the provision of transport in under-developed areas and the rapid improvement of existing facilities in other areas. Large quantities of raw materials have had to be transported up to distances of 1,500 miles to factories whose products have later needed distribution, while the dispersion has also led to increased inter-regional shipments. The growth of industry and mining in Central Asia and in Siberia has required the organization of food supplies from the countryside to the new towns, besides the maintenance of food supplies to the older towns of European Russia. Considering the great distances which separate consumers and producers in a difficult physical environment, it is not surprising that the Soviets have directed great attention to transport problems, and consequently no geography of their country can be complete without a study of different forms of transport and the relations to the physical background.

The greatest contrast with North American or West European countries is the primacy of the railways and the comparative under-development of road transport for anything other than local use. Another contrast with the United States is the slow development of pipelines, even allowing for disproportions between the petroleum and natural gas production of the two countries. Transport in Russia is foremost concerned in the movement of goods, with perhaps the exception of aircraft. Features of the Russian transport pattern reminiscent of the West in the earlier years of the century are due not so much to lacking technological standards as to the peculiar suitability of railways to Russian needs and conditions. The division of traffic between the prime hauliers shown in Table 21 brings out clearly the leading participation of railways in both goods and passenger haulage.

The railway share of long-distance passenger traffic is higher than the table suggests because these figures include urban

 DOI: 10.4324/9781003172048-9

commutation traffic. Road traffic accounts for a relatively small proportion of total traffic, although its share of the originating tonnage is high, a combination which reflects its widespread use as a feeder to the railways and as a distributor in town and country, where short hauls are most characteristic. The share of road transport has, however, risen considerably since 1945, and it is already claiming some of the short-haul railway traffic, particularly of passengers with the extension of town and country bus services. The predominant east-west orientation of movement in the Soviet Union is shown in the small share of total traffic of the extensive but largely north-south-oriented river system. The rise

TABLE 21

PARTICIPATION OF PRIME HAULIERS IN TRANSPORT IN U.S.S.R.

	Railways	Sea	Rivers	Road	Pipeline	Air
	As Percentage of Total					
Goods traffic in ton miles	83·0	6·3	5·4	3·7	1·6	n.a.
Passenger traffic in passenger miles	80·6	0·8	1·9	14·9	—	1·8
Goods tonnage	23·5	0·9	2·6	71·7	1·1	n.a.
Passengers	23·0	0·1	1·2	75·7	—	n.a.

All figures for 1956.

Source: *Transport i Sviyaz SSSR: Statisticheskiy Sbornik*, Moscow, 1957.
Narodnoye Khozyaystvo SSSR: Statisticheskiy Sbornik, Moscow, 1956.

of an industrial economy without marked seasonal traffic has tended to divert goods and passengers from rivers with seasonal restrictions on navigation; but the diversion of goods to the railways has not always been in accordance with Government policy. Slowness has also lost traffic which the rivers enjoyed in tsarist times: from Moscow to Volgograd takes 182 hours by express steamer but only 37·5 hours by the fastest train and less than two hours by air. Although the Soviet Union has an extremely long coast line, maritime traffic is a small fraction of the whole: coastwise traffic is limited by the nature of Soviet seas and the overwhelming distribution of economic activity in the continental heart of the country, while international sea traffic fluctuates with the volume of Soviet overseas trade, which at the best is never great. Despite inadequate statistical material, air traffic, particularly for passengers, appears to have developed remarkably since 1945, both internally and internationally. Pipeline transport, which accounted so far for little over 1 per cent equivalent total traffic, has been extended, notably by new lines in Siberia.

The steady increase in the share of total traffic on the railways between 1913 and 1950 has only recently been halted by transfer to motor vehicles and pipelines. Nevertheless, forty years after the Revolution, over three-quarters of the total is handled by this means peculiarly suited to Russian geographical and economic conditions. The vastness of the country, whose area is just over 8·6 million square miles, roughly three times the U.S.A., has created characteristically long hauls, notably along the 6,000-mile east-west axis, where the railway can best handle traffic in the absence of suitable river and sea routes, while the fortunate topography has helped to keep construction costs low and does not hinder operation. As discussed in Chapter 1, only the peripheral regions in the east and south offer substantial mountain barriers or dissected upland country which create serious engineering and operational problems. Tunnels have been few, but long bridges are needed to cross the great rivers, although the Russians have a recognized ability for bridging. The vast open lands, however, allow the penetration deep into the country of cold Arctic and arid Central Asian extremes of climate; but railways can usually be kept going under such conditions after other forms of transport have ceased. A further factor has been the demand for bulk long-distance haulage of heavy commodities by the Soviet concepts of regional planning and industrial location.

The densest part of the railway system is in European Russia, where to the west of the great trunk-line from Leningrad via the Central Industrial region to the Donbass, a true 'network', largely constructed before 1917, exists and is most highly developed in the industrial districts around Moscow and in the Donets coalfield. To the east of this main trunk-line, the system has a pronounced eastwards orientation towards the Volga and the gateways to Siberia; lines also fan out to Central Asia, Caucasia, and (of more recent date) the European North and North-east; and it is only since 1939 that a north-south railway link along the Volga has been completed. The small network of railways serving the mining and industrial towns of the central and southern Ural has been created since the Revolution. A strong physical control is seen in Caucasia where main lines skirt the mountain massifs, sending off short branches into the foothills, and coast-to-coast contact is made by the Suram-Kura route, involving sharp gradients, while the railway along the southern frontier has a primarily strategic rôle. Little importance now attaches to the lines which cross the Turkish frontier to Sarikamis and the Persian frontier to Tabriz. The former was built at a time when the territory of Kars was held by Russia after 1878, and extended in 1916 to its present terminus. The small railway system of the Central Asian oases, particularly in and around the Fergana valley, is linked to European Russia and Siberia by three trunk lines across poor steppe and semi-desert, two of which have been completed since 1917, and a further route to Europe is under construction from Kungrad. It is also joined to Caucasia by a train ferry across the Caspian between Baku and Krasnovodsk. Siberia and the Far East still depend fundamentally on the almost legendary Trans-Siberian railway, completed before the Soviet period, but increased in capacity and recently duplicated by the South Siberian railway, joining the southern Ural to the Kuzbass. The characteristic feature of the Soviet period has been to build feeders to these main lines from mines and outlying industrial centres. As an international link, the rôle of the Trans-Siberian has been increased by completion (1955) of the Trans-Mongolian railway, from Ulan Ude to Tsining in China via Ulan Bator, joining Russia directly to the developing Chinese industrial area between Paotow and Taiyuan.

HISTORICAL DEVELOPMENT OF THE RAILWAY SYSTEM

The reticule is compounded of elements closely related to the historical and economic development of Russia over the last century and a quarter. The first railway, opened in 1837, was of purely local significance, from Pushkin (Tsarskoye Selo) to Leningrad (S. Peterburg): but it was soon followed by main routes conceived for their political importance. The first was the remarkably straight and evenly-graded Moscow-Leningrad line (1843–1851) followed by the Leningrad-Warsaw railway (1848–1861), later extended to Vienna. Development, however, was

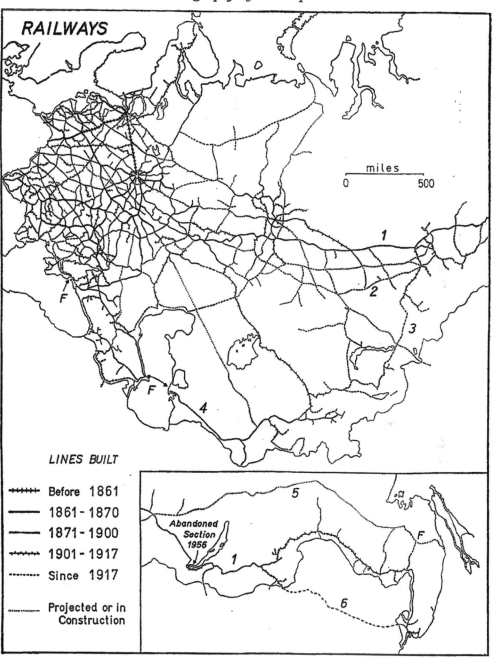

RAILWAYS

miles
0   500

LINES BUILT

┼┼┼┼┼ Before 1861
───── 1861 - 1870
───── 1871 - 1900
┼┼┼┼┼ 1901 - 1917
······· Since 1917

▬▬▬▬ Projected or in
          Construction

Abandoned
Section
1956

FIG. 52

Historical growth of railways. Based largely on Khachaturov, T. S., *Razmeshcheniye Transporta*, Moscow, 1939, and Nikolskiy, I. V., *Geografiya Transporta SSSR*, Moscow, 1960, augmented by sources listed in bibliography to Chapter 9. F: train ferry. 1 Trans-Siberian, 2 South Siberian, 3 Turksib (Turkestan-Siberian), 4 Trans-Caspian, 5 Baykal-Amur Trunk, 6 Chinese Eastern.

slow up to 1860, when building began to create an extensive local food-carrying system around Moscow and to link the city to other towns nearby (*e.g.*, to Gorkiy, 1861). Something of the slow start may be ascribed to Russian underestimation of the traffic likely to be attracted by railways: at the time of construction of the Moscow-Leningrad railway, it was thought the maximum traffic might reach 200,000 tons of freight and 120,000 passengers yearly. By the end of the first year, the line had carried 169,000 tons of freight and 780,000 passengers. During the 'sixties another element was added: lines to export grain to the ports of the Baltic and Black Sea, a development which reached the Volga by 1872. For a time after 1874, attention focused on Königsberg, for it was cheaper to ship grain to Finland and adjoining regions by the German port (the result of low German freight rates) than to send it through Russian territory. Lines were also built south to the Black Sea ports, notably Odessa, the door to Russia's black earth granary, so that during the 1870's, as the result of this building, Black Sea and Baltic ports accounted for over 30 per cent of Russian export trade; but with the flow of cheap extra-European grain in the last quarter of the century, and the ultimate separation of the Baltic ports after 1918, these railways declined, though they were still to play an important part in the changing economic geography in the Soviet period.

In 1877, war with Turkey intensified Russian interest in Caucasia, where the railway from Tbilisi to Poti had been finished in 1872, though it did not reach Baku until 1883. A final joining of the Caucasian railways to the European system remained until 1900. Several lines were built, like the Tbilisi-Dzhulfa line (1902–1913), primarily for military purposes. The slow construction in Caucasia, where Russian engineers had met difficult physical conditions for the first time, is a contrast to the rapid completion in the easier terrain of the Ukraine and Byelorussia. Industrialization spreading in Russia at the end of the nineteenth century added a further element: the first beginnings of local networks in industrial regions. The Perm-Chusovoy-Sverdlovsk railway (1878) was a start in the Ural, while its extension to Tyumen in 1882 opened a 'portage' between the Kama and Ob-Irtysh river systems. In 1876, a company had been formed to build railways in the Donets coalfield — completing the Yelenovka-Zhdanov line (1879–1881) — for the export of Donets coal by sea, and the railway from the Donets coalfield to the iron-ore mines of Krivoy Rog (1881–1884), which is still one of the economically most important in Russia. By 1904, the Donets coalfield was provided with direct communication via Dno to Leningrad.

The turn of the century brought the extension across the Volga and the start of the great pioneering railways into Siberia and Central Asia. The first move eastwards was made on completion of the Batraki bridge across the Volga; by 1888 Ufa had been reached, followed in 1890 by Zlatoust and in 1892 Chelyabinsk, the starting-point of the Trans-Siberian railway. Movement across the open steppe lands was easy, though progress slowed in the bitter winter, and Omsk was reached by 1894. The line followed the general direction of the old Siberian Tract and crossed the Irtysh on a bridge 2,100 feet long. Increased swampiness on the section to Novosibirsk reduced the speed of construction and raised building costs so that, because of the need for economy, the university town of Tomsk was by-passed, only later to have a branch connection. East of Novosibirsk, intensified winter conditions hampered work for about six months in the year, while the need to survey little-known country and to provide labour in a sparsely-populated area added to the difficulties. By 1897, Krasnoyarsk had been reached, with a long bridge over the Yenisey; but there was a temporary stop when the line reached Irkutsk in 1898. Building continued in Transbaykalia, where *permafrost* conditions were first met, and where the rough country raised new construction problems. At Sretensk the railway ended and river steamers sailed down the Shilka and Amur to Khabarovsk to connect with the Ussuri railway from Vladivostok (1891–1899), to which direct rail connection was first made across Manchuria via Harbin, opened in 1904 after delay arising from destruction in the Boxer Rebellion. The train ferry across Lake Baykal was replaced in 1905 when a 42-mile section with 38 tunnels was completed around the southern shore. The last section, from Kuenga to Khabarovsk, was finished between 1908 and 1916 to link Vladivostok finally to Moscow over Russian territory.

Railways into Central Asia arose from a Russian desire to consolidate politically a recently acquired territory. Physical conditions contrasted to those of the Trans-Siberian, but were nevertheless adverse. The first penetration was by the military Trans-Caspian railway, today a line mainly concerned with movement of oil, but designed to join the major cities (Ashkhabad, Tashkent, Bukhara, and Samarkand) and to serve the southern frontier, on which work started in 1881 at Uzun-Ade (later continued to the Caspian at Krasnovodsk) to reach Tashkent by 1899: in 1905, the Trans-Aral railway finally joined the Volga by land to these towns and hastened the carriage of cotton to the mills of Moscow and Ivanovo. Immediately before the end of the tsarist period, railways were also pushed into Northern European Russia with the lines to Arkhangelsk (1897–1898 as narrow gauge, converted to broad gauge in 1915) and from Petrozavodsk to Murmansk (1915), while in the western frontier districts a number of primarily military railways were also built.[1]

In 1917, the route length was about 43,686 miles, of which over 80 per cent was in European Russia. The lines kept conspicuously to low, open country, and in a few places crossed the 1,000-foot contour. The system was designed primarily to serve an agricultural country with mainly westward-oriented relations, but there were, however, already the signs of industrialization, reflected in the last years of the nineteenth century in the greater density of the railway system in the Donbass and around Moscow. The economy was well established on traditional sites west of the Volga and was starting to show marked locational tendencies, but interest was turning east, where in the last quarter of the nineteenth century new trunk-lines were built to join European Russia with Central Asia and Siberia. Although the railway system varied greatly in quality, it was not rudimentary or incomplete for contemporary requirements. The new concepts brought by the Revolution introduced three themes that have continuously influenced post-1917 transport policy with consequent geographical repercussions: first, the unequal territorial distribution of economic activity of tsarist times was to be removed, which was closely related to the second theme of developing 'backward' areas; and the third has been the need to give full consideration to strategic and logistical problems.

Until 1928, Soviet efforts concentrated on rehabilitation of the railway system, and a few selected lines were rapidly increased in capacity as inter-regional 'supertrunk-lines', which tended to canalize the flow of traffic and create a wide gulf between intensity of use of main and secondary railways, while new construction completed before the first five-year plan was largely of lines started before 1917. The 'gigantomania' in planning and the wish to create vast industrial complexes in the eastern districts, requiring large supply and market areas, tended to raise further the demands on transport in and between regions, and put a premium on the 'supertrunk-lines'; for example, the creation in the first plan (1928–1932) of the Ural-Kuznetsk Combine depended on linking into a single industrial unit two producing regions joined by one double-track railway across 1,500 miles of poor, open steppe country, while the siting of the great Magnitogorsk steelworks, as a part of the Combine, required extensive railway building to join it to the national system. The turn to the east reduced the importance of many railways in the European West as channels of Russian traffic, while former secondary lines became overnight the new arteries of communication : the well-developed 'grain' lines from the black earth lands to the Baltic and Black Sea ports lost their purpose, while the lightly constructed branch to the coal-mines of the Kuznetsk Basin, built originally to carry small supplies of coal for railway use, became the aorta of the new 'Second Metallurgical Base'. Some railways entirely changed their function : in the intense concentration on heavy industrial development, a third main line to carry coal and metal from the Donbass to the Moscow region was built by linking together sections of old grain-exporting lines by some new construction, undertaken during 1927 and 1928 and finished by double-tracking during 1932.

The first five-year plan, marking the start of the Soviet period of railway development, was characterized by construction in association with the development of the Ural-Kuznetsk industrial combine and the shift of railway development into areas of greater physical problems, needing more outlay on civil engineering works, while structures capable of carrying the heavier industrial loads envisaged took longer to create than light lines primarily carrying agricultural produce. The establishment of large new works at Magnitogorsk and Chelyabinsk brought the opening in 1930 of a railway between the two and the provision of an outlet to the Volga and Centre from Magnitogorsk by way of Orenburg and Orsk, providing a further southern link between the Volga and the Ural in addition to the Gorkiy-Kotelnich and Arzamas-Kazan-Sverdlovsk lines completed in the late 1920's. In expanding the Kuznetsk coalfield, better facilities were provided and the growing mining district around Karaganda was joined to the Trans-Siberian main line by a railway north to Petropavlovsk.

In European Russia, in contrast, efforts were devoted to double-tracking important lines in the Donbass and between that district and Moscow, but some short lines were built in the western frontier districts for purely military purposes to rectify the changed tactical conditions along the new frontier. Like many lines built in the Second World War, they improved the north to south links not well developed in the Russian railway system. The great achievement of this first plan was the building at immense human cost of the Turkestan-Siberian (Turksib) railway, projected and even surveyed in tsarist times. It joins Western Siberia with the Central Asian oases from Semipalatinsk to Lugovoy on the Frunze line (completed 1924) across the poor steppe country along the frontier with Sinkiang, over the rough upland country of the eastern Melkosopochnik and the desolate sands of the Semirechye, east of Lake Balkhash. The line requires several long rock cuttings and large bridges across deep, dry wadis. In the Chu-Ili mountains, there are gradients of 1 : 70. Apart from its obvious military value, it has made little contribution to the development of the country it traverses, but there is, however, a substantial traffic of wheat from the Altay country and Kuznetsk coal to the Central Asian oases and the Begovat steelworks in exchange for cotton for the mills of Barnaul and Novosibirsk. An example of railway development in the regional planning of 'complex economies' by *Gosplan*,[2] it has been a factor in the closer integration of the Central Asian republics into the Union by increasing their dependence on inter-regional exchange of goods and by using their food-producing irrigated land for industrial crops consumed elsewhere.

Slowness of railway construction in the second five-year plan appears to have been caused by a bottleneck in the supply of rails between 1932 and 1935 as new rails were used for replacements or to develop the selected supertrunks, but the main emphasis was on improving inter-regional communications. More significant was the third five-year plan, begun in 1938 but interrupted by the war, whose first intentions had been to improve inter-regional communications by completing a number of short railway links, though later, under pressure of world events, a number of military lines were undertaken; but the first serious attention to the development of Eastern Siberia was the construction during this plan of a number of short feeders, notably the line from Ulan Ude to the Mongolian frontier. The Russian railway system

increased rapidly during 1939 and 1940 by the inclusion of 12·7
thousand route miles in acquired territories, even though conversion of this track to broad gauge [3] drew away resources from
other uses. The pattern of inter-regional improvements was to
provide better connection between European Russia and the Ural
across the Volga, to improve railways to Western Siberia, and to
strengthen ties between Caucasia and other regions, while penetration of the forest and mineral wealth of the European North-
east also called for additional railway lines. On the Volga, with
the demand for more regular shipments, goods were moving
increasingly by railway in the months when river traffic ceased,
but they were forced to make a long, roundabout journey because
there was no suitable north-south route parallel to the river.
Provision of such a railway was a major project and was originally
to run along the east bank from Kazan to Saratov, where it would
cross to the Donbass; but under war conditions it was built on
the west bank, from Sviyazhk station to Ivolinskaya, near Volgograd, and came into operation in 1942. War also saw improvements in the lower Volga, by the completion of the Kizlyar-
Astrakhan railway, joining Caucasia to the Trans-Volga, and the
railway from Paromnaya (opposite Volgograd) to the Astrakhan-
Saratov line, both important carriers of Lend-Lease war materials
and Caucasian petroleum. The growth of petroleum production
in the Emba-Guryev oilfield and the increasing difficulties of
shipping on the shallow eastern Caspian, brought the construction
of a railway from Guryev to the refinery town of Orsk, opened in
1942, while a further link in the Caspian chain, between Astrakhan
and Guryev, will ultimately provide a direct communication
from Caucasia to the Ural. Loss to the Germans in 1941–1942
of the lower Don and the Stavropol Plateau hastened the building
of the western Caucasian coast-railway and train-ferry connections across the Kerch Strait between Caucasia and the Southern
Ukraine. Similarly, in the north, the German threat to the line
to Murmansk from around Lake Ladoga hastened the completion
of the Obozerskaya-Soroka railway, to join Murmansk to Arkhangelsk. At disregard to cost, the Russians also offset the loss
of Donbass coal by laying a railway to the Vorkuta coalfield,
which had formerly depended on summer shipments by sea, but
the line was not opened officially until 1950, after extensive reconstruction of the lightly-laid wartime track. It has been suggested
that the supplies of Vorkuta coal and Ukhta oil made available

by this railway have resulted in some railways from the Ukraine to the North-west, which formerly carried Donbass coal and Caucasian oil northwards, not being restored to their full pre-1941 capacity.

The gigantic 1,200-mile-long Baykal-Amur Trunk-line project announced in the third plan was to leave the Trans-Siberian at Tayshet and cross north of Lake Baykal through the roughly-dissected Aldan Plateau and Stanovoy Ranges to the Pacific near Soviet Harbour. Its prime purpose was to have been a northern relief in case the Japanese were to cut the Trans-Siberian where it runs near the Manchurian frontier, but it would certainly have opened rich metalliferous and coal-bearing country. Only two sections are known to be complete : the Tayshet-Bratsk-Ust Kut line, which carries traffic to the new hydro-electric town at Bratsk and to the Lena river for continuation to Yakutsk (opened 1954) and the section from the east bank of the Amur opposite Komsomolsk to Soviet Harbour (first revealed in 1946). Some authorities see the railway still under construction by vast compulsory labour gangs ; but in view of the extremely difficult topography and *permafrost* problems, it is more likely that it has been abandoned, if only temporarily. Soviet policy appears to prefer to build branch lines to particular areas from the main Trans-Siberian line, as in the Bureya coalfield and the suggested line extending from Bolshoy Never to Chulman in Yakutia where coal is to be mined. It is unlikely that other 'rumoured projects', such as a branch line to Anadyr in North-east Siberia, have been officially considered.[4]

The grossly over-ambitious fourth five-year plan (1946–1950) and the fifth five-year plan (1951–1955) were particularly concerned with the restoration of war-devastated western railways, but there are many indications that several pre-1941 lines have, however, not been fully restored. In Byelorussia, the route length is still 850 miles shorter than in the same area before 1939. A number of lines crossing the present Soviet frontier have certainly been dismantled. All destroyed lines were first restored to single track and as late as 1958 some second tracks had not been replaced. Interest has centred east of the Volga, principally in the Ural and Western Siberia, where the outstanding new construction has been the duplication of the Trans-Siberian railway, badly overloaded in the Omsk-Novosibirsk section, by a parallel South Siberian railway, some 250 miles to the south in the poorer

steppe country between Kartaly in the Ural and the Kuzbass at
Artyshta. About 30 per cent of the route existed before 1946, but
between 1952 and 1953, however, the missing sections, Akmolinsk-
Pavlodar and Kulunda-Barnaul-Artyshta, were completed. In
1953, a further link between Siberia and Central Asia was
completed south from Karaganda by the line from Mointy to the
Turksib at Chu, which had been first suggested in the third plan.
It involved considerable constructional problems, particularly the
supply of water, in the arid Bet-Pak-Dala, well-named 'Plain of
Misfortune'. Originally a line from Magnitogorsk to the Volga
via Sterlitamak and Buzuluk was envisaged in the immediate
postwar plans, but completion appears to have been postponed,
even though it would be an important link between the metal
producers of the eastern slope of the Ural, the oilfields of the
Bashkir A.S.S.R., and the Volga lands. The 1946 plan also
started work on a project discussed in 1928 and again in 1940,
a railway from Chardzhou (on the Amu Darya) across the Ust-
Urt Plateau to Aleksandrov Gay, the railhead on the edge of the
Kazakh steppe since 1885. Despite the problems of crossing the
desolate Ust-Urt, the line would serve to open up the mineral
wealth of the Mangyshlak Peninsula and join the Emba oilfields
to Central Asia, besides its immediate rôle of serving irrigation
projects in the abandoned Tertiary delta of the Amu Darya.
Progress was rapid until 1948, closely associated with the Turkmen Canal (later abandoned) and the Takhia Tash irrigation
scheme, after which a shortage of second-hand rails (which throws
an interesting light on the nature of some of these branches)
brought construction to a halt, although Kungrad had been
reached by 1955. A substantial and increasing part of postwar
construction has been devoted to building lines to aid operating
efficiency and to complete unfinished links. In the 1956–1960
plan a line was begun from Barnaul via Kamen-na-Obi and
Kulunda to Omsk to relieve the overloaded Omsk-Novosibirsk
section of westbound goods from the Kuzbass, but it will also help
to open to agricultural development the potentially rich Kulunda
Steppe after its amelioration by anti-flood work connected with
the Kamen Barrage has been completed. West of the Ural, an
important line to join the chemicals industry of the upper Kama
with the oil of the Second Baku and the Volga manufacturing
towns had been partly completed by 1958. The Glazov-Agryz
section is already open while the Agryz-Aktash-Bugulma section

is reported open for goods traffic and the final section (from Bugulma to Timashevo) under construction. The rapid expansion of industry and the search for new sources of raw materials has involved considerable railway building, notably in Kazakhstan, where the extension of the cultivated area has developed a large-scale narrow-gauge network; [5] the new West Karelian railway opens up large areas of forest and iron-ore deposits; and on the Pechora railway, branches are under construction into new forest areas, while an extension from Vorkuta to Labytnangi on the lower Ob is used to send out considerable shipments of coal by water. In Central Asia, the Frunze-Rybachye railway, completed in 1950, has begun to carry coal from the new Sovetskoye mines, sent by a narrow-gauge railway to Pristan Przhevalsk, then shipped across Lake Issyk-Kul to Rybachye. Another mountainous line now links Yerevan in Armenia with the shore of Lake Sevan.

No review can omit the two new international lines between Russia and China, for, when their full effect is felt, they may recast the whole geography of Inner Asia. A line completed to the Mongolian frontier from Ulan Ude before 1941 was extended to the Mongolian capital and through traffic started in 1949, after completion of the line with rails reputedly dismantled from railways in Eastern Europe. The second part of the line, built by Russian and Chinese engineers, runs from the coal-mines at Nalaykha, south of Ulan Bator, to Tsining in China, with Russian broad gauge as far as Erhlien. This 500-mile section was completed within two years and cuts the distance from Peking to Moscow by 950 miles. 'International' traffic started on 1st January 1956.[6] The second route is to cross the remote and comparatively little-known Sinkiang, to join the Chinese railways at Lanchow via Urumchi with the Soviet railways at Aktogay on the Turksib railway. The Russians have completed their section to the Soviet border near Ala-Kul and the Chinese have pressed westwards from Yumen which they had reached by 1956. The first route joins the rapidly developing Western Siberian and Transbaykalian districts to the growing Chinese heavy industrial centre between Paotow and Taiyuan, while the second line connects the coal and non-ferrous centres of Kazakhstan with the oilfields of Sinkiang and the Whang-Ho basin; and the Russians have also discussed the building of a Karaganda-Aktogay railway to complete this route.

TABLE 22

DISTRIBUTION OF RAILWAY SYSTEM IN RELATION TO AREA
AND POPULATION IN THE ECONOMIC REGIONS OF THE U.S.S.R.

Region	Track	Area	Population	Length of Track to	
				Area	Population
	'ooo km.	'ooo km.²	'ooo	km. per 'ooo km.²	km. per 'ooo pop.
North and North-west	9·7	1,600	9,900	6·0	0·9
Centre	21·6	1,000	43,400	21·6	0·5
Volga	5·8	391	10,000	14·8	0·6
N. Caucasus	5·6	430	11,000	13·0	0·5
Ural	9·1	752	15,700	12·1	0·6
W. Siberia	5·7	2,500	11,800	2·3	0·5
E. Siberia	4·9	7,200	6,500	0·7	0·7
Far East	5·7	3,100	4,300	1·8	1·3
South	21·7	635	43,300	34·0	0·5
West	12·7	397	14,400	31·9	0·8
Of which—					
Byelorussia	5·4	208	8,000	25·9	0·7
Lithuania	2·1	65	2,700	32·3	0·8
Latvia	3·1	64	2,000	48·4	1·5
Estonia	1·4	45	1,100	31·1	1·3
Kaliningrad	0·7	15	600	46·6	1·1
Transcaucasia	3·4	189	9,000	17·9	0·4
Of which—					
Georgia	1·3	72	4,000	18·0	0·3
Azerb'zhan	1·6	87	3,400	18·4	0·5
Armenia	0·5	30	1,600	16·6	0·3
Kazakhstan	9·6	2,700	8,500	3·6	1·1
Central Asia	5·1·	1,227	12,400	4·1	0·4
Of which—					
Uzbekistan	2·3	399	7,300	5·8	0·3
Kirgizia	0·4	198	1,900	2·1	0·2
Tadzhikia	0·3	142	1,800	2·1	0·2
Turkmenia	2·1	488	1,400	4·2	1·5
Average	120·7	22,121	200,200	5·4	0·6

Source: Sarantsev, P. L., *Geografiya Putey Soobshcheniya*, Moscow, 1957.
Figures for 1956.

In 1956, route length was 75·1 thousand miles, an increase of 31,317 miles since 1917, of which 12,700 miles were added by territorial incorporation. New construction amounted to 15,680 miles of railways to serve industry and agriculture, 9,075 miles as lines designed primarily for transit traffic, and 4,125 miles to join outlying towns, harbours, and mines to the main system. About 3,700 miles of route was narrow gauge, mostly 750 mm.[7] The bulk of new construction has been in the Ural and in Asiatic Russia: in 1956, about a quarter of the route length was in Asiatic Russia, and about a third of the total route length in the Ural and Asiatic Russia combined.

Roughly a further third of the route length was in the Centre, while the vast regions of the Far East and Eastern Siberia could claim only one-eleventh of the total. Despite the rapid expansion of railway mileage in the eastern regions, extending from the Volga to Western Siberia, the best-served districts still remain in European Russia, particularly in the small Baltic republics, the Centre, and the South (Ukraine). The high ratio of track length to population and the low ratio of track length to area is an index of the primarily transit function of the few railways in the Far East, in Kazakhstan, and in Turkmenistan. The bulk of railway construction has remained in the easier terrain, though lines have been pushed into exceptionally difficult country to reach specific objectives; examples are the line to the Pechora coalfield, the Frunze-Rybachye line, and the Mointy-Chu railway. Building has not been speculative, for each new railway has had a clearly defined objective, and some projects have been abandoned or postponed as soon as their objectives ceased or changed (*e.g.*, Baykal-Amur Trunk-line).

Railway development and operation are strongly influenced by local conditions of climate, terrain, and economy. Easy terrain, which has speeded construction and kept costs down, has characterized Russian railways by their low-ruling gradients: in 1940, over three-quarters of the system had gradients of 1 : 166 or less and only 1·8 per cent of the system had gradients over 1 : 66. There are few stretches which prohibit express working, though the Suram Pass route (1 : 50),[8] the Chu-Ili mountain section of the Turksib (1 : 70), and the Trans-Siberian Ulan Ude section (1 : 67) are still obstacles. The Suram Pass was one of the first routes to be electrified in the U.S.S.R. for this reason; other expedients, such as specially designed locomotives, have been used on the Ulan Ude section. Some mountain branches in Caucasia also have gradients too steep for adhesion working and special arrangements are made to work them. On the main routes, efforts have been made to ease gradients, either by engineering works or avoiding lines. About two-thirds of the Russian track length is straight track, though some lines through the Ural and on the rough shield terrain of Karelia are known for their bad curves, but the radius of curves has been increased to allow longer and heavier trains to be worked. Tunnels are the exception on Russian railways, though the route along the west coast of Caucasia has several, but the most difficult section is the 42-mile route along the southern shore of Lake Baykal which has 38 tunnels, and the new Novokuznetsk-Abakan railway (1958) has a long tunnel of unspecified length. Absence of tunnels and overhead bridges was undoubtedly a factor in the development of a loading gauge large even for the 5-foot gauge with a height of 17 feet 2⅝ inches (U.S.A. — 15 feet 6 inches; U.K. — 13 feet 6 inches) and a maximum width of 10 feet 8 inches. Broad rivers have provided the Russians with many problems in bridging, but all former wooden trestle bridges have been now replaced by steel (some of welded construction in European Russia) or concrete structures. Bridges over 60 yards in length account for 2 per cent of the total number and 22 per cent of the total length of all bridges. The longer bridges consist usually of a series of braced-girder metal spans, of box or hog-back type, resting on stone piers and abutments and constructed to withstand ice pressure besides extending across flood plains sufficiently to allow for spring floods. Typical bridges are the one across the Volga at Batraki, with thirteen 351-foot main spans and a length of 4,700 feet (completed in 1888); the Trans-Siberian railway bridges over the Irtysh (2,100 feet — 6 spans), the Ob (2,670 feet — 7 main spans) and the Yenisey (2,800-foot bridge with 6 main spans). The Russians have used train ferries in place of bridges on lines with light traffic or at particularly difficult obstacles, as for instance the train ferry used across Lake Baykal from Baykal to Mysovaya from 1899 to 1905 or the ferry still used to cross the Amur at Komsomolsk, while others are at work on the Volga, Dnepr, across the Strait of Kerch and across the Caspian between Baku and Krasnovodsk. In winter, rails are sometimes laid across frozen rivers and lakes and small groups of wagons propelled across, while at some places, specially prepared stakes in the river bed are used to give additional strength to the track. Such a track laid across the ice of Lake Ladoga was a vital supply channel to Leningrad during the German siege in the Second World War.

Shelter belts of trees along railways add a feature to the Russian landscape. As protection against snow, these were started in 1861 on the Moscow-Gorkiy railway. By 1900, there were over 2,475 miles of track so protected, and in 1950, shelter belts existed along 23,275 miles, representing some 250,000 acres of woodland. In arid areas, belts are used for protection of railway water supply or to anchor drifting sands, and in saline areas special bushes are used and belts of this type often extend up to 120 yards from the track. About 33,000 miles of track (425,000 acres) have been planted for protection against sand or snow, while a further 8,250 miles (125,000 acres) are so protected to prevent wind damage and soil erosion. Some 25,000 acres of trees have been planted to protect railway water supplies, providing a typical aspect of railway stations in the open steppe and semi-desert country.

Even in the open, level terrain, railway engineering has not been without its problems. In European Russia, railways avoid the marshy lands, such as those of the Polesye and around Lake Ilmen. In the north-west, railways are sited on the higher and drier interfluves, many of them terminal moraines, like the Moscow-Minsk railway, while the somewhat greater density of railways in the central black earth belt has arisen not only from the more intensive land use, but also from the ease of railway construction in this drier area. The railway from Petrozavodsk to Murmansk is an example of a line built quickly and at a low cost, which sacrificed distance to follow the firmer exposed parts of the Fenno-Scandian shield above the extensive muskeg-type swamp. In wet country, railways use a larger number of sleepers per mile to reduce the load on the poorly resistant sub-stratum and axle loads are kept down on most secondary lines to 18 tons or less. In the steppelands, areas of extensive gullying have influenced railway construction by forcing builders to locate lines above their possible extension, and many of the measures taken against gully-erosion have been designed primarily for transport. Lines across the gullies have to be carried on bridges or embankments with drainage culverts. Development of railway communications over the greater part of Siberia forces the Russians into areas of *permafrost* [9] of varying intensity which are a serious obstacle to construction and to continuous operations. The first serious problems caused by permanently frozen ground were met when building the Amur section of the Trans-Siberian railway. As far as Sretensk, the average cost of the Trans-Siberian railway was 101,550 rubles per mile, whereas between Kuenga and Khabarovsk, the average cost rose to 228,000 rubles because of trouble with *permafrost*. Work is limited to the summer, when the ground thaws for a few feet at the surface, but explosives have to be used in the frozen layers. Cuttings and embankments tend to disintegrate owing to solifluction, and the roadbed and engineering structures are thrown out of alignment by formation of ice nodules caused by water upwelling from below the frozen layers and then freezing. Five out of every eight culverts and bridges on the Amur railway have needed major repairs since their construction between 1908–1916. To keep traffic operating, particularly in the spring, considerable additional numbers of permanent-way workers are needed.

The principal hindrance to railway operation in Central Asia is the deficiency of water which has frequently to be piped over long distances for railway personnel and locomotives, but to some stations water is brought in tank wagons. During the construction of the Mointy-Chu railway across the arid Bet-Pak-Dala, supply of water was a major problem. Much of the water is limey or saline and produces additional scaling on locomotive boilers. Condenser tenders have been fitted to some locomotives to enable them to run further without taking on water, but there is some loss of efficiency for the same fuel consumption, so that availability of oil from Nebit-Dag introduced the use of a few diesel locomotives even before 1939 on the Ashkhabad railway, which have an advantage of a 95 per cent economy in water. [10] Shelter belts help to protect water resources, to arrest drifting sand, and even to reduce the effect of sudden torrential rains. In the foothills of the Tyan Shan, anti-seismic precautions have been necessary on some lines.

The geography of power resources has had some influence on the selection of railway motive power. With large resources of reasonably good coals scattered throughout the country, the steam locomotive has for long been an obvious choice and some locomotives have been fitted with fireboxes to burn the poorer fuels not sought by industry such as wood, lignite, and even anthracite dust. In 1955, steam traction accounted for 86 per cent of the freight ton miles and consumed roughly a quarter of the bituminous coal output, but it had fallen rapidly to about 40 per cent in 1962. Steam locomotives have, however, shown very low thermal efficiency in arid Central Asia and in the extremely low winter temperatures of Siberia, where it is reported that loss of heat raises fuel consumption by 70 per cent. Constantly increasing train weights have called for increasingly powerful locomotives without extensive strengthening of the track or bridges or raising axle loads. For more powerful steam locomotives this required the use of articulated wheel systems, of which some prototypes were introduced after 1945 on the Ulan Ude section of the Trans-Siberian railway but proved unduly costly to maintain, and better power to weight ratios for given axle loads could be obtained with other forms of traction. The improved petroleum situation since 1950, with the greatly expanded output from the Ural-Volga fields, has increased the use of diesel locomotives, which by 1955 already handled 5·5 per cent of the freight ton miles and had risen to about 22 per cent by 1960. Diesel locomotives are already operating in considerable numbers on the railways of the arid zone: in the Emba-Guryev oilfields and the adjacent Trans-Aral railway; on the lower Volga, using Ural-Volga oil and riverborne Caucasian oil; in Northern Caucasia (using local fuels); on the Ashkhabad railway and other lines supplied with Nebit Dag and Fergana petroleum; and a diesel locomotive, specially insulated for operation in cold climates, works on the Vorkuta branch. Experiments have also been made with gas generator units and the use of natural gas in containers at a pressure of 200 atmospheres! Plans for railway electrification have been a long-term part of the State Electrification Plan conceived in the early 1920's.[11] Despite the high initial capital cost, electrification is unlikely to be hindered by power shortage if the large potential water power can be harnessed for generation and already, by using only thermal generation, the same work can be performed by electric traction with a saving of a quarter to a third in coal consumption. The geography of present and past railway electrification should be studied in relation to power generation and the nature of traffic (Fig. 40). The first electric railway was the Surakhany-Baku suburban line opened in 1926, using petroleum-generated current; later, in 1929, the first Moscow suburban service was started, using thermal generation. The first electric main-line railway was the heavily-graded 39-mile section of the Suram Pass on the Batumi-Baku main line: here power was supplied from both a petroleum generator at Tbilisi and the hydro-electric *Zages* plant. Apart from extended development of suburban electric services around Leningrad and the initiation of the Moscow Metro, the opening in the early thirties of the *Dneproges* plant brought the electrification of a short but heavily laden goods railway between Dolgintsevo and Zaporozhye. Electrification between 1932 and 1939 of the Kizel-Chusovaya-Sverdlovsk line in the Urals, with gradients up to 1 : 56, was based on thermal generation, using local low-grade coal, and facilitated movement of heavy coal and ore trains. Cheap hydro-electric power in Caucasia has been used to electrify steep mountain lines such as Mineralnye Vody-Kislovodsk-Zhelez-novodsk. The Kandalaksha hydro-electric plant enabled electric traction to be used on the Loukhi-Kandalaksha-Murmansk section of the Kirov railway in the mining area of the Kola Peninsula. The supply of power from a number of the large hydro-electric projects completed since the Second World War and the anticipated supply from others nearing completion have started the extension of electric traction to selected 'supertrunk-lines', of which the main feature is the electrification of the route from Moscow to Vladivostok, already electrified in a number of sections, which was completed to Irkutsk by 1960 and will reach Vladivostok by 1970. Current will be supplied for the eastern sections from the large hydro-electric stations now under construction on the Angara and Yenisey. The Irkutsk station, completed in 1956, already supplies power for the new electrified line from Irkutsk to Slyudyanka via Andrianovskaya [12] which avoids a section of the Trans-Siberian railway now flooded by the rising waters behind the Irkutsk barrage at the Lake Baykal source of the Angara. Other main lines converted include Moscow-Donbass/Moscow-Leningrad and the lines from the Donbass along the Black Sea coast of Caucasia as far as Leninakan, and work proceeds on the line from Chop via Lvov to Kiev and Moscow. In 1955, electric traction hauled 8·5 per cent of the total goods traffic, saving 5·9 million tons of coal in comparison with the equivalent steam traction but consuming only 3·9 per cent of the national output of electric power.

Russian railway operations combine features from both American and European practice. Khachaturov [13] has claimed that goods train weights in the Soviet Union were comparable with those of America, but there was a much higher density of traffic, which could be compared with Europe. Whereas railways in America carry the occasional fast but very heavy freight train, Soviet railways have more frequent but lighter and slower trains. Fundamentally, the Soviet railways are designed to carry freight: for every passenger mile there is a traffic of 7 ton miles, unlike the virtual parity in the relationship in the United Kingdom. With such great distances, it is hardly surprising that hauls of goods in Russia are long: in 1950 the average haul of a ton of goods in Russia was 595 miles, compared with 552 miles in U.S.A. and 99 miles in the United Kingdom,[14] though the average hauls, which vary greatly between different commodities, have tended to increase as short-haul traffic has been transferred to motor vehicles. Similarly train weights have risen with the use of electric and diesel traction. Unfortunately, adequate statistical material is not available for a similar review of passenger traffic. Freight and passenger movement does not, however, take place with equal intensity all over the system and there is a concentration on a few 'supertrunk-lines', caused partly by the geographical conditions, partly by the distribution of the system which allows few alternative routes, and finally by Government policy which has directed resources into their development at the expense of secondary lines.

Examination of commodity flows shows this concentration on the main trunk routes (Fig. 53). In 1954, the quarter of the route length which was double track and included nearly all the main supertrunk routes carried 64 per cent of the total goods traffic at an average density of 10·8 million ton miles per mile, compared with the national average of 5·2 million ton miles per mile. Another Soviet figure claims that 46 per cent of the railways carry 86 per cent of the freight traffic and more than half the lines account for only 14 per cent.[15]

Although in every region the total tonnage of goods handled increased substantially, the share of receipts and dispatches has changed distinctly in favour of the eastern regions, with the very large increase in the share of the Ural, Western Siberia, and Kazakhstan outstanding. Relative declines have been shown in the North-west and the Volga region, while the South (which includes the Donets coalfield) has lost its former commanding position, though there has been much development during Soviet

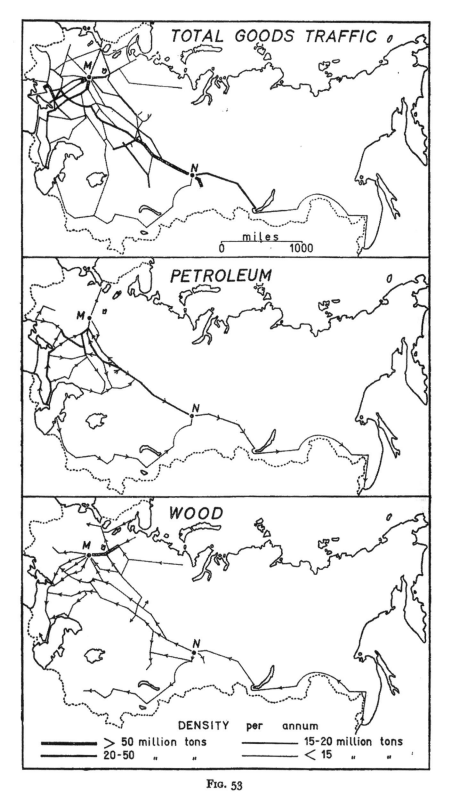

TOTAL GOODS TRAFFIC

miles
0   1000

PETROLEUM

WOOD

DENSITY   per   annum

⎯⎯⎯⎯ > 50 million tons   ⎯⎯⎯ 15-20 million tons
⎯⎯⎯⎯   20-50 „   „   ⎯⎯   < 15 „   „

FIG. 53

Goods traffic flows. Redrawn from Nikolskiy, I. V., *Geografiya Transporta SSSR*, Moscow, 1960, pl. 23, 25, 26. M: Moscow; N: Novosibirsk.

times in the European North, particularly associated with mining. Comparison between the proportions of goods dispatched and received gives some indication of the direction of flows. The regions developed since 1917, for example, Kazakhstan, European North and Western Siberia, as producers of raw materials all dispatch more than they receive and the South also still dispatches more than it receives. The Ural region, which consumes large

TABLE 23

PERCENTAGE DISTRIBUTION BETWEEN ECONOMIC REGIONS
OF GOODS DISPATCHED AND RECEIVED BY RAILWAY, 1913–1952

Economic Region	Dispatched Goods		Received Goods	
	1913	1952	1913	1952
North	0·2	3·3	0·5	1·6
North-west	6·1	3·5	9·4	4·9
West	2·0	3·7	1·1	4·3
Centre	20·9	15·1	23·5	19·6
South	40·0	27·0	34·6	23·6
Caucasia	12·0	9·0	14·5	7·7
Volga	9·1	4·3	6·4	5·3
Total West	90·3	65·9	90·0	67·0
Ural	5·3	13·7	4·7	14·6
Kazakhstan	0·6	4·2	0·4	3·3
Central Asia	1·3	2·3	1·8	2·5
West Siberia	1·6	7·6	1·8	6·2
East Siberia	} 0·9 {	3·6	} 1·3 {	3·1
Far East		2·7		3·3
Total East	9·7	34·1	10·0	33·0

Source: Khanukov, E. D., *Transport i Razmeshcheniye Proizvodstva*, Moscow, 1956.

tonnages of ores and coal, receives more than it dispatches, like the Centre which is the largest consumer of all types of goods and a large manufacturer, and which converts bulk low-grade shipments into a smaller volume of manufactured consignments. With industrialization, seasonal flows of goods have declined, since the railways depend less on agricultural freights, but industrialization has, however, brought a greater movement of goods between regions (Fig. 54) and over longer distances.[16]

Over 80 per cent of the traffic originates in the haulage of heavy, low-value goods such as coal (a third of total traffic),

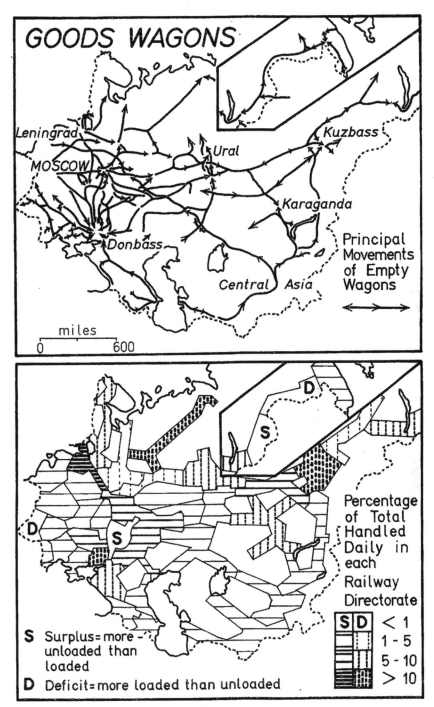

FIG. 54

Goods wagon movement. Railway directorates with a *deficit* load more wagons than they unload daily, consequently having to bring in empties from outside; a *surplus* indicates daily loadings less than unloadings, with consequent outflow of empty wagons. Khanukov, E. D., *Transport i Razmeshcheniye Proizvodstva*, Moscow, 1956, fig. 32 and table, p. 310.

mineral building materials (a fifth), wood and ores. Since 1913, the proportion of coal traffic has shown a substantial increase (from 19 per cent to 28 per cent) while the share of grains in the total traffic has fallen from 14 per cent to 5 per cent and timber, too, has declined. The coal supply of Western European Russia is dominantly from the Donbass but the high-grade Moscow lignite, which does not stand transport well, overlaps with the distribution of Donbass coal in the Central Industrial district (Fig. 55). The Donbass is also the primary supplier for Caucasia, with Tkvarcheli and Tkibuli output limited to local consumption. Central Asian mines also have a limited supply area and overlap throughout with Karaganda coals, which by high quality stand long hauls even to the western parts of the Turkmen S.S.R., and which are also sent to the Volga and the Ural, which both draw from many different fields. Kuzbass coals go predominantly to the Ural. The Volga, without coals of its own, is a zone where the supply of Donbass and Kuzbass coals overlaps. Far Eastern output is directed primarily to the *Amurstal* works at Komsomolsk,

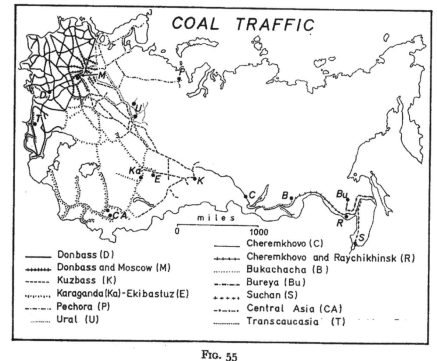

Fig. 55

Coal traffic. Compiled from numerous sources.

which also draw coal from the Cheremkhovo field, but this long haul will later be replaced by Chulman coals. Like other patterns, the proportion of consignments made by the eastern regions has risen substantially: in 1927, at the start of the five-year plans, 80·6 per cent of the coal dispatched originated at stations in the western districts, but by 1955 this had fallen to 54·8 per cent. Nevertheless, the region consigning and receiving the greatest quantities of coal was the South (Donbass coalfield), although its share has fallen steadily since 1940. The second most important dispatcher was Western Siberia, chiefly from the Kuzbass, closely followed by the Ural, which has become an increasingly important source. Since 1940, the Pechora coalfield has joined the suppliers, particularly to the European North. The Central Industrial district consumes large quantities of coal from the Donbass and most local lignite production, besides Kuzbass coals, which are used chiefly on the railways. The Ural metallurgical industry cannot satisfy its needs from local production, so that large quantities have to be obtained from the Kuzbass, Karaganda, and the new mines of Ekibastuz. The supply of eastern coals has risen in the Volga and Centre: in 1950, only 7 per cent of total consumption was received from eastern fields, but by 1955 this reached 17 per cent, when 10·6 million tons were received from distances up to 2,750 miles. An unusual feature has been shipment of lignites from Central Asia to the Volga and from Eastern Siberia to Western Siberia over distances of more than 800 miles.

Petroleum movement is more complicated, though it represents only 6 per cent of originating tonnage and 10 per cent of railway goods traffic. Large quantities of petroleum are sent by water and the proportion moved by pipeline is also increasing. Traffic originates at both wells and refineries, but because refineries have not responded to changes in the location of crude production, movement also includes haulage of crude petroleum from wells over long distances to refineries; for example, between Ural-Volga wells and Northern Caucasian refineries. Railways distribute most of the petroleum unloaded from Volga river boats for consumption in the Centre and North, while railway transport is also provided for petroleum for Western European Russia landed from Caucasia at Odessa and other Ukrainian ports. The rapid growth of the Ural-Volga fields has made the railways of the Volga region and western Ural the main dispatchers of petroleum, accounting in 1955 for 49·1 per cent of total shipments. The substantial increase in quantities shipped and the great increase in the proportion from these railways is shown by Table 24.

Almost all regions now draw the bulk of their supplies from the Ural-Volga fields (Fig. 53), while the proportion supplied by the Caucasian fields has dropped. Western and Eastern Siberia, for instance, have increased their imports from the Ural-Volga

TABLE 24

CHANGES IN PATTERN OF MOVEMENT OF PETROLEUM ON RAILWAYS IN THE U.S.S.R., 1940–1955

Railway	1940 Tonnage*	%	1950 Tonnage*	%	1955 Tonnage*	%
All railways Of which—	29,430	100	43,496	100	78,032	100
Volga lines	5,212	17·8	14,707	33·9	31,410	40·0
Ural lines	1,184	4·0	2,497	5·7	7,070	9·1
Total for— Volga and Ural lines	6,396	21·7	17,204	39·6	38,471	49·1
Central Asian lines	2,350	8·0	3,994	9·2	4,800	6·1
N. Caucasian lines	8,523	29·0	9,322	21·5	11,922	15·2
Transcaucasian lines	4,161	14·2	3,003	6·9	4,763	6·1
Total for— Caucasian lines	12,684	43·2	12,325	28·4	16,685	21·3

* In thousand metric tons.

Source: *Voprosy Ratsionalizatsii Perevozok Vazhneyshikh Gruzov*, Moscow, 1957.

fields and the Central Asian fields at the expense of Caucasia. The Far East since 1945 has increased the proportion of its requirements obtained from Sakhalin. Both water transport and pipeline shipment of petroleum are cheaper than railway carriage and will tend to displace and reorient the flows.

The movement of metal goods and engineering products is, by the wide range of articles involved and inevitable local specialisms, one of the most complicated patterns. Fig. 56 is an attempt to reconstruct some movements of selected metal goods based on

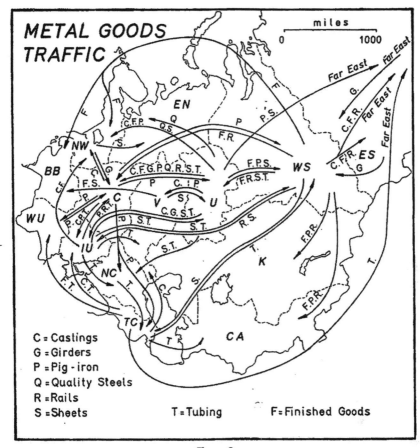

METAL GOODS TRAFFIC

C = Castings
G = Girders
P = Pig-iron
Q = Quality Steels
R = Rails
S = Sheets

T = Tubing   F = Finished Goods

FIG. 56

Metal goods traffic. Compiled from numerous sources. BB: Byelorussia and Baltic; C: Centre; CA: Central Asia; EN: European North; ES: Eastern Siberia; IU: Industrial Ukraine; K: Kazakhstan; NC: Northern Caucasia; NW: North-west; TC: Transcaucasia; U: Ural; V: Volga; WS: Western Siberia; WU: Western Ukraine.

various Soviet sources, usually for 1954–1956. Not only do the main iron- and steel-producing districts distribute to the rest of the country, they also ship goods between each other to cover deficiencies. In 1955, about 71·6 million tons [17] of ferrous metal goods (including scrap) were carried by the railways, of which 70 per cent was shipped from the South and the Ural. The Centre and Western Siberia accounted for a further 16·2 per cent. The South, the Ural, and Western Siberia dispatched a greater tonnage than they received, mostly to regions which consume metal in engineering and construction work and have only a limited basic

335

production of their own. The Centre has a production deficit and imports a wide range of metal goods, though it also exports some of its phosphoric pig in exchange for non-phosphoric types. Total shipment of pig in 1955 amounted to 11·2 million tons. Rails for new construction or replacement form a substantial part of the metal goods moving from the Ural and Western Siberia, notably to the Centre. Some movements are complicated : steel billets are sent from Kuzbass furnaces to Petrovsk-Zabaykalskiy rolling mills and ultimately return to Western Siberia as rolled products. A large traffic in tubes and pipes takes place between the main producing regions, the South, the Ural, and Transcaucasia, since they do not each produce all the types they require locally. Construction work in Siberia since 1945 has created a strong flow of pipes eastwards amounting to 590 thousand tons in 1955. These complex flows take place nevertheless predominantly along the main 'supertrunks'.

A substantial traffic in building materials has been created by the development of large civil engineering projects, but the pattern of flow changes as specific projects are completed. One of the most important components is cement, of which 85 per cent moves by railway wagons, some specially designed for bulk handling; the movement is, however, distinguished by shipments from European producers to consumers in the eastern regions and in Central Asia. Timber is used not only in construction but also in manufacture and in the mines, which consume about a quarter of the annual production. The railways carry much of the manufactured timber goods, such as houses and furniture. Logs are floated, where possible, for long distances, though some may have to be carried by the railways to their final destinations. In general, the movement is from the producing regions of Northern European Russia to the south, the Ukraine and lower Volga. The movement of foodstuffs is also carried out by railway, though much short-distance traffic is now handled by motor lorry, which has also taken over the movement of sugar beet to refineries formerly carried by railway. Special refrigerator trains operate for long-distance transport, mostly of fresh meat and fish. Comparison between the patterns of movement of the different goods (Fig. 53) suggests that the bulk of traffic is on two main axes : one west to east between the Centre, the Ural, and Western Siberia, closely followed by north-south traffic between the North-west, the Centre, and the South. These two axes do in fact join together the major industrial districts of the country served by the 'supertrunk-lines' while their projections link the secondary districts, such as the Far East, Transbaykalia, Central Asia, and Caucasia. In new construction, Soviet plans have concentrated on improving communications between the industrial districts served by these principal routes, and despite development of the eastern regions, the traffic nucleus remains around the Centre, the Volga, the Ural, and the South.

Passenger traffic is secondary to goods traffic : between 1913 and 1956, the density of passengers per mile of track increased by almost three times, whereas the density of goods rose by eight times.[18] In 1956, railways carried 1,658 million passengers, a ratio of 8 passengers per head of the population compared to 20 passengers per head of the population in the United Kingdom, and total passenger traffic amounted to 117,480 million passenger miles. By 1962, passengers carried exceeded 2,000 million. About one-quarter of the traffic and 85 per cent of the passengers were suburban, with an average length of journey of 19 miles. The average length of all types of passenger journey fell from 112 miles in 1913 to 51 miles in 1956, representing more frequent but shorter journeys than before the Revolution : a reflection of the great suburban development discussed in Chapter 4. Fig. 57 illustrates the strong flow of the passenger trains over the trunk-lines and the poor services on secondary routes, and, in comparison with freight traffic, reveals a similar intensive use of the main east-west railways and the Leningrad-Moscow-Donbass routes, with services tending to radiate from Moscow which has through trains or carriages to all the capitals of republics and other principal towns. Over such great distances most principal connections may be several days on their journey, so sleeping-cars are a very common feature, particularly as services are not particularly fast. The best train on the almost level line from Moscow to Leningrad takes six hours twenty minutes for 404 miles (1961), but an ordinary passenger train takes over twelve hours. From Moscow by the Trans-Siberian Express to Vladivostok remains a nine-day journey for 5,800 miles. Speed is restricted by the light roadbed on secondary routes. Ample allowance is made for recovery time and for passing trains on single-track routes. To run one train a day over the principal long-distance routes involves the use of large quantities of motive power and rolling stock; for instance, the present service three times a week from

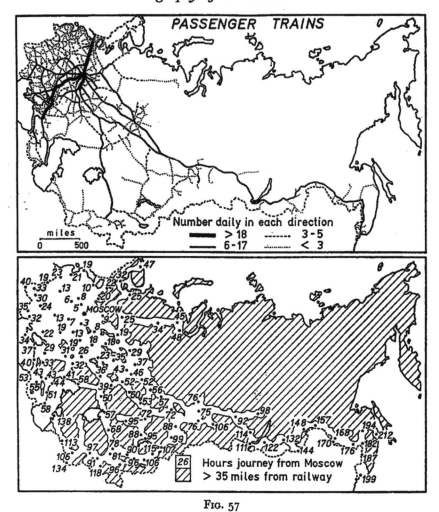

FIG. 57

Passenger trains. Based on official Soviet state railways timetables (summer schedules) and *Atlas Shkem Zheleznikh Dorog SSSR*, Moscow, 1963.

Moscow to Vladivostok requires thirty sets of coaches. Using special coaches on which the bogies can be changed from Russian broad gauge to European standard gauge, services are now run to Vienna, Berlin, other Eastern European capitals, and to Peking and Pyongyang. Moscow has the largest suburban passenger traffic of any Russian city and is served by an extensive system of electrified railways on which a frequent service of trains runs up to fifty miles from the capital and carries about half a million people each day, comprising 95 per cent of Moscow's commuters.

The Moscow underground railway, opened in 1935 and one of the showpieces of the Soviet Union, consists of three radial lines and one circular route, which join the main railway stations and serve the principal points of the city, carrying about 2·5 million people daily. Other electric suburban services are provided in Leningrad (which also has an underground railway first operated in 1955), in Riga, Tallin, Kiev,[19] and Kharkov. The electric railways of the Apsheron Peninsula (Baku) were among the first of their kind in the country.

WATER TRANSPORT

Measured in terms of traffic, water transport is the second most important haulier; but despite substantial increases in total traffic, the share of water transport has fallen greatly since 1913. Rivers have shown the largest decline in the proportion of total traffic, though they have been the great historical routeways of the country, offering easy routes for trade and goods moved along them and across the low-watershed portages between (Fig. 58). The way from 'the Varangians to the Greeks', noted by Nestor, was via the Dnepr and the Western Dvina or Volkhov and the Volga has acted both as a routeway and as a barrier, while in Siberia the rivers served as the lines of penetration for Russian traders and later settlers. The physical characteristics of the rivers have been described in Chapter 1; they flow north or south to the seas that border Russia (though the Volga, the Amu Darya, and Syr Darya have inland drainage basins), a pattern, however, which does not fit well with the patterns of movement, industry, and settlement in modern Russia, which are predominantly east-west. Even without seasonal fluctuations in traffic, rivers are, however, limited in their period of navigability, either by low water in summer or ice in winter, and those which flow north discharge mostly into seas frozen for considerable periods each year. About 95,000 miles are suitable for navigation, some 68,400 miles by large steam and diesel river-boats, while 330,000 miles are suitable for rafting. Of 146·8 million tons of goods carried by rivers in 1956, 49 per cent consisted of timber and firewood, 23 per cent was composed of mineral building materials, petroleum goods accounted for 10 per cent, 4 per cent was grain, and 1 per cent, salt. Since the first five-year plan,

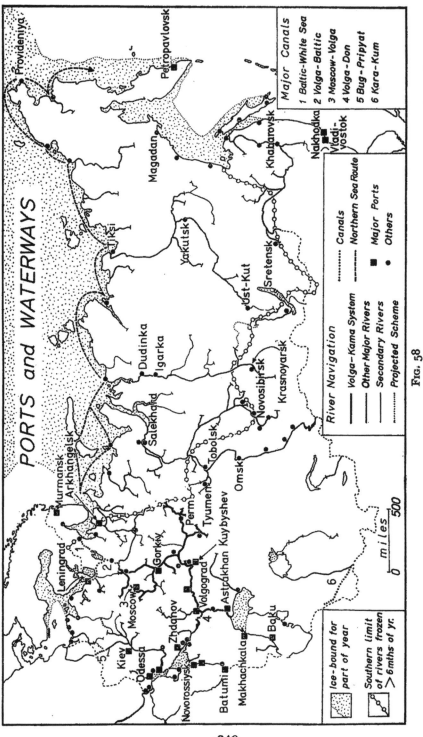

Fig. 58

Ports and waterways. Plotted from Soviet atlas maps and from Admiralty Pilot handbooks for the various seas. River classification based on Nikolskiy, I. V., *Geografiya Transporta SSSR*, Moscow, 1960, and *Transport i Svyaz*, Moscow, 1957.

diversion of certain goods to railways, despite increasing yearly tonnages, has resulted in a rise in the share of timber and mineral building materials. The number of passengers has risen from 11·2 million in 1913 to over 80 million in 1956, but the length of passenger journey has dropped from 103 miles in 1913 to only 35 miles in 1956.[20] Like the railway system, certain stretches of river are used far more intensively than others.

The greatest and most intensively used system in European Russia is that of the Volga, with its tributaries, the Kama, the Oka, and the Belaya, which rise in the forest lands and drain across open steppe in the south to the inland sea of the Caspian. Within their drainage area live about 70 million people. Steamers navigate the Volga upstream as far as Rzhev or Kalinin and downstream to the delta at Astrakhan; on the Kama, navigation by river-boats is from Berezniki to the confluence with the Volga, while the Belaya is used upstream to the oil town of Sterlitamak. The Oka is used to above Kaluga and also to Moscow, though the capital has its principal connection through the Moscow Canal which runs north to join the Volga at Ivankovo. By artificial construction the Volga has been joined to other river systems: to the Svir and Leningrad by the Mari Canal system (first completed in 1810) [21] and to the Sukhona-Northern Dvina rivers by the somewhat antiquated Northern Dvina Canal. By these canals, Volga traffic may reach the White Sea and the Baltic, and the link to the Baltic is presently (1962) being enlarged to take sea-going vessels by rebuilding the old Mari Canal system. In the south, the completion (1952) of the Volga-Don Canal and the Tsimlyansk reservoir corrected one of nature's errors by linking the Volga to the Don across the Yergeni Hills by a canal, which follows a depression in the hills, crossing the watershed at a level of 132 feet above the surface of the Don and 264 feet above the Volga; four locks act on the Don slope and nine on the Volga slope. These canals allow Moscow to be described as the 'Port of the Five Seas'. Regulation of the Volga channel has been carried out at many points, and further improvements will arise after completion of the hydro-electric projects now under construction, so that it is hoped the summer low water will be eliminated, though there has been considerable improvement since the completion of the Moscow Canal and the Rybinsk reservoir (1941).[22] Navigation is usually limited in late summer in several reaches by lack of water; while in spring, melt water raises the level of the river, spreads over the low-lying banks and interrupts navigation. Ice cover on the river lasts 164 days at Kalinin, 172 days at Ulyanovsk, and 109 days at Astrakhan; at the break-up, ships often have to seek shelter behind the spits (*zatony*) along the river. On the upper and middle reaches, navigation lasts 6½ months, on the lower reaches, 7 to 8 months; on the Kama, navigation lasts 6 months.

The first steamship was operated in 1817, but it was not until the middle of the century that steam navigation developed, replacing the traditional sailing vessels with their famous hauliers (*burlaki*), and traffic increased greatly after 1862 when the 'portage' railway from Volgograd to Kalach-na-Donu, across the Volga-Don watershed, enabled the easier export of goods to the seaport of Rostov-na-Donu. Expansion of Baku oil production in the 1880's further increased traffic upstream. The Volga-Kama system, over 14,850 miles of waterway suitable for river-boats and a further 36,300 miles on which rafts can be floated, now claims about 65 per cent of Soviet river traffic. Shipping comprises river freighters, barges towed singly or in groups by tugs, depending on their size, and special vessels such as tankers. Some of the modern self-propelled tankers are of 3,300 tons deadweight, while oil barges of larger size are drawn by tugs, but about a third of the traffic is composed of timber rafts drawn by tugs, though semi-finished and finished wooden goods also move in barges. Petroleum is an important upstream freight along with salt and grain countering the downstream flow of timber, most of which originates on the Kama, and cement forms a substantial part of cargoes originating in the middle reaches, with increasing quantities of phosphorite and fertilizer shipped downstream from the Kama. A recent Soviet textbook claims Volga traffic is equivalent to that of ten to twelve main-line railways and freights are attracted to the river from distances of over 150 miles away.[23]

Petroleum, one of the main upstream cargoes, is brought to the river from Baku by Caspian ships; but because of the falling level of the Caspian Sea, these large boats cannot reach Astrakhan, and cargo is transhipped to river barges at the artificial port of 'Twelve-foot Roadstead', connected to the town by a 150-mile-long dredged canal, while southwards through this artificial port passes timber on its way to Central Asia and Transcaucasia. Astrakhan is the centre of large-scale fish-preserving, whose produce is also shipped upstream to feed the industrial towns to

the north, and has large shipyards. Between the delta and Volgograd, the river, in a braided channel, flows across a flood plain
over thirty miles wide where navigation is hindered by shifting
channels and sand-banks. On the long eastern distributary, the
Akhtuba, lies Vladimirovka, which exports salt brought to it by
railway from the Baskunchak deposits. At the bend of the Volga,
between the steppe and semi-desert, where the watershed between
the Volga and Don narrows, lies the strategically and commerci-
ally important port of Volgograd, built on a series of Volga
terraces with the harbour installations on the lower levels, which
acts as a transhipment point between the river and railway and
between the Volga and the Don. Its importance increased after
the completion in 1862 of the railway across the watershed to the
Don at Kalach, and later with the growth of the industrial Donbass. Timber, oil, salt, and fish are received from the Volga,
while coal, metals, and grains are loaded on to river-boats, and
sawmilling and petroleum refining in the town depend for raw
materials on the Volga cargoes. Krasnoarmeysk, an outport in
a well-protected harbour in a *zaton*, lies at the entrance to the
new Volga-Don Canal and has a large shipyard. Upstream, on
the west bank, Kamyshin transfers Volga freights to the railway
and sends out grain and glassware. There are several grain
elevators at the port, which is being enlarged to handle cotton
brought by water from Central Asia to its large new textile plant,
and ship repairs are also undertaken. The first railway bridge
across the river above Astrakhan is at Saratov, another west-bank
port, equipped with special quays to handle a wide range of
cargoes, including cotton from Central Asia. One of the few
ports on the low east bank, Engels, despite its liability to inundation, is a growing industrial centre, which was helped at the end
of the nineteenth century by improvements of the fairway in this
reach of the river. Another east-bank port is Pugachev (on the
Great Irgiz) which can be reached in spring by steamboats. The
first important port of the middle Volga, Volsk, built on a lower
west-bank terrace, exports large quantities of cement, but the
principal ports of the middle reaches are in the Samara bend, a
difficult section for navigation. Batraki, on the west bank, where
the railway crosses the river, the outport for the industrial and
railway centre of Syzran, is chiefly a transfer point for goods from
rail to river. On the east bank, at the confluence of the Volga
and Samara rivers, Kuybyshev, occupying a commanding position

above the river between the mixed forest belt, the forest steppe, and the true steppe, deals with a wide range of cargoes, and has become a point for the shipment of petroleum from the Ural-Volga fields. The river current against the left bank here deepens the channel along the foot of the Sokol Hills and ice tends to break up earlier than elsewhere on the middle Volga. Above the bend, Ulyanovsk, on a watershed little more than a mile wide between the Volga and Sviyaga, is to be expanded by completion of new port facilities and the filling of the Kuybyshev reservoir, recently completed. The confluence of the Kama is served by the ship-repairing settlement at Kuybyshevskiy Zaton. To the north, on the east bank, Kazan, the port for the central and northern Ural, is in a hilly situation, separated from the Volga by a wide, sandy terrace which will be submerged under the waters of the Kuybyshev reservoir. The town has a considerable river trade, lying between the coniferous and mixed forest belt near to where the river is crossed by the main railway to Sverdlovsk, and highly developed market gardening accounts for shipments of vegetables. Other nearby ports are Cheboksary and Vasilsursk at the mouth of the Sura. One of the greatest of all Soviet river ports is Gorkiy, at the confluence of the Oka and Volga, which as Nizhniy Novgorod was the site of one of the biggest mediaeval European fairs, and today is the administrative centre for Volga traffic. The port lies near the ancient Kremlin overlooking the Oka, while a large harbour has been constructed on the right bank of the Volga, and port facilities are provided at some of the suburbs of Gorkiy. Cargoes handled comprise petroleum (both from Caucasia and the Ural-Volga fields), timber, manufactures, and food goods. Other ports on the upper reaches are Kineshma, Kostroma, Yaroslavl, Rybinsk, and Kalinin, all primarily concerned with transhipment from river to rail. By the canal, built before the Second World War, Moscow receives a wide range of goods, most of which are unshipped at quays in the southern part of the city. Passenger boats, from a terminal in the northern part of the city near Tushino, regularly sail as far as Rostov-na-Donu during the summer, needing eleven days downstream and thirteen days upstream. On the Kama, over 80 per cent of the traffic is timber, mostly in rafts, but growing amounts of chemicals and wood by-products, such as cellulose and paper, are sent downstream. Upstream freights include coal, petroleum, and salt. Perm, on the high right bank, is the chief port, while others are

Krasnokamsk, Kambarka, Solikamsk, and Berezniki. On the Belaya, Ufa, on the right bank, is the main port.

The second most important river is the Dnepr, 1,425 miles long, whose navigation has increased since the completion of the Dneproges dam in 1932 made it possible to navigate the 53-mile-long rapids between Dnepropetrovsk and Zaporozhye, while movement in the lower reaches has been eased by completion of the Kakhovka barrage. The river is joined to the Bug by a canal which can take only small river-boats, and a canal of limited capacity also joins it to the Western Dvina. The Dnepr has a navigable season of about 265 days at Kiev, though this may be restricted by low water in dry summers. The principal freights upstream are petroleum from Kherson (brought to the port across the Black Sea from Batumi), coal from the Donbass, and some metal ores. Downstream, grain is sent to the elevators at Kherson and timber to Dnepropetrovsk, from where it is carried by rail to the Donbass. Enlargement of the Dnepr-Bug canal would allow the carriage by water of ores and coal from the Ukraine to the new Polish steelworks at Nowa Huta on the Vistula a few miles east of Kraków. The Don is navigable from the estuary to Liski, 1,100 miles, and its tributary, the Northern Donets, also carries considerable traffic. Coal and grain together form 70 per cent of the traffic, while timber brought from the Volga is an important item. Formerly, goods from the Volga were received at Kalach-na-Donu by railway, but since completion of the Volga-Don Canal, Volga boats can sail to Rostov through the Tsimlyansk reservoir, which has raised the level of the water in the middle Don by over 70 feet. Traffic is expected to increase on completion of the harbour at Ust-Donets Port (near Kamensk), particularly shipments of coal to the Volga basin.

Rivers in Byelorussia and the Baltic republics are unimportant, principally used for timber rafts, though there are some small river-boats and tugboats with barges. The Neman is navigated by river-boats on the reach between Kaunas and Klaypeda and plans suggest linking it for 5,000-ton vessels to the Dnepr by a system of canals. The Western Dvina is linked to the Dnepr system by a short canal and the Berezina river. The rivers of the European North-west and North acted in the past as the arteries along which Slavonic colonization took place, but are important in modern times for the movement of timber, particularly the Northern Dvina, with the Sukhona and Vychegda, and by means

of the Northern Dvina Canal these rivers are joined to the Volga.
The greatest limitation is the short shipping season: on the
Northern Dvina it lasts for only 170 days and the lower reaches of
the Pechora may be ice-bound for 220 days. A connection between
the White Sea and the Baltic was made before the Second World
War by canals joining together the many lakes of Karelia, now
used to ship fish to Leningrad and the Volga towns and also
consignments of apatite and other minerals from the Kola
Peninsula. The principal freight on the Northern Dvina is timber
floated downstream to Arkhangelsk, the largest sawmilling centre
in Russia, and some coal and petroleum move down the Pechora
to service ships in the Kara Sea, but it is likely that use of the river
has declined since the opening of the Kotlas-Vorkuta railway.
There is a close relation between development of forest resources
and river systems in the north, since the rivers provide the easiest
means of access and the simplest way of exporting cut timber.

The great rivers of Siberia have been relatively little developed
for transport, although some are so large that sea-going ships can
sail up them. Siberian rivers suffer from a short navigation
season: on the Ob at Salekhard, navigation lasts for 152 days,
increasing to 189 days at Tobolsk on the Irtysh; on the Lena,
navigation lasts for 88 days in the north and some 145 days in
south; at Krasnoyarsk, on the Yenisey, navigation can be done
on 197 days. When the ice breaks there is serious flooding, which
hampers shipping, also halted by frequent fog in summer on the
lower reaches. River-boats sail up the Irtysh to Semipalatinsk,
and on the Ob, usually to Novosibirsk, but some local navigation
is undertaken further upstream on both rivers. The main upstream freights are timber and, since 1953, coal from the Vorkuta
mines. The Yenisey is used in the 450 miles below Igarka by sea-
going ships to carry away timber and the produce of the Taymyr
Polymetal Combine, which also move upstream to the railway
port at Krasnoyarsk, carried in barges and river-boats of up to
3,000 tons deadweight. Local traffic continues upstream to
Minusinsk. Navigation on the Angara is hindered by rapids:
from the confluence with the Yenisey ships may sail to the
Murskiy rapids; and from Lake Baykal navigation is possible
downstream to the Pyarniy rapids at Bratsk, but completion of the
Angara hydro-electric 'cascade' will make navigation possible
between the Yenisey and Lake Baykal. Tugs and barges are used
on Lake Baykal. Passenger boats from Baykal quay take about

six days to reach Nizhniy Angarsk. The main routeway from Southern Siberia to Yakutia is along the Lena, which is usually navigated upstream to the railhead of the branch line from Tayshet at Osetrovo (Lena), and some ships sail to Kachuga in the high-water season. Downstream freights, including petroleum, manufactured goods, and food, seem to exceed upstream freight of timber and ores. Some coal moves on the middle Lena since the opening of the Sangar and Kangalasskiye Kopi mines. Besides being ice-bound for seven months in the year, the lower Lena is reported to be inadequately charted, so that ships frequently ground on sandbanks, while extreme difficulty in navigating the delta and estuarine bar causes sea-going ships to call at Tiksi Bay to transfer cargoes.[24]

In the Far East, the Amur is usually navigated upstream and along its tributary, the Shilka, to Sretensk. The average depth is about 10 feet, though for the last 340 miles to the estuary there is a depth of 20 feet. Sea-going ships, however, cannot cross the shallow Amur estuarine bar. The river is ice-free for 155 days at Sretensk, 185 days at Blagoveshchensk, and 160 days at the mouth, but it suffers from low water in the summer, when boulders and sand bars are a danger to ships, while sudden dangerous rises in level occur after heavy rain. The Amur was used as the main route to the Far East until the completion of the Amur section of the Trans-Siberian railway to Khabarovsk. The main harbours on the lower reaches are Khabarovsk, Komsomolsk, and Nikolayevsk.

Some navigation is undertaken on the Syr and Amu Darya in Central Asia, and goods also move across the Aral Sea to the railway station at Aralsk. Freight consists of cotton, wool, and minerals for European Russia, with return cargoes of grain and some manufactured goods. Navigation is restricted because of shallows, particularly during the summer months, and throughout the year the Syr Darya is used only in certain sections because of shallowness.

ROAD TRANSPORT

The contribution of road transport to total traffic is small but it deals with a large proportion of the originating tonnage, being essentially a distributor in towns and country districts and a feeder to long-distance hauliers, particularly the railways, and it is only since 1945 that inter-city transport has begun to develop on an appreciable scale. Road transport is not particularly suited to the extreme climatic conditions and the long distances characteristic of movement in Russia, and it may never occupy the strong position held in the United Kingdom and in the United States. Primitive roads have, however, an important historical part in Russian development: examples are the many portages between rivers in the mediaeval period and the Great Siberian Tract in the colonization of Siberia in the nineteenth century, while the caravan routes through such towns as Merv (Mary) and Samarkand were the slender link between Europe and Asia for many centuries, and the Silk Road had its terminus within what is now Soviet territory.

Roads with a hard, prepared surface comprise about a fifth of the total length; the remainder are dirt roads, for the greater part unimproved and ungraded. Hardly 1 per cent of the length is asphalted or concreted, only 4 per cent is cobbled, and about a tenth has a prepared gravel surface. European Russia claims most of the surfaced roads (Fig. 59). Whatever the surface, use by road vehicles is likely to be interrupted for long periods in winter by ice and snow, and often by flooding in spring. Unsurfaced roads become impassable quagmires in spring and all forms of surface suffer badly from frost. In some respects, winter is the best time for road transport, since the frozen ground can carry heavy vehicles and even rivers frozen in winter are used for lorries in areas where roads are few. With constant heavy traffic, frozen roads become uneven and are more suited to heavy vehicles than light cars. It is significant that Russian statistics for 1956 regard 88 per cent of the roads as of local importance and only 12 per cent as important for inter-regional connections. A guide to the function of roads in Siberia and Central Asia is given by Soviet maps marking 'routes without rails', roads which run into areas not yet penetrated by railways. Although a great deal of the traffic is by motor lorry, horses are still used (particularly in the countryside), while camels and other draught animals have regional importance.

Regional examination shows substantial development of road transport in European Russia, where a skeletal network of main surfaced highways exists. The system links together Moscow, Kiev, Kharkov, and Minsk, important regional capitals; and the long, straight, easily-graded highways with fairly good surfaces

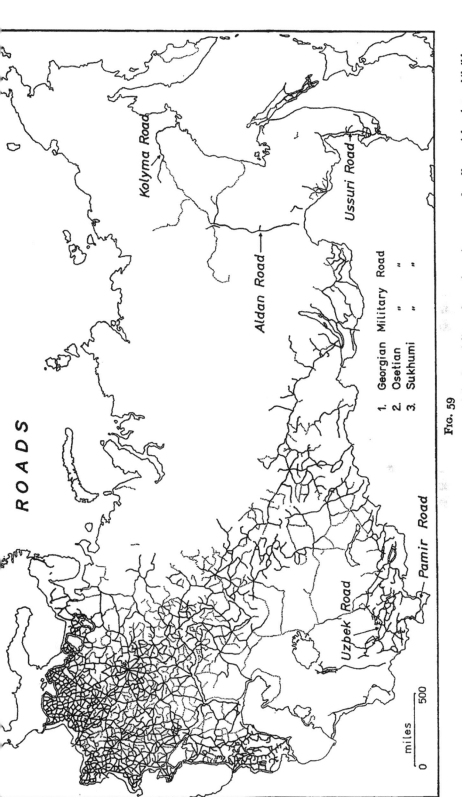

ROADS

Kolyma Road

Ussuri Road

Aldan Road

Pamir Road

Uzbek Road

1. Georgian Military Road
2. Osetian   "   "
3. Sukhumi   "   "

Fig. 59

miles
0   500

Roads. Solid lines mark roads of national and regional importance, while dotted lines show those important locally. *Atlas Avtomobilnikh Dorog SSSR*, Moscow, 1960, which gives no indication of surface type.

have some of the engineering features of railways. Concrete is not regarded as a particularly satisfactory surface, since it suffers more from frost than other forms of surface, and asphalt or grit are commonly used. Most major highways are scheduled for reconstruction and a number are planned or are already under construction. Some of these roads were started before 1941: German military intelligence on several occasions spoke with surprise at finding good new roads which were not shown on available maps. In Northern European Russia, use of roads is limited by climatic conditions, though in places such as the Kola Peninsula roads have been built from mines to the railways, like the road carrying ore from the Lovozero mine to the railway, while a road serves the Pechenga mines at the end of the former Finnish Arctic Highway, now partly controlled by Russia.[25] In the Kola Peninsula, 'winter roads' are used only during the long frozen period and are frequently impassable in summer. Elsewhere, roads, sometimes built of wood, penetrate into the forests and several important roads join Syktyvkar, capital of the Komi A.S.S.R., to the railway system.

Mountain roads are characteristic of Caucasia and Central Asia. Three highways, built with considerable skill, cross the Great Caucasus; the Georgian Military Highway is the most famous and may be crossed by tourists in Russia, but the Sukhumi and Osetian Military Highways are also important, even though their military rôle has now largely gone. The greatest of the many mountain roads in Central Asia is the Pamir Highway, serving the difficult country of Badakhshan, said to be wide enough for two-way traffic and to have a reasonable surface used by lorries and buses. A number of roads cross the mountains into China, including a branch from Sary Tash on the Pamir Highway to Kashgar. Many are traditional caravan routes; but the Ayaguz-Chuguchak-Urumchi road carried large motor-lorry convoys between China and Russia during the Second World War and has since carried equipment from Russia to the developing industries of Sinkiang. In the Central Asian deserts, some routes carry not only motor lorries but also regular bus services, serving outlying settlements and mines.

In Siberia, roads have been important in the development of the remoter north and north-east where, much cheaper in initial cost and maintenance than railway tracks, they act as feeders to railways and rivers, though some are complementary to the river system, which may be used as roads in winter. Several roads cross from the Trans-Siberian railway into the Mongol People's Republic: most important are the Chuya Highway, which runs from the railhead at Biysk via Inya into the western cattle-raising districts, and the Usa Highway, from Minusinsk via Kyzyl (Tuva A.O.) to the important wool centre at Khadkhal. The road from Ulan Ude to Ulan Bator in Mongolia has been duplicated by the railway since 1949. Another railway has duplicated the road from Tulun via Zagorsk to the Lena river. A road of great importance runs from the Trans-Siberian railway at Bolshoy Never via the Aldan mining settlements to Yakutsk, but in the future it may be converted into a railway with the planned expansion of mining activity. The Ussuri Highway from Vladivostok to Khabarovsk is reported to be a good motor road, and another reasonably maintained road is the Kolyma Highway from the port of Magadan to the settlements on the upper Kolyma. Roads suffer badly from rains and floods because most have no proper foundations and are frequently swamped, while *permafrost* is a common problem. The vast northern part of Siberia is generally roadless, depending on tracks or paths for the usual means of communications; but during the spring and summer, overland movement is restricted by flooding and water-logging. Native sledges drawn by reindeer or dogs are used extensively and there are considerable structural differences between sledges in Eastern Siberia and in the western districts, where they are heavier and clumsier. Horses are not common, though they are used in mountainous parts as pack animals and Russians have introduced them into the Far East coastal areas, such as Kamchatka and Sakhalin.[26]

On the roads, traffic is chiefly composed of motor lorries, some cars, and buses, which are increasing in number. Since 1945, the Russians have intensified the use of motor lorries between towns (including large articulated vehicles) and the runs have steadily risen in length: in 1958, motor-lorry transport was initiated between Moscow and Riga, a distance of several hundred miles. In the countryside, the motor lorry makes runs of 30 to 40 miles and it is planned to transfer the short-haul traffic up to 50 miles to motor vehicles. Private motor cars are less frequent on main roads than other vehicles, though they are more plentiful in towns where transport is still provided mostly by trams or trolley buses. Only half the towns have motor-bus services, which account for 35 per cent of the passengers, and buses are most important in towns in Central Asia and least important in Western European Russia. There are many inter-town bus services operating regular runs one or more times daily: Riga has over thirty daily services to other towns and has four services daily to Leningrad. The average length of journey by country bus is about 23 miles and by town bus, 2 miles.

SEA TRANSPORT

Physical characteristics of Soviet seas have been described in Chapter 1. The western seas are separated from those of the east by a long and roundabout journey beyond Russia's borders, and the long northern coastal route is little used by ships because of its unusually difficult climate. The Baltic, parts of whose eastern shores are held by Russia, is restricted in use in the east by freezing in winter. The Black Sea, with the expansion of Russian influence into the Balkans, is virtually a Russian lake, though its exit is held by a foreign power, and the Caspian is an inland sea, in area greater than the United Kingdom. The Far Eastern seas suffer from long frozen winters, which close the harbours and narrow waters to ships. Except in the far west, the Northern Sea Route is also closed for the greater part of the year, while at all times navigation is hazardous. The problem of keeping ports open in the winter and the risks of navigation in icy waters have been considered major factors in the Russian political search for warm water ports, but the fluctuations in Russian foreign trade and the emphasis on the continental aspects of the economy since 1917 have left ports with facilities frequently under-employed, yet goods shipped by sea transport have increased from 15 million tons in 1940 to 54 million tons in 1955. Coastal shipping between Russian ports in the same sea is the leading movement, followed by international traffic, and third place is occupied by shipments between Russian ports on different seas. The greatest traffic, 36·8 per cent of the total, is on the Black Sea and Sea of Azov; Caspian traffic amounts to almost another third, while only a sixth is handled by the Far Eastern ports. Baltic ports account for 8·9 per cent and the long northern sea route handles 7·1 per cent of the traffic.

Navigation of the Black Sea and Sea of Azov is reasonably

easy; but care is needed on the northern coast because of shallows, long spits, and sandbanks. Strong currents off the south-western Crimea in some wind conditions make navigation difficult for small boats, and strong winds frequently develop in summer in association with local low-pressure systems over the Black Sea, while on the Caucasian coast in winter, strong north-eastern winds close harbours. Wind in some of the estuaries also makes movement difficult for big ships, as may reduced visibility caused by spring sea mist or summer dust storms. The Russian coast is reasonably well provided with aids to navigation. The Black Sea and its adjacent waters have a vigorous coasting traffic. Oil is shipped from Tuapse, Batumi, and Novorossiysk to other ports, notably those west of the Crimea. Ores are sent from Kerch to Zhdanov in return for coal shipments. Ores and coal are also moved between Caucasian and Ukrainian ports. Other cargoes include cement, salt, fruit, fish, and grain. The principal place in traffic is held, however, by shipments of Caucasian and Ukrainian goods to ports on other Russian seas: principal cargoes are cement, grains, oil, sugar, and ores. International trade includes the importation of foodstuffs, rubber, jute, and industrial equipment. Exports include shipments of coal and ores to the Danube for use in the industrial development of Eastern Europe. Manganese ore is an important item in the foreign export trade.[27]

The principal port, Odessa, has a short season with ice but can be kept open most years by icebreakers. Originally the chief export port for Ukrainian grain, it is now one of the principal Russian ports in international trade, well equipped with quays to handle special freights, including refrigerators, oil storage, large warehouses, and grain elevators; and it has developed passenger traffic. Odessa is the base for Soviet research ships sailing to the Antarctic. East, in the drowned valley of the Dnepr and Southern Bug, lie Kherson and Nikolayev. Kherson, approached by a dredged channel in the Dnepr delta opened in 1901, handles oil and grain, for which there are several elevators, and deals in cargoes moving up and down the Dnepr. Nikolayev, 60 miles from the sea on a sandbank in the deep-water estuary of the Southern Bug is a better port and one of Russia's principal ship-building centres, with a training college devoted to the subject. It is equipped to deal with grain and petroleum. The Crimean ports are less important: Port Khorl and Skadovsk are difficult to approach and are frozen for long periods; Yevpatoriya, which exports salt, has a restricted harbour; the naval base of Sevastopol, opened as a 'window on the world' in 1784, has a safe natural harbour sheltered from all winds. Yalta, chiefly a health resort, is a harbour of refuge, and the small harbour at Feodosiya exports grain. Passage to the Sea of Azov is through the Kerch Strait, which freezes in winter for about 36 days but can be kept open by icebreaker and through which navigation is difficult in certain conditions, though the Kerch-Yenikale channel has 23 feet of water. Since 1955, a train ferry has operated from Kerch to the Kuban shore, but the port is mainly concerned with shipping iron-ore to Zhdanov in return for coal. Ice lasts longer (up to 180 days) in the Sea of Azov, and more icebreaker assistance is required. At Taganrog, ice lasts about 113 days and navigation for 252 days. Zhdanov has a season of 288 days: it is the main port, shipping in ore and exporting coal and metal. Rostov-na-Donu, 40 miles up-river, lying between the heavy industrial belt to the west and the rich grain lands to the east, well equipped and approached by a dredged canal, is difficult for large sea-going ships to reach because of the narrowness and curves of the channel. At the quays, water is 12 feet deep, but constant dredging against silting is needed. Taganrog and Azov act as its outports. Other small ports are Yeysk, Osipenko, and Temryuk, which has a river-boat service up the Kuban. The ports of the Caucasian coast are not generally troubled with ice. The best equipped is Novorossiysk, which lies in a bay sheltered from the west, but the harbour is difficult to approach and may close in autumn and winter when the strong north-east *bora* blows. It ships out large quantities of cement and petroleum. Tuapse is principally an exporter of petroleum. The principal port in the south, Batumi, which has rather limited accommodation, dispatches petroleum from the terminus of the pipeline from Baku, and has a wide range of imports for consumption in Transcaucasia. Poti, which silts easily in north-west gales, exports citrus fruits, wines, and manganese from Chiatura.

The Baltic has more severe icing conditions than the Black Sea, particularly in the Gulf of Finland, but ice does not greatly affect navigation west of 22° E. Wind is important in winter navigation, driving the ice inshore or out to sea, particularly in Riga Bay. Ice lasts from 130 to 170 day in the eastern parts of the Gulf of Finland and about 90 days at Tallin. Coastwise trade is less important than in the Black Sea. Leningrad, the most important Soviet port, established on the low mud-flats in the Neva estuary, had its original commercial port on Kotlin Island, the site of the naval base and fortress of Kronshtadt, but a deep-water channel some 25 miles long was built during 1875–93 to enable large ships to sail into Leningrad port. The harbour, with about 200 berths, several refrigerated warehouses, and elevators, can usually be kept open by icebreakers, but it may close any time between 22nd January and 5th May. Motor vehicles are sometimes run over the ice from Kronshtadt to enable cargoes to be landed. Depth of water in the Neva estuary is much influenced by the prevailing wind, which often blows from the west and dams back water in the river. The port handles a wide range of cargoes, notably outgoing timber, and it has imported coal from Silesia via the Polish ports since 1945. Westwards, Tallin is the next large port and is usually kept open in winter by icebreakers and is often regarded as a winter port for Leningrad. Navigation in the Bay of Riga is difficult in winter because of the presence of ice, particularly if approached from the north, and even in the bay itself icebreaker assistance is needed by most ships. Ice is present at Riga between the end of December and mid-April. The port is restricted in size, though the depth is between 26 feet and 29 feet, but constant dredging is needed to keep the roadsteads clear of silt. The entrance of the Gulf of Finland was covered from 1940 to 1941 by the Russian concession at the Finnish port of Hankö, while between 1945 and 1955 the Russians held a similar concession at Porkkala, and the Finnish Åland islands are a demilitarized zone. Liepaya has a depth of 27 feet and ice is present between early December to mid-March. Klaypeda, formerly Memel, offers 23 feet of water, though owing to drift ice it may close between 2 and 19 days each year unless icebreaker assistance can be obtained. Kaliningrad (Königsberg), approached by a dredged ship canal, is closed to steamships for 15 to 91 days, but traffic is maintained at the outport of Baltiysk (Pillau), the winter port of the Russian Baltic Fleet since 1945. The south-eastern ports all suffer from an eastwards drift of sand and bad silting. In general, all the Baltic ports west of Leningrad have still not recovered their pre-war pattern of trade as food exports from the Baltic republics have been negligible since 1945, and the ports now function mainly as entrepôts for their republics and the Byelorussian hinterland.

The Northern Sea Route, one of the enigmas of Soviet Arctic policy, was first navigated by Nordenskiöld in 1878-1879, forming a 3,000-mile-long route of great strategic significance as an 'All-Red' route to the Far East from European Russia. Physical conditions are hard: ice persists for at least seven months in the year and in some parts for ten months, with a good deal experienced even in the open season, when navigation is made more difficult by frequent fogs. Running the route has called for several ancillary services: bunkering and stores, icebreakers, accurate weather forecasting, study of ice movement, charting, and erecting navigational aids. Since 1932, the operation of the route and all auxiliary services have been controlled by *Glavsevmorput* (the Chief Administration of the Northern Sea Route) with headquarters at Leningrad. Improvements in traffic and in navigating conditions have arisen not only from better technique, particularly in forecasting, but also from a suggested improvement in ice conditions. Some Soviet authorities believe, however, that a warm period ended about 1940 and since then ice has tended to worsen and the open season to shorten. Year-to-year operation is entirely dependent on ice conditions. Possibly a shortage of icebreakers has limited the annual number of convoys, but the atom-powered icebreaker, *Lenin,* has eased its use since 1960 and a number of icebreakers delivered to the Soviet Union after 1945 as reparations from Finland have also doubtless simplified the problem of providing escorts for convoys.

The season first opens in the Kara Sea, where warm water brought down by the great Siberian rivers helps an early break-up. Entry is made by the Kara Gates or the Matochkin Shar, which is an easy if narrow passage in clear weather. Ships sail to the lower Ob, mostly to Novyy Port, as shallow waters in the river prevent larger vessels sailing beyond Labytnangi. On the Yenisey delta, Dikson is principally a bunkering and operating base, but ships (up to 8,000 tons) sail up to Igarka, where the channel is four miles wide, the principal lumber port established in 1927 at a point where the current becomes too difficult for tugs to take timber rafts further downstream. Development of the Taymyr Polymetal Combine at Norilsk has expanded cargoes to and from the port of Dudinka. At Ust-Yeniseysk, navigation, lasting from 15th June to 28th October, though ice appears by 15th October in normal years, requires care over the bar at the mouth. Navigation eastwards becomes increasingly difficult along the Laptev Coast, where shoals extend out to sea. The most difficult section is the shallow Vilkitskiy Strait, which is hard even for icebreakers to navigate, where a narrow channel for convoys is available during August and September, but it may be blocked all summer by grounded icebergs. Most ships passing through this section are 3,500-4,000 tons deadweight. To the east, ships call at Nordvik at the mouth of the Khatanga river, where a base is maintained and where much geological prospecting has taken place. Tiksi Bay acts as the roadstead for goods to and from the Lena, which are here transferred to river-boats as sea-going vessels have difficulty in navigating the delta because there is a shallow bar. Ice brought down by the river is a danger in the delta and blocks some of the channels throughout the summer. Other calls are at Ambarchik and Nizhne Kolymsk, which serve the goldfields of Kolyma, and at Chaun Bay, ports usually served from Vladivostok and ships unload into lighters for carriage to the shore. The first Pacific port is Provideniya Bay, modernized and mechanized during the Second World War to deal with Lend-Lease cargoes.

At the western end of the Northern Sea Route lie two major Russian ports. Murmansk, developed during the First World War, acts as a well-equipped outport for Leningrad during the winter, though it has a large trade of its own. Terminus of the Kirov railway, the harbour lies on the south bank of the deep-water Kola Inlet, 25 miles from the Barents Sea, which under the influence of the North Atlantic Drift does not freeze, while the harbour is easily kept open all the winter with icebreaker assistance if necessary. There is a large fishing and canning industry in the town, also a shipyard at which major repairs can be undertaken. Cargoes include ores imported from the ice-free port of Pechenga (Petsamo) to the west and cargoes of coal are received in the summer from Russian mining concessions on Spitsbergen, where Russian interests have been influenced by the strategic need to keep open the sea approaches to this warm-water port. Arkhangelsk is less fortunate as the shipping season lasts only from 15th May to 25th November. It has been claimed that if necessary it can be kept open all the year by constant icebreaker attention.[28] The most difficult period to keep the port open is during the break-up of the ice. Navigation in the dredged channels of the delta is restricted, but above Arkhangelsk the river widens and dredging is constantly required in the delta channels, particularly the northern ones used by sea-going vessels. The port lies on the north bank alongside the town and on the south bank adjacent to the railway terminus. Besides ship repairing and fishing, the principal trade of Arkhangelsk and its suburbs is in timber for it is the largest timber port in the Soviet Union.

The long coast of the Far Eastern region accounts for a small proportion of Soviet sea traffic, and shipping, except in the south, is composed largely of internal Russian services. Ports suffer badly from ice, particularly those north of 44° N. and in sheltered bays and inlets. Okhotsk Sea ports may be ice-bound for up to 200 days: in some places ice persists until July. The principal port, Vladivostok, founded in 1860, and the base for the Russian Pacific Fleet after 1872, developed quickly after railway construction at the end of the nineteenth century. At the southern end of the Muravyev Peninsula, the town lies on the deep and well-protected bay of the Golden Horn, which surrounding hills shelter from the cold west and north-west winds. Ice persists for three to four months in the winter, but the port is kept open. The port, which was modernized and expanded between the wars, has a wide general traffic but exports coal, timber, and canned fish, though it is sometimes closed to non-Russian vessels. More recently, Nakhodka, a good natural harbour which remains ice-free longer than Vladivostok, has been developed. Ice becomes worse northwards along the Sikhote Alin coast. Soviet Harbour and Vananino have been developed to replace Nikolayevsk as a port for the lower Amur, with which they have been connected by railway to Komsomolsk since 1946, and export chiefly timber. Use of Nikolayevsk is limited not only by ice but also by the bar at the mouth of the Amur, as vessels drawing over 12 feet of water cannot pass, and navigation in the channel is hazardous due to shoals and limited to an ice-free season of only 160 days. Attempts to improve the entrance have failed. Movement through the Straits of Tartary is prevented by ice between December and March. Sakhalin ports handle exports of petroleum and coal and import goods for consumption on the island. The ports of Kamchatka and the Okhotsk Sea are used only during the short summer season, though they are the principal means of intra-regional communication. Magadan, on Nagayevo Bay, developed since 1931, and the port for the Kolyma district, is a large settlement for such latitudes (population *c.*

60,000), and is kept open all the year by icebreakers. It has replaced the old port of Okhotsk. At some places along this coast during the winter, ships are offloaded at the end of the fast shore ice and cargoes carried ashore by sledges or motor lorries. Petropavlovsk-Kamchatskiy, the principal port of Kamchatka, lies in the sheltered Avacha Bay, seldom ice-bound for long, and is well equipped to handle ships up to 5,000 tons. This is the winter harbour for boats working on the Northern Sea Route. For the west-coast fishing industry, a new port at Ozernovskiy has been under construction in the sixth five-year plan.

The immense inland Caspian Sea is in some respects an extension of the Volga waterway. Navigation is, however, comparable with truly sea-going conditions: storms which sweep off the steppe and desert force ships to seek safety in the many small harbours of refuge, and ice forms on the northern part of the Sea in winter. Most ports lie on the western shores as the eastern shores are largely desert. In the southern basin, Baku, lying in a broad bay, is the chief port, but constant dredging is needed to make it available to the largest boats, and it is difficult to approach during frequent north-east or south-east winds. With the nearby Port Apsheron, constructed during the 1930's, it exports petroleum. A train ferry crosses to the eastern shore at Krasnovodsk, where navigation has been made difficult by a growing sandbar and dredging is now necessary. At the terminus of the Trans-Caspian railway, Krasnovodsk exports Nebit-Dag petroleum and Central Asian cotton in exchange for timber and manufactured goods. Lenkoran and Astara, on the western shore, are roadsteads: at the former, ships can lie inshore only in calm weather, but north-east and south-east winds are frequent. Astara is not properly equipped as a port and ships anchor a mile offshore. Port Ilyich is no longer used because of silting. Bekdash, on the eastern shore, is the anchorage for ships serving the chemicals industry of Kara-Bogaz-Gol. In the northern basin, Makhachkala, a large port, is being extended to replace ports silted by the falling level of the Caspian, which has caused the greatest inconvenience in the northern basin. A product of silting of the Volga estuary, the artificial port of 'Twelve-Foot Roadstead' is used to transfer goods from Caspian boats to river-boats, which reach it from Astrakhan by a dredged channel with a minimum depth of 5 feet. Astrakhan, although more exactly a Volga river port, is the focus of movement on the northern Caspian. Guryev, now well away from the sea, reached by a dredged canal through the Ural estuary, is declining as railways replace water transport, and the once important roadstead at Prorva is silted and deserted. Northbound traffic consists of petroleum from Baku, Makhachkala, and Krasnovodsk, mostly destined for Astrakhan. Makhachkala exports North-Caucasian petroleum and petroleum received by rail from Baku, besides grain from the Northern Caucasus. Also northbound are fish and foodstuffs from Caucasia and Central Asia, besides rice, wool, salt, chemicals, and some ores, sent to the Volga delta. Southbound freights are timber, wooden goods, sugar, grains, coal, and some manufactured goods.

AIR TRANSPORT

Over the vast distances common in Russia, the speed and directness of flight are a great advantage: aviation has made an important contribution to communication over the immense, empty lands of Siberia, the Far East, and Central Asia. The majority of air services radiate from Moscow, the centre of both internal and international flights, to nearly all principal cities. Civil air services are provided to such remote centres as Yakutsk, Magadan, and Norilsk, although the densest system of air services is provided between the Centre, the South, the Volga, and Caucasia. Moscow has three airports, all well outside the town.

The *Glavsevmorput* operates large numbers of aircraft to transport personnel and supplies between the Arctic bases and other northern centres. Aircraft are also used extensively for weather forecasting and ice observation. Low-flying aircraft have also been used to track and shoot wolf packs. In Central Asia, aircraft have been employed in finding pasture for flocks in drought years and maintaining contact with remote herding communities.[29]

The Soviet Union releases little information about the use of air transport; but despite recent Russian advances in aircraft technology, the medium does not appear to be as widely applied as might be expected, though there are many local internal services, and in 1963 about 24 million passengers were probably carried. By its geographical location, the Soviet Union could play a major part in international air transport, since many great circle routes between North America and Eurasia, Europe and the Far East, Africa and Asia, and future trans-polar routes, can be made across its territory.

NOTES ON CHAPTER 9

1. Useful discussion from the historical angle of Russian railway development is found in Ames, E., 'A Century of Russian Railroad Construction: 1837–1936', *American Slavic and East European Review*, 1947. More recent years are discussed by Hunter, H., 'Soviet Railroads since 1940', *Bulletins on Soviet Economic Development*, Series I, 1950.

2. The rôle of railways in regional planning is discussed in Kibalchich, O. A., and Stepanov, M. N., 'Present Problems of the Economic Organization of the U.S.S.R.', translated from the Russian in *Petermanns Mitteilungen*, vol. 99, 1955.

3. Russian main-line railways are 5' gauge: in most other European countries, railways are 4' 8½" gauge. It is unlikely that the Russians have left in operation any European standard-gauge lines in the incorporated territories, except for operational needs in frontier areas. Russian broad gauge is used in Finland and also extends for some distance across the Romanian boundary. There is no truth in rumours that a 5'-gauge line had been laid to Berlin. Narrow-gauge railways (750 mm. and metre) are common in the Soviet Union.

4. Such rumoured projects are frequent in the Soviet Union and belong to a modern type of 'traveller's tale'.

5. Discussed further in Mellor, R. E. H., 'Narrow-Gauge Railways in Russia's Virgin Lands', *Geography*, 1956.

6. The first recorded passenger service was in the summer of 1957, when through trains from Moscow to Peking via Tsining took 10 days. Ulan Bator to Peking required 51 hours. Judging from scanty descriptions and photographical material, the line appears lightly built.

7. Discussed in *Voprosy Razvitiya Zheleznodorozhnogo Transporta*. Moscow, 1957. Taaffe, R. N., *Rail Transportation and the Economic Development of Soviet Central Asia*. Chicago, 1960.

8. Lickey Incline on the Bristol-Birmingham line is 1 : 37¾.

9. *Permafrost* is discussed at length in Chapter 2. It is one of the most difficult physical problems to overcome in constructional work in many parts of Siberia.

10. Technical discussion of motive power is treated in Rakov, A., *Lokomotivy Zheleznikh Dorog Sovetskogo Soyuz*. Moscow, 1955.

11. Rakov, A., *op. cit.*, also Blackman, J. H., *Transport Development and Locomotive Technology in the Soviet Union*. Columbia, U.S.A., 1957.
Mellor, R. E. H., 'Motive Power and Its Problems on Soviet Railways'. *Locomotive*, vol. 65, 1959.

12. Evidence in the Russian Railway Passenger Timetable (Summer 1958) indicates that trains now use the new line via Andrianovskaya. No trains run between Irkutsk and Baykal, but the latter is served by two daily local trains from Slyudyanka. This is clearly related to changes in water level created by the Angara Barrage.

13. Khachaturov, T. S., 'The Organization and Development of Transport in the U.S.S.R.', *International Affairs*. London, 1945.
    Ökonomik des Transportwesens. Berlin, 1962.
14. Khanukov, E. D., *Transport i Razmeshcheniye Proizvodstva*. Moscow, 1956.
    Williams, E. W., *Freight Transportation in the Soviet Union*. Princeton (N.J.), 1962.
15. *Voprosy Razvitiya Zheleznodorozhnogo Transporta*.
    Nikolskiy, I. V., *Geografiya Transporta SSSR*. Moscow, 1960.
16. According to Khanukov, *op. cit.*, over 80 per cent of the empty wagons on Soviet railways are filled daily by six railway directorates: Donets, Tomsk, Pechora, Karaganda, Sverdlovsk, and Kirov. Further, 60 per cent of the empty wagons are used for coal. As a result of this distribution, empty running is a serious item: this averaged 38·3 per cent of the total distance run for all types of wagon in 1953. Of all types of wagons running empty, 44 per cent were open coal wagons.
    Nikolskiy, I. V., *op. cit.*
17. Based on figures in *Voprosy Ratsionalizatsii Perevozok Vazhneyshikh Gruzov*. Moscow, 1957. Of the 71·6 million tons of ferrous metals moved by rail in 1955, pig iron formed 11·2 million tons, bars 2·7 million tons, ingots over 6·0 million tons, girders, 1·6 million tons, pipes 4·8 million tons, and rails 4·6 million tons.
18. *Transport i Sviyaz SSSR: Statisticheskiy Sbornik*. Moscow, 1957.
19. An underground railway is under construction in Kiev.
20. Sarantsev, P. L., *Geografiya Putey Soobshcheniya*, Moscow, 1957, and *Transport i Sviyaz*, Moscow, 1957.
21. *Pravda* reported in June 1958 that the Mari Canal system was being enlarged to take bigger river boats.
22. Improvement in navigation is only one aspect of the Great Volga project: others include hydro-electric power and irrigation.
23. Baranskiy, N. N., *Ekonomicheskaya Geografiya SSSR*. Moscow, 1956.
24. Armstrong, T., *The Northern Sea Route*. Cambridge, 1952.
    Krypton, C., *The Northern Sea Route*. London, 1956.
25. It was reported in 1955 that Pechenga was served by a railway from Murmansk. In the 1958 Summer Railway Passenger Timetable this line has been included.
26. Thiel, E., *The Soviet Far East*, London, 1957, discussed native transport.
    Also Mikhaylov, N. I., *Sibir: Fiziko-Geograficheskiy Ocherk*. Moscow, 1956.
27. Navigational data based on the various *Admiralty Pilots* and the *Russian Atlas of Oceanography*. Moscow, 1950–1953. Other material drawn from Sarantsev, P. L., *op. cit.* Also Taskin, G. A., 'Falling Level of the Caspian Sea in Relation to Soviet Economic Development', *Geographical Review*, vol. 44, 1954.
28. The ability of the Russians to keep Arkhangelsk open all the year round never seems to have been fully tested. The claim revolves around remarks by I. D. Papanin, Head of *Glavsevmorput* (1934–1947), during the Second World War. See Armstrong, T., 'The Soviet Northern Sea Route', *Geographical Journal*, vol. 120, 1955.
29. Frequent reference is made to aircraft in Russian geographical studies, but there is no standard work on the modern employment of aircraft. Soviet statistics on air transport are scanty.
    Kish, G., 'Soviet Air Transport', *Geographical Review*, vol. 48, 1958.

APPENDIX

Population of Towns with over 50,000 People at the 1959 Census

Town	Former Name §	Population (thousands)	Increase 1939–1959 (1939 = 100)
NORTH			
Arkhangelsk	Archangel*	256	102
Vologda		138	145
Cherepovets		92	284
Severodvinsk	Molotovsk	79	368
Syktyvkar	Ust-Sysolsk	64	263
Vorkuta		55	—
NORTH-WEST			
Leningrad	S. Peterburg		
	St. Petersburg*		
	Petrograd	2,888	96
	(with suburbs)	(3,300)	97
Murmansk	Romanov-na-Murmane	226	189
Kaliningrad	Königsberg (Ger.)	202	...
Petrozavodsk	Kalininsk	135	194
Pskov	Pleskov		
	Pleskau (Ger.)	81	134
Novgorod		61	153
Velikiye Luki		59	168
Vyborg	Viipuri (Fin.)	51	...
CENTRE			
Moskva	Moscow*	5,032†	120
Gorkiy	Nizhniy Novgorod	942**	146
Voronezh		454	132
Yaroslavl		406	132
Tula		345	121
Ivanovo	Ivanovo-Voznesensk	332	116
Kalinin	Tver	261	121
Penza		254	159
Kirov	Khlynov, Vyatka	252	175
Ryazan		213	224

Town	Former Name §	Population (thousands)	Increase 1939–1959 (1939 = 100)
Bryansk	Debryansk	206	118
Kursk		203	169
Rybinsk	Shcherbakov	181	126
Kostroma		171	141
Tambov		170	160
Dzerzhinsk	Chernoye, Rastyapino	163	158
Lipetsk		156	234
Vladimir		154	230
Orel		152	138
Smolensk		146	93
Perovo		143	227
Kaluga		133	149
Kuntsevo		128	210
Podolsk		124	171
Babushkin	Losinoostrovskaya	112	158
Orekhovo-Zuyevo		108	109
Novomoskovsk	Bobriki, Stalinogorsk	107	140
Serpukhov		105	116
Kolomna		100	133
Kovrov		100	148
Mytishchi		99	164
Elektrostal		97	227
Lyubertsy		93	200
Noginsk	Bogorodsk	93	114
Saransk		90	221
Tushino		90	226
Yoshkar-Ola	Tsarevgorod Tsarevokokshaisk Krasnokokshaisk	88	324
Lyublino		86	172
Kineshma		84	112
Cheboksary		83	278
Michurinsk	Kozlov	80	112
Yelets		78	153
Zagorsk	Sergiyevskiy Posad Sergiyev	73	165
Murom		73	181
Belgorod		71	208
Vyshniy Volochek		66	104
Balashov		64	135
Shuya		64	110
Yegoryevsk		59	105
Balashikha		58	200
Kuznetsk		57	150
Pavlovskiy Posad		55	129

Appendix

Town	Former Name §	Population (thousands)	Increase 1939–1959 (1939 ≑ 100)
Borisoglebsk		54	102
Uzlovaya		54	299
Klin		53	192
Gus-Khrustalnyy		53	133
Vichuga		51	109

VOLGA

Town	Former Name §	Population (thousands)	Increase 1939–1959 (1939 ≑ 100)
Kuybyshev	Samara	806	206
Kazan		643	162
Volgograd	Tsaritsyn Stalingrad	591	133
Saratov		581	156
Astrakhan		294	116
Ulyanovsk	Simbirsk	205	209
Syzran		148	177
Engels	Pokrovsk	90	131
Chapayevsk	Ivashchenkovo, Troitsk	83	143
Volzhskiy	Lopatino	67	—
Novokuybyshevsk		63	—
Volsk		62	110
Bugulma		61	245
Zelenodolsk	Zeleniy Dol	60	199
Kamyshin	Dmitriyevsk	55	230
Chistopol		51	161
Melekess		51	156

NORTH CAUCASUS

Town	Former Name §	Population (thousands)	Increase 1939–1959 (1939 ≑ 100)
Rostov-na-Donu		597	117
Krasnodar	Yekaterinodar	312	162
Groznyy		240	139
Taganrog		201	107
Shakhty	Aleksandrovsk-Grushevskiy	196	145
Ordzhonikidze	Vladikavkaz Dzaudzhikau	164	125
Stavropol	Voroshilovsk	140	164
Makhachkala	Petrovsk	119	137
Armavir		111	133
Novoshakhtinsk	Imeni III Internatsionala Komintern	104	216
Sochi		95	153
Novocherkassk		94	124
Novorossiysk		93	98
Nalchik		87	181

Town	Former Name §	Population (thousands)	Increase 1939–1959 (1939=100)
Maykop		82	147
Kislovodsk		79	155
Pyatigorsk		69	112
Kamensk-Shakhtinskiy		58	135
Yeysk		55	122
Kropotkin	Romanovskiy Khutor	54	130
Gukovo		53	598
Bataysk		52	107

URAL

Sverdlovsk	Yekaterinburg	777	184
Chelyabinsk		688	252
Perm	Molotov	628	205
Ufa		546	212
Nizhniy Tagil		338	212
Magnitogorsk		311	213
Izhevsk		283	161
Orenburg	Chkalov	260	151
Orsk		176	265
Zlatoust		161	162
Kopeysk		160	267
Kamensk-Uralskiy		141	278
Sterlitamak		111	287
Berezniki		106	207
Serov	Nadezhinsk, Kabakovsk	98	151
Pervouralsk		90	205
Korkino		85	739
Troitsk		76	164
Lysva		73	142
Sarapul		68	161
Kungur		65	179
Oktyabrskiy		65	—
Krasnoturinsk		62	647
Chusovoy		60	140
Salavat		60	—
Asbest		60	207
Kizel	Uglegorsk	60	152
Votkinsk		59	153
Beloretsk		59	145
Glazov		59	358
Novo-Troitsk		57	22 times
Revda		55	170
Buzuluk		55	129
Krasnokamsk		54	183

Appendix

Town	Former Name §	Population (thousands)	Increase 1939–1959 (1939 = 100)
WESTERN SIBERIA			
Novosibirsk	Novo-Nikolayevsk	887**	219
Omsk		579	201
Novokuznetsk	Kuznetsk, Stalinsk	377	228
Barnaul		320	216
Prokopyevsk		282	263
Kemerovo	Shcheglovsk	277	209
Tomsk		249	171
Tyumen		150	189
Biysk		146	182
Kurgan		145	272
Leninsk-Kuznetskiy	Kolchugino, Lenino	132	160
Kiselevsk		130	296
Anzhero-Sudzhensk		116	168
Rubtsovsk		111	294
Belovo		107	247
Osinniki		68	268
Mezhdurechensk		55	—
Shadrinsk		52	168
EASTERN SIBERIA			
Krasnoyarsk		409	215
Irkutsk		365	146
Ulan Ude	Udinsk, Verkhneudinsk	174	138
Chita		171	142
Angarsk		134	—
Cheremkhovo		123	220
Norilsk		108	778
Yakutsk		74	140
Kansk		74	177
Abakan	Ust-Abakanskoye	56	154
Chernogorsk		51	294
Bratsk		51	—
FAR EAST			
Khabarovsk		322	155
Vladivostok		283	137
Komsomolsk-na-Amure		177	250
Ussuriysk	Nikolsk-Ussuriyskiy, Voroshilov	104	144
Blagoveshchensk		94	159
Yuzhno-Sakhalinsk	Vladimirovka, Toyohara (Jap.)	86	...

Town	Former Name §	Population (thousands)	Increase 1939–1959 (1939 = 100)
Petropavlovsk-Kamchatskiy		86	242
Nakhodka		63	—
Magadan		62	227
Svobodnyy	Alekseyevsk	57	129
Artem		55	157

UKRAINE

Kiev		1,102	130
Kharkov		930**	112
Donetsk	Yuzovka, Stalino	701	150
Odessa		667	111
Dnepropetrovsk	Yekaterinoslav	658	125
Zaporozhye		435	154
Lvov	Lwów (Pol.), Lemberg (Ger.)	410	121
Krivoy Rog		386	204
Makeyevka	Dmitriyevsk	358	148
Gorlovka		293	161
Zhdanov	Mariupol	284	128
Lugansk	Voroshilovgrad	274	128
Nikolayev	Vernoleninsk	224	133
Dneprodzerzhinsk	Kamenskoye	194	131
Simferopol		189	132
Kadiyevka	Sergo	180	133
Kherson		157	162
Sevastopol		148	130
Chernovtsy	Cernăuţi (Rom.) Czernowitz (Ger.)	145	137
Poltava		141	110
Kirovograd	Yelisavetgrad, Zinovyevsk, Kirovo	127	127
Vinnitsa		121	130
Kramatorsk		115	122
Zhitomir		105	110
Kerch		99	95
Kommunarsk	Alchevsk, Voroshilovsk	98	179
Sumy		97	151
Melitopol		95	125
Krasnyy Luch		94	159
Yenakiyevo	Rykovo, Ordzhonikidze	92	104
Chistyakovo		92	185
Chernigov		89	130
Konstantinovka		89	93
Kremenchug		86	95

Appendix

Town	Former Name §	Population (thousands)	Increase 1939–1959 (1939 = 100)
Slavyansk		83	106
Cherkassy		83	160
Nikopol		81	140
Belaya Tserkov		71	149
Ivano-Frankovsk	Stanislav, Stanisławów (Pol.), Stanislau (Ger.)	66	102
Berdyansk	Osipenko	65	126
Khmelnitskiy	Proskurov	62	165
Sverdlovsk		62	166
Artemovsk	Bakhmut	61	110
Yevpatoriya	Kozlov, Eupatoria **	57	122
Rovno	Równe (Pol.)	57	132
Konotop		53	117
Berdichev		53	85
Ternopol		52	104

BYELORUSSIA

Town	Former Name §	Population (thousands)	Increase 1939–1959 (1939 = 100)
Minsk		509	214
Gomel		166	120
Vitebsk		148	88
Mogilev		121	122
Bobruysk		97	115
Brest	Brest-Litovsk, Brześć nad Bugiem (Pol.)	73	176
Grodno	Gardinas (Lith.)	72	146
Orsha		64	119
Borisov		59	121
Baranovichi	Baranowicze (Pol.)	58	211

UZBEKISTAN

Town	Former Name §	Population (thousands)	Increase 1939–1959 (1939 = 100)
Tashkent		911**	166
Samarkand	Afrosiab, Maracanda	195	143
Andizhan		129	153
Namangan		122	154
Kokand		105	124
Fergana	Novyy Margelan, Skobelev	80	224
Bukhara	Bokhara *	69	137
Margelan	Staryy Margelan	68	147
Chirchik	Kirgiz-Kulak	65	444
Angren		55	—

2 B

Geography of the U.S.S.R.

Town	Former Name §	Population (thousands)	Increase 1939–1959 (1939 = 100)
KAZAKHSTAN			
Alma Ata	Vernyy	455	206
Karaganda		398	255
Semipalatinsk		155	141
Chimkent	Isfidzhab	153	205
Petropavlovsk		131	143
Ust-Kamenogorsk	Zashchita	117	580
Uralsk	Yaitskiy Gorodok	105	156
Tselinograd	Akmolinsk	101	314
Aktyubinsk		97	199
Pavlodar		90	314
Kustanay		86	256
Guryev		78	188
Dzhambul	Yangi, Aulia-Ata, Mirzoyan	67	177
Leninogorsk	Ridder	67	133
Kzyl-Orda	Ak-Mechet, Petrovsk	66	141
Zyryanovsk		54	335
Temir-Tau	Samarkand	54	11 times
Balkhash		53	164
GEORGIA			
Tbilisi	Tiflis	694	134
Kutaisi		128	165
Batumi		82	117
Sukhumi		64	145
Rustavi		62	—
AZERBAYDZHAN			
Baku		636	111
	(with suburbs)	968**	125
Kirovabad	Gandzha, Yelizavetpol	116	117
Sumgait		52	822
LITHUANIA			
Vilnyus	Vilnius (Lith.), Wilno (Pol.), Vilno	235	109

Appendix

Town	Former Name §	Population (thousands)	Increase 1939–1959 (1939 = 100)
Kaunas	Kowno (Pol.), Kauen (Ger.), Kovno	214	140
Klaypeda	Klaipeda (Lith.), Memel (Ger.)	89	...
Shyaulyay	Šiauliai (Lith.), Schaulen (Ger.), Shavli	60	190
MOLDAVIA			
Kishinev	Chişinău (Rom.)	214	191
Beltsy	Bălţi (Rom.)	67	218
Tiraspol		62	165
LATVIA			
Riga		605	170
Liepaya	Liepaja (Lat.), Libau (Ger.), Libava	71	134
Daugavpils	Dünaburg (Ger.), Dvinsk	65	126
ESTONIA			
Tallin	Tallinn (Est.), Revel	280	175
Tartu	Dorpat (Ger.), Derpt, Yuryev	74	130
KIRGIZIA			
Frunze	Pishpek	217	234
Osh		65	194
TADZHIKSTAN			
Dushanbe	Dyushambe, Stalinabad	224	271
Leninabad	Khodzhent	77	169

Town	Former Name §	Population (thousands)	Increase 1939–1959 (1939 = 100)
ARMENIA			
Yerevan	Erivan	509	250
Leninakan	Aleksandropol	108	160
TURKMENISTAN			
Ashkhabad	Ashabad, Poltoratsk	170	134
Chardzhou	Leninsk-Turkmenskiy	66	120

Notes

§ Former names also include English conventionals (*) and foreign forms (Est.: Estonian; Fin.: Finnish; Ger.: German; Jap.: Japanese; Lat.: Latvian; Lith.: Lithuanian; Pol.: Polish; Rom.: Romanian).
** Towns so marked have reached a million inhabitants since the census.
† Excluding suburbs.
— Not recorded in last census (1939).
... Not in Soviet territory at last census (1939) or no comparable figure for that date.

GLOSSARY

Adyr	Mountain foothills in Central Asia.
Afganets	Dusty SW or WSW wind blowing in Central Asia (40-70 days p.a.). Associated with intrusions of cold air masses from N or NW.
Agrogorod	New agricultural town to replace scattered villages of collective or state farms.
Bakhcha (Baksha)	Garden cultivation (often irrigated) of crops such as melons and cucumbers, etc.
Balka	Usually broad valley in steppe developed from *ovragi* (q.v.). Usage varies slightly throughout Russia.
Basseyn	Basin. Usually applied to coalfields (e.g. Donets Basin) in abbreviated form -*bass* (e.g. Donbass).
Belka	Snow covered peak. Usually applied in Siberia.
Barkhan	Crescentic sand dune.
Bor	Hillock. Usually a term to describe pine groves on drier sandy hillocks.
Bogarni	Non-irrigated cultivation (mostly cereals) in dry Central Asian foothills.
Boyar	Alloidal landlord of old Russia.
Bulgun(n)yakh	A hydrolaccolith, ice blister or pingo-like form developed in areas of *permafrost*.
Chernozem	Black earth. A fertile dark soil of the steppe.
Chink	Escarpment in Central Asia.
Comecon	Council for Mutual Economic Assistance. Founded 1949 to further economic integration among Communist *bloc* states.
Dolina	Valley.
Glavsevmorput	Chief Administration of Northern Sea Route. H.Q. in Leningrad.
Glint	North-facing escarpment in Northern European Russia and Baltic littoral marking where ancient shield rocks dip beneath Palaeozoic deposits.
Golets	Bare rounded summits. Usually in Siberia.
Gora	Mountain. Sometimes replaced by *Pik* (peak) or by Turkic *Tau* (mountain).
Graben	German: ditch. A long narrow depressed tract, usually a fault-sided rift valley.
Ibe (Ebe)	Warm föhn-like wind blowing in cold season in Dzungarian Gates.
Ilmen	Elongated water-filled depression of Volga delta and northern Caspian littoral.

373

Izba	Hut. Usually wooden or log hut of Northern and Central Russia.
Kebir	Puffy *solonchak*, a highly friable soil.
Khata	Hut. Usually the frame house of Southern Russia.
Khrebet	Mountain range.
Kir	Sand bound with petroleum seepage and wind-eroded into fantastic shapes.
Kolkhoz	Collective farm worked by members (*kolkhozniki*).
Krasnozem	Red earth — semi-lateritic soil of humid Caucasia.
Kray	Country. Usually applied to large, sparsely-settled administrative areas of Siberia.
Krotovina	Earth-filled abandoned animal burrow of steppe.
Kulak	Well-to-do peasant. Literally 'a tightly gripped fist'.
Kultbaza	Culture base. Settlement (mostly in Siberia) primarily used to introduce Soviet ideas to natives.
Kum	Desert.
Kurgan	Tumulus (mostly in Southern Russia and Central Asia).
Kurum	Rock-strewn slopes in the Sayan mountains.
Kryazh	Hilly ridge.
Liman	(1) long narrow bay or estuary along northern Black Sea coast; (2) cultivated terraces in arid areas.
Loess (Löss)	Fine yellow-grey loam.
Makhorka	Inferior type of tobacco.
Mar	Wooded swamp covered with small hillocks, usually in *permafrost* areas. Eastern Siberia, Far East.
Melkosopochnik	Low hills with gentle slopes rising from plain. Result of erosion under continental conditions.
More	Sea.
Mys	Cape.
Naled	Sheets of ice formed by water outflowing on the surface of ground in *permafrost* areas. May be several hundred square yards in extent and from three to twelve feet deep, formed of successive sheets.
Nizhmennost	Plains or low undulating country.
Oblast	Primary division of Soviet territorial-administrative system. Usually named after the industrial centre that administers it.
Odnodvorets	Peasant freeholder of old Russia.
Okrug	A low rank of administrative district, usually a small reservation of a minor national group.
Ostrog	Palisaded settlement of Siberia: often a place of banishment.
Ostrov	Island. *Poluostrov*: peninsula.
Ovrag	Deep erosion gully in steppe; often water-logged floor.
Ozero	Lake. In Turkic areas replaced by *Kul*.
Padina	Shallow hollow, notably in steppe.
Pereval	Mountain pass.
Perevozka	Portage between rivers.

374

Plaven	Floodplain of Southern Russian river.
Ploskogorye	Plateau, upland, tableland. *Plato* also used.
Podzol	Poor acid soil with ashen-coloured layer in boreal forest.
Polonina	High mountain pasture in Carpathia.
Pomestie	Landed estate based on feudal principles, held by *pomeshchik*.
Pomory	Coastal dwellers, usually around the northern seas.
Proliv	Strait or channel.
Purga	Violent arctic blizzard.
Rasputitsa	Spring thaw, also applied to autumn floods.
Ravnina	Plain or lowland.
Rayon	Small administrative-territorial unit.
Serozem	Grey soil of desert and semi-desert.
Shor (Sor)	Swampy salt covered area or *solonchak*.
Skladchataya Strana	Folded country.
Solod	Leached saline soil.
Solonchak	Saline soil without structure.
Solonets	Formerly saline soil from which salts leached. Surface salt on soil.
Sopka	Round-topped hill, often volcanic.
Sovkhoz	State farm.
Sovnarkhoz	Regional economic administrative commission.
Stanitsa	Cossack village.
Step (Steppe)	Dry grassland, where interspersed by trees, wooded steppe (*Lesostep*).
Sukhovey	Dry hot SE or S wind in steppe.
Syrt	Bare flat-topped hills or plateau in Southern Russia and Central Asia.
Takyr	Clay flats. Hard barren crust in dry weather, bog in wet weather. Barren alkaline soil in heavy clay.
Talik	Unfrozen area amid *permafrost*.
Taryn	Type of *naled* (q.v.) formation on rivers, notably when completely frozen.
Tayga	Boreal coniferous forest.
Toltry	Limestone ridges in Ukraine, remains of coral reefs of a Miocene sea.
Tugay (Togay)	Forest and rank meadow association of Central Asian rivers.
Tundra	Treeless subpolar moor. Marshy, mossy or stony. Where stunted trees occur, wooded tundra (*Lesotundra*).
Urman	Dense coniferous forest in Western Siberia.
Urstromtal	River valley used by swollen stream carrying Quaternary glacial meltwater. *Pradolina* also used.
Uvaly	Low hillocky country (also *Uvalistaya Strana*).
Vechnaya Merzlota	Permanently frozen ground : *permafrost*.
Vodokhranilishche	Reservoir.
Vorota	Gate: used to denote major passes or gaps.
Vozvyshennost	Upland or low plateau.
Vpadina	Depression.

Yayla	Summer pasture, notably in limestone plateau of S. Crimea.
Yelan	Coarse river meadow subject to spring floods.
Yurt	Felt tent stretched over a lattice frame used by Asiatic nomads.
Zaliv	Bay (also *Guba*).
Zamor	Fall in oxygen content in slow moving river water, stagnating Siberian rivers. Discomforts fish.
Zamorozki	First autumn frosts, agriculturally important in Siberia.
Zaton	River backwater used as refuge harbour for ships in winter.
Zemlya	Country or land.

BIBLIOGRAPHY

SELECTED GENERAL TEXTBOOKS

BALZAK, S. S., VASYUTIN, F. E., FEIGIN, Y. G., *Economic Geography of the U.S.S.R.*, New York, 1949. Translation of a now dated but still useful Soviet textbook.

BARANSKIY, N. N., *Ekonomicheskaya Geografiya SSSR*, Moscow, 1954, or English translation, *Economic Geography of the U.S.S.R.*, Moscow, 1956, with revisions.

BERG, L. S., *Priroda SSSR*, Moscow, 1938. English translation: *The Natural Regions of the U.S.S.R.*, New York, 1950.

BERG, L. S., *Geograficheskiye Landshaftnyye Zony Sovetskogo Soyuza*, Moscow, Vol. I, 1947, Vol. II, 1952.

CAMPBELL, R. W., *Soviet Economic Power: Its Organisation, Growth and Challenge*, Cambridge (U.S.A.), 1960.

CHERDANTSEV, G. N. (ed), *Ekonomicheskaya Geografiya SSSR*, Moscow, R.S.F.S.R., 1956, other union republics, 1957, systematic review, 1958.

COLE, J. P., GERMAN, F. C., *A Geography of the U.S.S.R.*, London, 1961.

CRANKSHAW, E., *Khrushchev's Russia*, Harmondsworth, 1959.

CRESSEY, G. B., *Soviet Potentials: A Geographic Appraisal*, Syracuse, 1962.

CRESSEY, G. B., *The Basis of Soviet Strength*, New York, 1950.

FICHELLE, P., *L'U.R.S.S.*, Paris, 1962.

GEORGE, P., *L'U.R.S.S.: Haute Asie-Iran*, Paris, 1947.

GREGORY, J. S., SHAVE, D. W., *The U.S.S.R.: A Geographical Survey*, London, 1947.

HARRIS, C. D. (ed.), *Soviet Geography: Accomplishments and Tasks*, New York, 1962.

JORRÉ, G., *The Soviet Union: The Land and its People*, 2nd ed., London, 1961.

KOLTUN, M. T., *Prirodnoye Rayonirovanniye Territorii SSSR*, Moscow, 1962.

KOSTENNIKOV, V. M., *Ekonomicheskiye Rayony SSSR*, Moscow, 1958.

KRUGER, K., *Unser Wissen über die UdSSR*, Berlin, 1956.

LEIMBACH, W., *Die Sowjetunion: Natur, Volk und Wirtschaft*, Stuttgart, 1950. Written by one who has travelled far in the Soviet Union.

LYALIKOV, N. I. (ed.), *Geografiya SSSR*, Moscow, 1955.

LYALIKOV, N. I., *Ekonomicheskaya Geografiya SSSR*, Moscow, 1960.

PAVLOV, M. Y., *Geografiya SSSR*, Moscow, 1954.

RAUS, O., *Die Sowjetunion: Politisch-ökonomisch-geographische Übersicht*, Berlin, 1961.

SAUSHKIN, Y. G., *Economic Geography of the Soviet Union*, Oslo, 1959.

SCHWARTZ, H., *Russia's Soviet Economy*, Englewood Cliffs, 1958.

SEGER, G., *Physische Geographie der Sowjetunion*, Berlin, 1960.

SHABAD, T., *The Geography of the U.S.S.R.*, New York, 1951.

Geography of the U.S.S.R.

STROYEV, K. F., *Fizicheskaya Geografiya SSSR*, Moscow, 1959.

UdSSR, Leipzig, 1959 — a translation into German of Vol. 50 of the Great Soviet Encyclopedia (Moscow, 1952–) devoted solely to the U.S.S.R.

UTECHIN, S. V., *Everyman's Concise Guide to Russia*, London, 1961.

SELECTED REGIONAL TEXTBOOKS

ANUCHIN, V. A., *Geografiya Sovetskogo Zakarpatya*, Moscow, 1956.

BELYUKAS, K-K. (ed.), *Litovskaya SSR*, Moscow, 1955.

DIBROVA, A. T., *Geografiya Ukrainskoy SSR*, Moscow-Kiev, 1961.

DOBRYNIN, B. F., *Fizicheskaya Geografiya SSSR: Evropeyskaya Chast i Kavkaz*, Moscow, 1948.

DOLGOPOLOV, K. V. (ed.), *Povolzhye: Ekonomiko-geografcheskaya Kharakteristika*, Moscow, 1957.

DZHAVAKHISHVILI, A. (ed.), *Gruzinskaya SSR: Ekonomiko-geografcheskaya Kharakteristika*, Moscow, 1956.

FREIKIN, Z. G., *Turkmenskaya SSR: Ekonomiko-geografcheskaya Kharakteristika*, Moscow, 1957.

KOMAR, I. V., *Ural: Ekonomiko-geografcheskaya Kharakteristika*, Moscow, 1959.

KOVALEVSKIY, G. T. (ed.), *Belorusskaya SSR*, Moscow, 1957.

KUTAFYEV, S. A., SHASTNEV, P. N. (eds.), *Rossiyskaya Federatsiya*, Moscow, 1959.

LEWYTZSKIY, B., *Die Sowjet-Ukraine 1944–1963*, Cologne, 1963.

MARYUKHAN, A. O., *Armyanskaya SSR*, Moscow, 1955.

MIKHAYLOV, N. N., *Sibir: Fiziko-geografcheskiy Ocherk*, Moscow, 1956.

MIKHAYLOV, N. N., *Gorny Yuzhnoy Sibiri*, Moscow, 1961.

MILKOV, F. M., *Fiziko-geografcheskoye Rayonirovanniye Tsentralnikh Chernozemnikh Oblastey*, Voronezh, 1961.

MILKOV, F. M., *Sredneye Povolzhye: Fiziko-geografcheskoye Opisaniye*, Moscow, 1953.

MINTS, A., *Podmoskovye*, Moscow, 1961.

MIRCHUK, I. (ed.), *The Ukraine and its People*, Munich, 1949.

NAZAREVSKIY, O. R. (ed.), *Kazakhstanskaya SSR: Ekonomiko-geografcheskaya Kharakteristika*, Moscow, 1957.

NARZIKULOV, I. K., Ryazantsev, S. N., *Tadzhikskaya SSR*, Moscow, 1956.

NESTERENKO, A. A., *Ukrainskaya SSR*, 2 vols., Kiev, 1957.

PAVLENKO, V., Ryazantsev, S. N., *Kirgizskaya SSR: Ekonomiko-geografcheskaya Kharakteristika*, Moscow, 1961.

POKSHISHEVSKIY, V. V., *Voprosy Geografii Naseleniya Vostochnoy Sibiri*, Moscow, 1962.

ROMANENKO, I. N., *Silskohospodarski Zoni Ukrainskoi RSR*, Kiev, 1961.

SKOSYREV, V., *Turkmenistan*, Moscow, 1955.

SUSLOV, S. P., *Fizicheskaya Geografiya SSSR: Aziatskaya Chast*, Moscow, 1956.

TARMISTO, V. Y., Rostovtsev, M. I., *Estonskaya SSR*, Moscow, 1957.

THIEL, E., *Sowjet-Fernost*, Munich, 1953.

VEIS, E. E., PURIN, V. P., *Latviyskaya SSR*, Moscow, 1957.

Bibliography

SHORT LIST OF SELECTED TEXTBOOKS
ON SPECIAL SUBJECTS

BERG, L. S., *Die Geschichte der russischen geographischen Entdeckungen*, Berlin, 1958.

BERG, L. S., *Ekonomicheskaya Geografiya-Ekonomicheskaya Kartografiya*, Moscow, 1960.

BONDARCHUK, V., *Osnovy Geomorfologiya*, Moscow, 1949.

CHOMBART DE LAUWE, P., *Les Paysans Soviétiques*, Paris, 1961.

DAVYDKIN, P. K. (ed.), *Khrestomatiya po Fizicheskoy Geografii SSSR*, Moscow, 1959.

GVOZDETSKIY, N. A., GLAZOVSKAYA, M. A., *Voprosy Fizicheskoy Geografii SSSR*, Moscow, 1959.

Hodgkins, J., *Soviet Power: Energy, Resources, Production and Potentials*, Englewood Cliffs, 1961.

HUNTER, H., *Soviet Transportation Policy*, Cambridge (U.S.A.), 1957.

JASNY, N., *The Socialized Agriculture of the U.S.S.R.*, Oxford, 1949.

KOCHAN, L., *The Making of Modern Russia*, Harmondsworth, 1963.

Khrestomatiya po Ekonomicheskoy Geografii SSSR, Moscow, 1961.

LENIN, V. I., *Ob Elektrifikatsii*, Moscow, 1958 (reissue).

LORIMER, F., *The Population of the Soviet Union: History and Prospects*, Geneva 1946.

LYASHCHENKO, P., *History of the National Economy of Russia*, New York, 1949.

LYUBOVNYY, I., *Voprosy Geografii Naseleniya SSSR*, Moscow, 1961.

MAVOR, J., *An Economic History of Russia*, 2 vols., London, 1914.

MONGAIT, A. L., *Archaeology in the U.S.S.R.*, Harmondsworth, 1961.

NECHKINA, M. V., RYBAKOV, B. A. (eds.), *Istoriya SSSR*, 2 vols., Moscow, 1952.

PARES, B., *The Russian Peasant and other Studies*, London, 1942.

SHCHUKIN, I. S., *Osnovy Geomorfologiya*, Moscow, 1960.

SHIMKIN, D. B., *Minerals: A Key to Soviet Power*, Cambridge (U.S.A.), 1953.

SLEZAK, J. O., *Breite Spur und Weite Strecken*, Berlin, 1963.

STEPANOV, P., *Geografiya tyazheloy Promyshlennosti SSSR*, Moscow, 1961.

SUMNER, B., *A Survey of Russian History*, London, 1944.

TOKAREV, S., *Etnografiya Narodov SSSR*, Moscow, 1958.

VERNADSKY, G., *A History of Russia*, Princeton, 1944.

WALLACE, D. M., *Russia*, 2 vols., London, 1912. An excellent account of Russia before the Revolution.

ZIEBER, P., *Die Sowjetische Erdölwirtschaft*, Hamburg, 1962.

ATLASES

Bolshoy Sovetskiy Atlas Mira, Moscow, 1937–1939. Vol. I is easily obtained in libraries, but Vol. II is very rare.

Atlas Mira, Moscow, 1954. This is a large atlas (50″ × 14″) with a separate gazetteer volume. Maps of the Soviet Union mostly at scales of 1 : 2.5 M and 1 : 1.5 M.

Morskoy Atlas, Moscow, 1955. Vol. I is mainly maps of oceans and seas, but Vol. II contains thematic maps.

Geograficheskiy Atlas dlya Uchiteley Sredney Shkoly, Moscow, 1954, 2nd revised edition, 1959.

Atlas SSSR dlya Sredney Shkoly — Kurs Ekonomicheskov Geografii, Moscow, 1959.

Atlas SSSR, Moscow, 1962. Contains a valuable collection of detailed thematic and topographic maps.

Atlas Istorii SSSR, Moscow, 1954–1960, 3 vols.

Atlas Istorii Geograficheskikh Issledovanii i Otkrytii, Moscow, 1959.

Atlas Selskogo Khozyaystva SSSR, Moscow, 1960.

Atlas Avtomobilnikh Dorog SSSR, Moscow, 1960.

Atlas Shkem Zheleznikh Dorog SSSR, Moscow, 1963.

Atlas Narodov Mira, Moscow, 1962. A great deal of valuable information is contained on the plates covering the Soviet Union.

Atlas Armyanskoy SSR, Moscow-Yerevan, 1961.

Atlas Belorusskoy SSR, Moscow-Minsk, 1958.

Atlas Ukrainskoy i Moldavskoy SSR, Moscow-Kiev, 1961.

Atlas Silskoho Hospodarstva Ukrainskoi RSR, Kiev, 1961.

The Times Atlas of the World, Vol. 2, London, 1959.

Oxford Regional Economic Atlas — U.S.S.R. and Eastern Europe, Oxford, 1960 (revised edition).

An Economic Atlas of the Soviet Union, ed. G. Kish, Ann Arbor, 1960. The rather poor maps conceal a wealth of scholarship.

PERIODICALS

The Soviet Press forms a useful source of information on new developments. *Pravda* and *Izvestiya* carry a wide range of material and some appears in English translation and abstract (along with material from other Soviet papers) in the daily broadsheet, *Soviet News*, published by the Soviet Embassy in London.

Articles on Russian problems can be found now in a wide selection of journals, though a growing number is designed specially to cover Soviet subjects. This list contains a selection of the more useful sources.

Soviet Geography: Review and Translation (New York).

Information U.S.S.R. (Oxford).

Central Asian Review (London).

Royal Central Asian Journal (London).

Slavonic and East European Review (London).

American Slavic and East European Review (New York).

Polar Record (Cambridge).

Soviet Studies (Glasgow).

Geographical Journal (London).

Geographical Review (New York).

Annals of Association of American Geographers (Chicago).

Economic Geography (Worcester, U.S.A.).

Bibliography

Quarterly Journal of Economics (Cambridge, U.S.A.).
International Affairs (London).
Foreign Affairs (New York).
Survey of Europe (Paris).
Petermanns Mitteilungen (Gotha).
Annales de Géographie (Paris).
Der Europäische Osten (Munich).
Osteuropa (Stuttgart).
Russischer Digest (Munich).
    Soviet periodicals, many with summaries in English, French or German:
Izvestiya Akademii Nauk SSSR (Geographical Series).
Izvestiya Vsesoyuznogo Geograficheskogo Obshchestva.
Geografiya v Shkole.
Geograficheskiy Sbornik.
Voprosy Geografii.
Geografiya i Khozyaystvo.
Vestnik Leningradskogo Universiteta (Series Geology and Geography).
Planovoye Khozyaystvo.

STATISTICAL HANDBOOKS

Since 1956 a number of Soviet statistical handbooks have appeared, but unfortunately some of these are very rare in the West. The contents are sometimes superficial and often leave unfortunate gaps, though they nevertheless provide an invaluable source of statistics to any student of the U.S.S.R. As far as they go, the statistics they contain are probably reliable, though some have been questioned by Western scholars. A complete list of the handbooks known in the West has been published in parts in *Soviet Studies* since 1960. Among the most useful publications are:

Narodnoye Khozyaystvo SSSR (1956) — an English translation was published in London (*The U.S.S.R. Economy. A Statistical Abstract, 1957*).
Narodnoye Khozyaystvo SSSR v 1956 godu — and subsequent years.
Promyshlennost SSSR 1957.
Ugolnaya Promyshlennost SSSR 1957.
Transport i Svyaz SSSR 1957.
Selskoye Khozyaystvo SSSR 1960. Early volumes had covered the sown area and distribution of crops, numbers of cattle and other livestock keeping, but the statistics are all gathered in conveniently concise form in this volume.

ILLUSTRATIONS

1*a*. The Kremlin forms the core of Moscow, which owes its importance to a central position in the European Russian plains

1*b*. The geographical diversity of the Soviet Union is reflected by the semi-desert Fergana valley, an area to be irrigated for cotton-growing

2*a*. Much of Moscow's real character is masked by façades of modern buildings along new thoroughfares

2*b*. Like all towns, Moscow is replacing the old wooden houses by modern flats. Housing is one of the most serious Soviet problems

3*a*. The Krasnoyarsk hydro-electric barrage in the Sayan foothills typifies major civil engineering projects in Siberia

3*b*. Lake Onega in the Karelian tayga. Late glacial ice-rafted boulders in the foreground

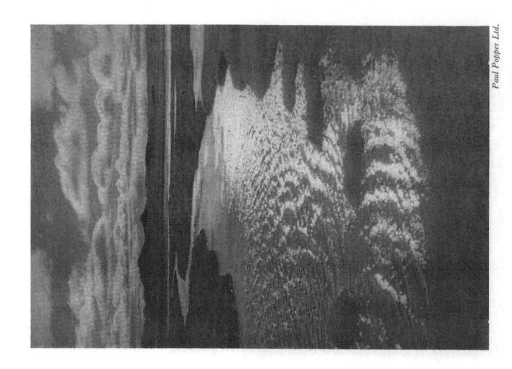

5a. A Svanetian village with its typical watchtower beneath the twin crystalline peaks of Mt. Ushba (15,400 ft.) in the Great Caucasus

5b. The bare, relatively plateau-like summits of the Crimean mountains are used as summer pasture. Patches of sunflower surround houses in the foreground

Souter-Smith

6a. View north across industrial Kiev and the Dnepr at the break up of the ice (April 1963). In the far distance is the grey outline of the high right river bank

Mellor

6b. A main street in Smolensk, with flats built after wartime destruction and the beautifully restored cathedral

7*a*. The church reflects the German Baltic influence in the Riga skyline. The photograph taken in 1949 shows remains of wartime damage

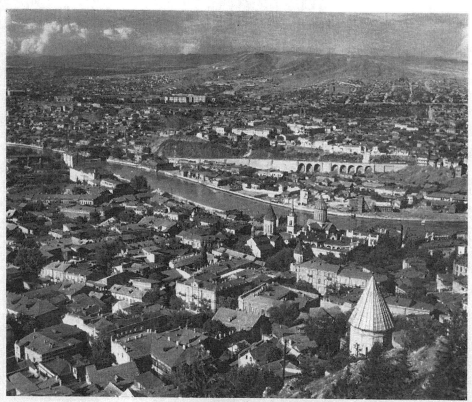

7*b*. Tbilisi, the Georgian capital, is named after local warm springs and stretches along the steep-sided valley of the Kura

8*a*. The Novokuznetsk iron and steel works were one of the early heavy industrial plants built in Siberia after the Revolution

8*b*. Search for oil takes place up to several miles offshore in the Caspian, notably around the Apsheron Peninsula

J. Allan Cash

9a. Coastal terraces and foothills in Western Caucasia are outstandingly fertile farming areas, as seen from this picture taken near Gagra

Camera Press Ltd.

9b. The arid, loessic piedmont belt in Central Asia can be very fertile when irrigated by canals like the Great Fergana shown here

Paul Popper Ltd.

10*a*. An Ukrainian village scene fast disappearing with rural improvements. Note the wandering animals and the primitive well

Paul Popper Ltd.

10*b*. Meadowland and forest in the rolling glaciated country west of Moscow. A typical farming scene in the northern forest belt

Paul Popper Ltd.

11*a*. Drying grain outdoors in the new Kazakh wheatlands near Tselinograd

Paul Popper Ltd.

11*b*. Light traffic allows main roads to be used for crop drying. A picture taken on the main road from Moscow to Minsk

12*a*. One of the many prosperous villages established before the Revolution in the forest fringe of Southern Siberia

12*b*. Heavy machinery in use on the immense flatness of the North Caucasian grainlands near Stavropol

Paul Popper Ltd.

13a. Bukhara, with the Kalan mosque in the foreground and the remains of the old wall in the trees in the background. Typical native houses can be seen on the left

J. Allan Cash

13b. Intense dissection of the soft rocks of the Tertiary Great Caucasus has cut gorges like this on the road to Ritsa

14*a*. Soviet guards on the frontier with Afghanistan beside the silt-laden Amu Darya in its braided channel

14*b*. The main Tashkent-Samarkand railway follows the tectonic Tamurlane Gate at the eastern end of the Nura Tau

15*a*. A suction dredger cutting the Kara Kum irrigation canal through shifting sand

15*b*. The Moscow-Volga Canal is designed primarily for navigation. Note the suburban electric train on the bridge

16*a*. The main Moscow-Leningrad highway is typical of new arterial roads in the U.S.S.R. The poster (*right*) calls for recruits for the Kazakh virgin lands

16*b*. The railway still dominates the Soviet communications pattern. A mainline express at Ternopol in the Ukraine

INDEX

Pages in italics indicate a figure in the text. No distinction is made between administrative units and towns of the same name. Space forbids the listing of all towns with alternative name forms: the principal towns with alternatives have been listed, but for smaller places see the Appendix.

Index

Index

THE END

PRINTED BY R. & R. CLARK, LTD., EDINBURGH

Printed and bound by CPI Group (UK) Ltd, Croydon, CR0 4YY

24/10/2024

01778282-0015